Routledge R

Nuclear Imperatives
and Public Trust

This title, first published in 1987, examines the topic of nuclear waste management, and the way in which the public reacts to this issue. Part 1 explores the sources of public unease, such as the way in which nuclear waste had failed to be properly contained in the past. Part 2 looks at the search for a waste policy and the introduction of The Nuclear Waste Policy Act. Part 3 examines the waste problem from the standpoint of it being an international issue, and finally, Part 4 looks to the future and the lessons that we can learn from past nuclear waste management failures. This book will be of interest to students of environmental management.

Nuclear Imperatives and Public Trust

Dealing with Radioactive Waste

Luther J. Carter

RFF PRESS
RESOURCES FOR THE FUTURE

First published in 1987
by Resources for the Future, Inc.

This edition first published in 2016 by Routledge
2 Park Square, Milton Park, Abingdon, Oxon, OX14 4RN
and by Routledge
711 Third Avenue, New York, NY 10017

Routledge is an imprint of the Taylor & Francis Group, an informa business

Publisher's Note
The publisher has gone to great lengths to ensure the quality of this reprint but points out that some imperfections in the original copies may be apparent.

Disclaimer
The publisher has made every effort to trace copyright holders and welcomes correspondence from those they have been unable to contact.

A Library of Congress record exists under LC control number: 87004706

ISBN 13: 978-1-138-94122-9 (hbk)
ISBN 13: 978-1-315-67348-6 (ebk)
ISBN 13: 978-1-138-94184-7 (pbk)

Nuclear Imperatives and Public Trust

Dealing with Radioactive Waste

LUTHER J. CARTER

Resources for the Future, Inc. -Washington, D.C.

Printed in the United States of America

Published by Resources for the Future, Inc.
1616 P Street, N.W., Washington, D.C. 20036
Books from Resources for the Future are distributed worldwide by The Johns Hopkins University Press

Library of Congress Cataloging-in-Publication Data

Carter, Luther J.
  Nuclear imperatives and public trust.

  Bibliography: p.
  Includes index.
  1. Radioactive waste disposal—Government policy—United States. 2. Radioactive waste disposal—Government policy—Case studies. I. Title.
HC110.E5C358   1987   363.7′28   87-4706
ISBN 0-915707-29-2

⊗ The paper in this book meets the guidelines for permanence and durability of the Committee on Production Guidelines for Book Longevity of the Council on Library Resources.

RESOURCES FOR THE FUTURE (RFF) is an independent nonprofit organization that advances research and public education in the development, conservation, and use of natural resources and in the quality of the environment. Established in 1952 with the cooperation of the Ford Foundation, it is supported by an endowment and by grants from foundations, government agencies, and corporations. Grants are accepted on the condition that RFF is solely responsible for the conduct of its research and the dissemination of its work to the public. The organization does not perform proprietary research.

RFF research is primarily social scientific, especially economic, and is concerned with the relationship of people to the natural environment—the basic resources of land, water, and air; the products and services derived from them; and the effects of production and consumption on environmental quality and human health and well-being. Grouped into three research divisions—Energy and Materials, Quality of the Environment, and Renewable Resources—staff members pursue a wide variety of interests, including food and agricultural policy, forest economics, natural gas policy, multiple use of public lands, mineral economics, air and water pollution, energy and national security, hazardous wastes, and the economics of outer space. Resident staff members conduct most of the organization's work; a few others carry out research elsewhere under grants from RFF.

Resources for the Future takes responsibility for the selection of subjects for study and for the appointment of fellows, as well as for their freedom of inquiry. The views of RFF staff members and the interpretations and conclusions of RFF publications should not be attributed to Resources for the Future, its directors, or its officers. As an organization, RFF does not take positions on laws, policies, or events, nor does it lobby.

This book is a product of the Energy and Materials Division of Resources for the Future, Joel Darmstadter, director. Luther J. Carter is a writer on environmental issues.

The book was edited by Samuel Allen and designed by Martha Ann Bari. The index was prepared by Baehr Publishing Services.

Contents

PART 2: SEARCHING FOR A WASTE POLICY

PART 3: EUROPE, JAPAN, AND THE INTERNATIONAL WASTE PROBLEM

Preface

Nuclear waste disposal has become such a sticky tangle of political and technical issues that anyone who picks up this problem may find it hard to put down. I first had to do with the nuclear waste problem in the late 1970s as a reporter for *Science* magazine, but this was a passing encounter that gave me little idea of what to expect from the much more intimate involvement to come.

This book, on which I began work in the fall of 1980, was originally planned to take a year and a half, a highly optimistic schedule shaped more by the limited availability of funds than by anything else. The fact is, I was embarking on a fascinating but grueling endeavor that was to last six full years.

The first year was taken up largely with field research in Europe and the United States. This included full body immersion in the history of nuclear waste management and in the details of how radioactive wastes are generated throughout the nuclear fuel cycle and of the fate of those wastes. This immersion was as necessary as it was time consuming. Knowing something in the abstract is far different from knowing it in detail; as the saying goes, "The devil is in the details."

The second year was given over mainly to preparing the initial drafts of the book's European, Japanese, and international chapters. Meanwhile, Congress had passed the Nuclear Waste Policy Act of 1982, so the third year was devoted to reconstructing that experience and describing how American nuclear waste policy-making began back in the Ford and the Carter administrations. A chapter examining part of the American experience in nuclear fuel reprocessing also was prepared during this period.

By the end of 1983, a first draft of the entire book was near completion. But I had the sense of being still engaged with a messy and confusing set of issues that I was not yet ready or able to leave or put down. What remained was to pull things together according to several integrative themes which were only then beginning to emerge. So over the next two years the entire book was rewritten, with new events taken into account as they occurred.

Not long after the manuscript was accepted for publication by Resources for the Future in the early spring of 1986, the waste repository siting process mandated by the Nuclear Waste Policy Act reached a point of deep political distress. By all the signs, Congress was not yet done with the waste issue; the 1982 act would almost surely have to be revisited and amended. Portents of this crisis in repository siting had been visible for some time. My analysis and conclusions were revised to take the new developments into account. Final revisions for the manuscript as a whole were completed by the end of October 1986.

My research, which included extensive interviewing (both in person and by telephone), continued throughout the entire book project. Further inquiries often arose from questions encountered in writing. I found I was able to keep abreast of developments in Europe and Japan through correspondence, trade publications, and interviews with nuclear attachés at foreign embassies and with visitors from abroad. The record of legislative hearings and government proceedings of one kind or another were important sources of information, at home and abroad, as were industry conferences.

Funding for the project has been entirely from foundations and other private sources. Support has come from Resources for the Future, the Rockefeller Foundation (in the form of an International Relations Fellowship), the Berlin Science Center (Wissenschaftszentrum Berlin), the Ruth Mott Foundation, the A.W. Mellon Foundation, the Fund for Investigative Journalism, and from a foundation which makes its grants anonymously.

I bear a heavy debt to the many who have helped me in my research. In the matter of repository siting and nuclear waste policy, I must mention in particular Critz George, Colin A. Heath, Thomas A. Cotton, Roger E. Kasperson, Kai N. Lee, and Thomas A. Moss. They served as continuing sources of information and advice. Among others who deserve mention as reviewers of parts of the manuscript (or who have gone out of their way to help in other ways) are Sam A. Carnes, Sandra Fucigna, Roger Gale, William A. Holmes, Michael J. Lacey, Robert Loux, Robert H. Neill, Peter Myers, Elizabeth Peele, Joel Snow, and Mason Willrich.

Harry W. Smedes has given generously of himself as a field geologist with a special gift for lucid description of rock formations and their origins. Other geologists from whom I have received much help include Lokesh Chaturvedi, Joseph D. Martinez, Eugene Roseboom, Newell Trask, Isaac Winograd, and Wendell D. Weart. In fuel reprocessing and related waste management issues, J. O. Blomeke has been a patient teacher and an exceptionally valuable and objective source of information. Among others who have helped in this field have been Clint Bastin, Joseph Leiberman, Goetz Oertel, Marvin Resnikoff, Walton Rodger, and Donald B. Trauger. On the health effects of low-level radiation Thomas B. Cochran, Dan Egan, Merril Eisenbud, John H. Harley, and Arthur C. Upton have been particularly helpful.

Many members of the House and Senate staff have helped me piece together how congressional policy has been made on nuclear waste and fuel cycle issues. I wish to mention in particular Gerald Brubaker, Ben Cooper, and Andrea Dravo. People in the academic and environmental communities and in environmental law who have gone out of their way to help and advise have included Frederick R. Anderson, Janet Bearden, David Berick, Joseph Browder, Christopher Duerksen, Christopher Flavin, William Futrell, John Holdren, James J. MacKenzie, Michael McCloskey, James Moorman, J. G. Speth, and Guy Martin.

Charles Van Doren and Warren Donnelly have been key sources of information and advice on the international and nuclear nonproliferation issues. Other reviewers of one or more of the chapters on the international and foreign experience are: on the United Kingdom, Donald G. Avery, Sir Douglas Black, H. John Dunster, Frank S. Feates, Thomas B. Stoel, Jr., and Vaughan T. Bowen; on Germany, David Albright, Meinolf Dierkes, Hermann Graf Hatzfeldt, Carsten Salander, Jeffrey M. Schevitz, Ulrich Steger, and Horst W. zur Horst; on Sweden, Thomas B. Johansson, Mans Lönnroth, and Erik Svenke; on France, Bertrand Barré, Dorothy Nelkin, and Michael Pollak (Nelkin and Pollak also reviewed the chapter on Germany); on Japan and the Pacific basin, Justin Bloom, Susan J. Pharr, Richard Scribner, and Richard P. Suttmeier. Allison Platt and Kent Harmon were among reviewers of the chapter on the international experience.

Several persons have been generous in sharing with me their own research materials or unpublished work. They include William P. Bishop and Richard A. Watson, Robert Leslie Cohen, and Randall F. Smith. Numerous government and industry public information specialists have assisted me, but I want to mention in particular William V. Merriman for his exceptional efforts.

Editors and journalist colleagues who have assisted me include Edwin L. Dale, Constance Holden, William Lanouette, Leslie Roberts, Jane Scully, Robert D. Selin, John Walsh, and Sam Allen. Allen, a gifted editor with a fine sense for the structure of a long and complex manuscript, edited the book as it was finally submitted.

Colleagues at Resources for the Future who have read all or parts of the manuscript and served as advisers are Philip H. Abelson, John F.

Ahearne, James M. Banner, Jr., Marion Clawson, Joel Darmstadter, Robert W. Fri, James Jasper, Hans H. Landsberg, Robert C. Mitchell, Bryan Norton, Clifford S. Russell, and Milton Russell.

Fisher Howe, institutional relations counselor at RFF at the start of the project, was tireless in his efforts to raise funds for it. As administrative assistant for the energy and materials division, Helen-Marie Streich did much to facilitate my work. Angela Blake, Peggy Kroder, John Mankin, and Mimi W. Rubin have done cheerfully and accurately the monumental job of word processing.

Finally, I dedicate this book with love to my wife, Marsha.

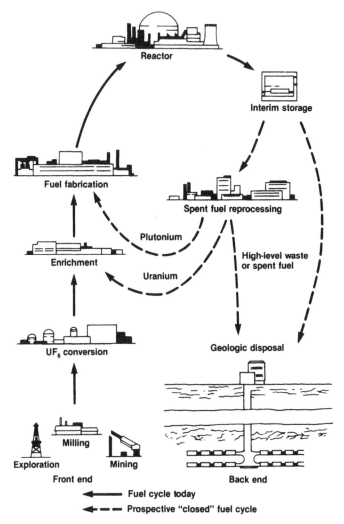

Reactor

Interim storage

Fuel fabrication

Spent fuel reprocessing

Plutonium

Enrichment

Uranium

High-level waste
or spent fuel

UF₆ conversion

Geologic disposal

Milling

Exploration   Mining

Front end   Back end

◄——— Fuel cycle today
◄— — — Prospective "closed" fuel cycle

The commercial nuclear fuel cycle includes activities for preparing and
using reactor fuel and for managing spent fuel and other radioactive
wastes produced in the process. It was originally intended that spent fuel
be stored for 6 months in water-filled basins at reactor sites to dissipate
thermal heat and allow decay of short-lived fission products. The spent
fuel would then be reprocessed and the resultant liquid high-level waste
solidified and disposed of in a federal repository. Since no repository has
been developed and no commercial reprocessing is being done, spent
fuel will remain in storage until repositories are available to close the
nuclear fuel cycle. *Source:* Council on Environmental Quality data in
Office of Technology Assessment, *Managing the Nation's Commercial
High-Level Radioactive Waste*, OTA-0-171 (Washington, D.C., U.S. Congress, March 1985).

Introduction

Nuclear power is subject to two indisputable imperatives. One is to safeguard potential nuclear explosives, particularly the plutonium created as an inevitable by-product of the fission process. The other is the imperative to contain radioactivity in reactor and nuclear fuel cycle operations.

These imperatives have to do with matters of perception as well as matters of fact, and partly for this reason success in satisfying them is still far from being achieved. Nuclear power has remained controversial and, uniquely among energy sources, has lacked broad public acceptance. Coal, oil, natural gas, and hydropower all have their problems, but no other energy source provokes anything like the controversy and emotional response that "nuclear" does.

Other important new technologies of the twentieth century, such as those that have given us the private automobile and commercial aviation, have been allowed to solve their problems as they have gone along. But for more than a decade now, and long before Chernobyl, nuclear power has been in political trouble because its problems were not solved convincingly—and some were hardly solved at all—before the technology was introduced on a commercial scale. That the public should be demanding more of nuclear power may suggest a double standard. But as I shall be noting time and again in this book, the exceptional demands imposed on nuclear power are understandable in light of both its origins as a postwar spin-off from the nuclear weapons program and the mysteries which this still unfamiliar technology represent for the public. They are understandable too in light of the often troubled history and flawed performance of the nuclear enterprise in the United States and other countries.

Radioactive waste disposal and management of the irradiated or "spent" fuel that is removed from the reactor are only part of the overall problem of safeguards and containment, but that part is very important and is

1

what particularly concerns us here. One starts with the fact that spent fuel contains both plutonium and the "fission products" (or fragments of split atoms) which are created in the reactor. Plutonium is the stuff dreams as well as bombs are made of. The "breeding" of this man-made fissionable element, which results from the transmutation of the fertile and abundant uranium-238 atom, would be the basis of the so-called breeder economy in which a limited resource of fissionable uranium would be expanded manyfold and indefinitely extended in time. But because plutonium is a nuclear explosive as well as a fuel, recovering it from the spent fuel (where it is in a form too diluted to be either an explosive or a fuel) has always represented a dilemma, albeit a dilemma that many have refused to recognize. The fission products are the source of most of the spent fuel's radioactivity, and, while some small part may find practical uses, these materials constitute the bulk of the high-level waste. Moreover, it is the buildup of the fission products—which inter- fere with the fission reaction by capturing neutrons—that makes it necessary to remove the fuel from the reactor at a point when significant amounts of fissionable uranium and plutonium remain.

The chemical step by which the unconsumed plutonium and uranium are separated from the fission products and from one another is known as reprocessing. Officially, reprocessing is sanctioned and encouraged in the United States, but because it is not now economic and has been denied government subsidies, commercial reprocessing in this country no longer exists and will not be revived for the foreseeable future. Accordingly, present U.S. policy is to regard spent fuel as waste and to dispose of it in geologic repositories, or deep underground labyrinths created by conventional methods of deep mining. No such repositories for spent fuel now exist, but a far-flung, multibillion-dollar effort has been in process to establish one repository by 1998 and possibly a second during the following decade. But today this effort is in political distress and is foundering. Whether a repository can be sited and built by the turn of the century is likely to depend on the adoption of innovative approaches to overcome the resistance of potential repository host states and to encourage a sense of urgency and cooperation among a variety of interests across the nation that have a stake in the nuclear waste issue.

In a number of European countries and in Japan official policy still calls for reprocessing of spent fuel as a key step in closing the nuclear fuel cycle and disposing of radioactive wastes. The economics of repro- cessing is no better there than in the United States. But the mix of official and ideological commitments to reprocessing will carry this undertaking forward at least for the next decade or two. The strength of these commitments has impelled countries where commercial reprocessing capacity is either lacking or insufficient to plan to have their reprocessing done abroad, in France and the United Kingdom; for French and British reprocessors, the willingness of foreign customers to pay premium prices has served as a kind of surrogate for public subsidy and encouraged what otherwise would be a financially troubled endeavor to go forward. Pending development of their own national reprocessing facilities, utilities of have-not countries (particularly Germany and Japan, where reactors are licensed under the condition that spent fuel will be reprocessed) have gone to the French and the British and entered into contracts extraordinarily favorable to the reprocessors. Indeed, these contracts are enabling the reprocessors to expand their installed capacity and offer a service which conditions in the national and international nuclear fuel markets ordinarily would never support.

How the high-level waste from fuel that is reprocessed is to be managed or disposed of over the foreseeable future may vary greatly from country to country. In principle, all nations appear to favor geologic disposal as the ultimate solution. But typically, as in the United States, there is political resistance to be overcome in siting repositories, and whereas French government leaders plainly have the stomach (and the tradition) for overriding any local protests that stand in their way, their British counterparts have shown no stomach for this at all.

Also, none of the countries of western Europe has the geologic and climatic diversity and the large desert areas found in the United States. Some of them will surely be found to lack the conditions appropriate to the siting of radioactive waste repositories even if it should make economic sense (it does not) for each of them to spend hundreds of millions and perhaps billions to establish such a facility. In Japan, as in France, Germany, and the United Kingdom, the nuclear program is large enough to justify establishing a national repository, but in this island country of earthquakes, volcanoes, and dense human settlement the constraints on repository siting are all too obvious. Certainly for some of the European countries and probably for Japan the waste disposal problem cries out for solutions that depend on one country's taking another's wastes or on the establishment of international repositories. But up to this time such arrangements, with one or two exceptions, have been regarded as politically impossible. In all countries the prospect is for both high-level waste and most of the spent fuel to remain for at least some few decades in interim surface storage. One major country, the United Kingdom, has sought to raise this practical necessity to a virtue and has abandoned its efforts to find a geologic disposal site for high-level waste.

As I shall be discussing in succeeding chapters, whether to reprocess spent fuel and recycle the plutonium or, alternatively, whether to treat the fuel as waste and dispose of it in a geologic formation is a choice

4

of crucial significance. This choice bears importantly on the problems of safeguards and containment. It also bears on the overall complexity of the nuclear enterprise and the potential for regulatory and political conflict and controversy.

An essential argument of this book is that the safeguards and containment imperatives are more likely to be satisfied in an effective and convincing manner if spent fuel is disposed of as waste. Furthermore, direct geologic disposal of spent fuel need not necessarily mean that use of the plutonium shall be forever forgone. Disposal to deep repositories in hard rock can be done retrievably, for up to fifty years or more. This would allow eventual recovery of the fuel if reprocessing and recycling ultimately can be justified in light of technical and economic viability, improved political conditions around the world, and the promise of safeguards adequate to the need. Retrievable disposal has the virtue, too, of fostering broader public support of geologic disposal inasmuch as the spent fuel can be removed from the repository in the event that unexpected problems arise.

I shall further argue that radioactive waste disposal and spent fuel management are best understood and dealt with if viewed in the larger perspective of the overall containment and safeguards problems. This perspective is one which, as I see it, permits a better assessment of the risks associated with waste isolation as compared to those associated with such problems as reactor safety, nuclear terrorism, and the proliferation of nuclear weapons. As I shall show, the more technically sophisticated nuclear critics recognize that, in terms of real danger to the public, the possibility of radioactivity escaping in harmful concentrations from a geologic repository is easily the least of the nuclear hazards.

But the public at large ordinarily seems not to discriminate clearly between one nuclear hazard and another. Why this is the case calls for close examination. The transport and disposal of spent fuel (or of any radioactive waste) can in many circumstances arouse such public fear and resistance as to make difficult what might otherwise be a timely and effective solution to the problem of isolating both fission products and plutonium. I shall take up this question, especially in chapter 2, where I explore the history of the nuclear enterprise in the United States and the likely sources of the widespread public fear of things nuclear.

An argument implicit throughout this book is that if nuclear power is ever to gain widespread public trust the nuclear enterprise should begin moving away from ever greater complexity to greater simplicity. Fuel reprocessing and plutonium recycling, this argument goes in part, move the enterprise in the wrong direction, for in this fuel cycle all the problems relevant to the containment and safeguard imperatives are made more difficult and complex. For instance, contrary to what many leaders of the enterprise have maintained, waste management and disposal are not made easier by fuel reprocessing. Reprocessing, a difficult industrial technology, makes for a proliferation of wastes and contaminated facilities and equipment. This theme of complexity and how it is affected by the choice of nuclear fuel cycles will be discussed in the chapters on containment, the reprocessing dilemma, and nuclear fuel cycle issues in Europe (particularly the chapters on France and the United Kingdom). It will be argued that in the light of both foreign and American experience the once-through cycle with direct geologic disposal of the spent fuel is for now, at least, the simplest, the most economic, and the most politically sensible and viable fuel cycle alternative.

In principle, the United States should be able to arrive at ideal solutions, within its own national borders, to its domestic geologic disposal problem, whereas for a number of foreign countries solutions may be possible only through international cooperation or joint undertakings. In this connection, the repository siting problem within the U.S. federal-state system bears a rough analogy to the international problem. I shall be discussing the merits of possible solutions in this country which, should they deserve and gain public trust and cooperation, could be an encouraging example for the rest of the world and perhaps even an indispensable precedent. But progress is not likely to be made here, I shall argue, without a much more widespread recognition of what I see as the realities:

One reality is that to survive the trials of real-world politics, any new repository-siting policy and strategy must be crafted with explicit and openly acknowledged concern for what is workable and with an uncompromising concern for technical soundness and public safety. As I shall point out, this reality has implications both for planning the overall national siting effort and for dealing with particular host states. The public's fear of radioactive waste, its alarm at proposals for waste "dumps" of any kind, and its distrust of the federal government are not likely to be overcome unless care is taken to avoid major environmental conflicts in choosing repository sites; unless the prospective host states and localities are assured that if the repository is built they will receive rewards as well as waste; and unless means are found to convince officials of the host state and locality that the siting investigation will be carried out competently and with integrity and the repository system proposed will leave little or no uncertainty about waste containment.

A second reality, as obvious as it is important, is that establishing the nation's first nuclear waste repositories and related fuel storage and transport systems demands political accommodations on the part of interested parties. But if all the major players—the electric utilities, the state and local governments, the environmental groups, and those who speak for these parties in Congress—are to confront the waste disposal problem more constructively than in the past, all must see a clear need and urgency to arrive at a solution. Such a perception must come in part from a recognition of the importance that nuclear power already has assumed in the energy mix of the United States and the world. Like it or not, nuclear power will not go away, not for decades, perhaps not for generations. So the real question has to do not with shutting down nuclear power but with making it safer and more publicly acceptable. To reemphasize the point, the proposition argued here is that the once-through fuel cycle, with disposal of the spent fuel, is the most straight-forward, easily managed of any, and, for this generation at least, likely to be found most serviceable. The sooner the effectiveness of this fuel cycle is demonstrated in a robust and practical manner, the more likely the nuclear imperatives of waste containment and plutonium safeguards will be convincingly answered.

Part 1
Sources of Public Unease

1
Containment

The terminal isolation of radioactive wastes, while an important aspect of containment, is not the most difficult aspect. The containment problems involved in operating reactors and fuel reprocessing and recycling facilities will almost certainly prove to be demonstrably more difficult. The point is an important one, because to exaggerate the difficulties and hazards associated with waste repositories skews the priorities for dealing with nuclear power's manifold problems, and can frustrate efforts to begin the early geologic disposal of spent fuel—the best means to safeguard plutonium.

The "Real" Versus the "Imaginary" Problems

Several years ago in London I talked with Sir John Hill, who at the time was still chairman of the United Kingdom's Atomic Energy Authority. "I've never come across any industry where the public perception of the problems is so totally different from the problems as seen by those of us in the industry who are actually dealing with them," Hill told me. In Hill's view, the problem of radioactive waste disposal was, in a technical sense, comparatively easy. The more demanding problem, as Hill saw it, was the everyday operational task of running the industry, in particular the fuel reprocessing plants, and containing the radioactivity in keeping with the standards adopted by the operators and imposed by regulators for the protection of workers and the public. "It's a difficult industry to run to high standards," Hill said. "The industry is dealing with very real [and] practical engineering, management, [and] operating problems. The greatest care has got to be taken to keep the plant going— to make sure that the operators behave meticulously; that they don't get irradiated, don't spread activity, don't get activity on their boots and carry it through the change rooms and over the roads. The public are worried about [things] which, in many ways, are nonproblems. They've been encouraged to see the industry in a different way. They have been frightened by talk of a quarter of a million years before activity dies away, which is scientifically true but in practice has little meaning or consequence."

John Holdren, a physicist and energy resources specialist at the University of California at Berkeley, is an academic of high standing who is widely respected as a responsible and knowledgeable nuclear critic. Holdren recognizes that bringing about safe terminal disposal of radioactive waste is a significant problem, but, in relative terms, he ranks it last among the major problems of nuclear power. "First comes the proliferation of nuclear weapons," he told me. "Second, I put the diversion of nuclear explosives by criminals and terrorists. Third comes accidents and sabotage at nuclear reactors and possibly at reprocessing plants. Only in fourth place do I rank the long-term management of radioactive waste. I base these rankings on real risks to human beings as measured by the probability of things going wrong and the consequences if they do go wrong. In the case of radioactive waste, the probability of massive failure of a repository is smaller than the probability of a major release of radiation in the event of a reactor core meltdown. And, if a massive failure did take place, the consequences would be smaller than the potentially catastrophic consequences of a reactor accident or an act of war or terrorism involving the use of nuclear explosives."[1]

I was to find that Holdren's ranking of the nuclear risks is concurred in by a number of other scientific critics or opponents of nuclear power, such as Amory Lovins of Friends of the Earth, Thomas B. Cochran of the Natural Resources Defense Council (NRDC), and Terry R. Lash, formerly of NRDC and now director of the Illinois Department of Nuclear Safety. In sum, Holdren, a knowledgeable American critic, and John Hill, a leader of the nuclear enterprise in Britain, both saw the public making too much of the waste problem and failing to achieve a clear sense of perspective on the nuclear priorities.

The Wastes

Radioactive wastes arise in every part of the nuclear power program from reactor operations to the running of reprocessing plants and other fuel cycle facilities. They arise in the facilities devoted to waste management themselves, for in handling waste, more waste is generated. John

[1] Interview by the author with John Holdren, November 19, 1981.

Hill has described the waste management problem that arises when spent fuel from a reactor is first received at a reprocessing plant, in what is perhaps the simplest step carried out at such a plant. The transport cask is opened while immersed in the receiving pond (which provides both shielding and cooling) and is emptied of the fuel elements. But before the cask can be returned, it must be thoroughly cleaned, with all measurable radiation removed from its exterior.

So you get hoses and scrubbers and clean it up. Then the [radio]activity is in the water. So you put the water in a tank, and what do you do with that? Wherever you go, nobody will let you throw any activity away ... You have got to take the wash water and purify that before you can throw it away. Then you've got the purification sludges ... They won't let you throw those away either. So you concentrate those in another plant and cast them in concrete [for disposal]. You're moving this activity around the whole time, which is fine while everything works well. But then ... the crane breaks, the pipe breaks, you drop it ... These things are always happening in factories. Equipment fails, or one of the fuel pins is broken and a lot of pellets fall out on the bottom of the pond. Then you've got to get them up. How? You can't just dive in and pick them up with your fingers. You've got to pick these things up with remote tongs and tools, and it takes you a week to do five minutes' work. And during that time, the pond doesn't work, nobody can get at it. You've got a problem on your hands.[2]

Waste management problems become a great deal more numerous and complex in the later steps in reprocessing plant operations, for at the step described above the great bulk of the radioactivity is still contained in the irradiated fuel rods and pellets. But this gets ahead of our story. Let's go back and look at the waste and containment problems that arise from the beginning of the very simplest of fuel cycles, the once-through cycle in which spent fuel is disposed of as waste.

Mill Tailings

The place to start is at the uranium mill. For every ton of uranium ore that is milled in the United States, not more than 2 kilograms (or about 5 pounds) of uranium is extracted, leaving the rest to be discharged as finely ground, sandy tailings. The tailings contain other naturally radioactive substances (such as radium) which are responsible for more than three-fourths of the radioactivity that was originally in the ore. From the mill they go as a slurry into a tailings pond, but in drying they form a large spreading delta around the pond and in this way huge tailing piles

---

[2]Interview by the author with Sir John Hill, London, September 1980.

have been created. About 90,000 cubic meters of tailings are created in producing the yellowcake, or U_3O_8, needed for each gigawatt year of electricity generation,[3] or for the annual power output of a 1,250-megawatt reactor operating at 80 percent of capacity. Should these tailings ever have to be moved (typically tailings are left near the uranium mill), they would fill ten one-mile-long trains of hopper cars. The amount of tailings generated in fueling a large nuclear power plant is almost a third again greater than the total amount of fly ash, bottom ash, and scrubber sludge left from the operation of a coal-fired electric plant of the same size.[4]

In the United States some 121 million cubic meters of tailings had accumulated by the end of 1983, distributed among piles at twenty-four active or recently active mill sites and at more than a score of sites shut down some years ago.[5] Most of these sites are in New Mexico and Wyoming, but a number are in Utah, Colorado, Texas, and Washington. The largest pile is at the Kerr-McGee mill near Grants, New Mexico; it covers 250 acres and rises to a height of about one hundred feet. The piles are not well enough stabilized to be left unattended, and if abandoned they would eventually be widely dispersed by forces of wind and water.

Failures of containment have occurred at mill tailings impoundments. This happened in July 1979 when a tailings dam gave way at the United Nuclear Corporation mill near Churchrock, New Mexico. The escape of some 100 million gallons of tailings solution left sixty miles or more of the Rio Puerco contaminated along its meandering course through Navajo grazing lands in New Mexico and Arizona. The actual hazard was found to be slight, but the Navajos suffered much inconvenience and anxiety. In particular, stock watering became a major problem, with water having to be hauled in from distant wells for sheep and cattle that normally would have been watered nearby on the river.

An especially troublesome and costly problem associated with abandoned mill tailings piles has been the tendency of unwitting or irresponsible individuals to use them as sources of landfill in the construction of homes, commercial buildings, and even schools. This has happened in numerous communities near uranium mining and milling sites in the

---

[3]A gigawatt year is the continuous generation of 1 billion watts of electric power over the period of one year.

[4]Health Physics Society, *Proceedings of the Low-Level Radioactive Waste Management Conference* of May 1979, Environmental Protection Agency document EPA/520/3-79-002 (Washington, D.C., 1979).

[5]Oak Ridge National Laboratory, *Spent Fuel and Radioactive Waste Inventories, Projections, and Characteristics*, DOE/NE-0017/2 (Washington, D.C., Government Printing Office, 1983) table 5.1, p. 174, and fig. 7.1, p. 197.

West. The worst situation by far has been the one at Grand Junction, Colorado, where tailings from the Climax uranium mill pile there were commonly diverted to construction sites before health authorities learned of the practice in 1966 and stopped it. Some 33,000 sites were surveyed in and around Grand Junction under a federal and state remedial program. Several thousand of them were found to be sufficiently contaminated to justify excavation of the tailings fill. The concern was the gaseous alpha-emitter radon-222 and its daughters, known causes of lung cancer in uranium miners exposed to them at high concentrations.[6] Although the concentrations were not at the levels experienced in poorly ventilated uranium mines, the radon emanations into dwellings and other buildings at the designated Grand Junction sites were deemed high enough to put the occupants of those structures at a cancer risk significantly greater than that of the general population.[7]

Excavating the tailings fill from beneath thousands of structures is a huge job, as I noted in visiting several of the Grand Junction sites one afternoon in late 1982. This remedial work had been in progress since the early 1970s, yet still had another ten years to go. At a former Safeway supermarket some workers were breaking up the concrete flooring with pneumatic drills while others were shoveling out the underlying fill to a depth of several feet. The fill was moving by conveyor belt to trucks waiting to carry it back to the still-exposed Climax pile from which the material was diverted some years before. This particular job was to cost more than $100,000, and similar large sites awaited their turn for remedial treatment, among them a bowling alley and a large downtown furniture store. The initial effort at Grand Junction has involved removing the tailings fill from beneath some 600 dwellings and other occupied structures. Still to come is a much larger effort to remove the tailings from beneath a variety of other occupied and unoccupied structures, from vacant land, and from beneath or around such public facilities as sidewalks and water and sewer mains. Also, the Climax tailings pile must be stabilized and probably relocated. Altogether, the total remedial effort at this one Colorado city will have taken twenty years and cost on the order of $222 million.[8]

The accumulation of huge tailings piles represents what is now widely recognized among regulators as an unacceptable and unnecessary waste

---

[6]The least penetrating form of radiation, alpha particles are more densely ionizing than beta or gamma rays. Ionization affects atoms or molecules by adding or removing electrons, and can damage tissue.

[7]Interview by the author with Jake Jacobi of the Radiation Control division, Colorado Department of Health, May 31, 1985.

[8]Interview by the author with Donald Groelsema, director of the Department of Energy's division of mill tailings projects, December 17, 1985.

disposal practice. The best answer to the problem is to require burial of the tailings well below grade, or below the surface of the adjacent terrain. The effect of such burial is to reduce radon emanations to levels close to natural background and, in many cases, to make isolation of the tailings no less effective than was the isolation of the original uranium ore. Indeed, the isolation can be more effective than that of natural ore bodies found within local or regional aquifers, such as those in ground-water-bearing sandstone in the Southwest. Thus, for a properly regulated industry, containment of tailings need not be a major problem. The major difficulty today is to remedy the problems left from the past.

Reactor Wastes

Radioactive waste is generated at other points along the front end of the nuclear fuel cycle, as at the uranium conversion and enrichment plants. But the waste volumes generated at these facilities are comparatively modest and the radioactivity levels tend to be low. The next point after the uranium mill where large amounts of waste are generated is at the reactor. Up to this point, all of the radioactivity in the wastes generated has been naturally occurring, with no radionuclides created yet by man's intervention and the splitting of atoms.[9] The reactor operations generate, as a result of the fission process, three kinds of radioactive waste: highly radioactive spent fuel; low-level waste resulting from routine, day-to-day efforts to prevent a buildup of radioactivity in the reactor cooling system, the spent fuel pool, and elsewhere in the plant; and decommissioning waste, mostly low-level, which results from the final decontamination and dismantlement of the plant. In addition to the wastes that accumulate continuously and routinely, there is the waste from the major decontamination of the cooling system which appears necessary at least once in the life history of every reactor, and the waste from the cleanup of any accident that may occur, as at Three Mile Island.

Spent fuel. To take the most common example, a 1,000-megawatt pressurized water reactor, of the kind first developed by Westinghouse and now dominant worldwide, is loaded with about 100 metric tons of

[9]A radionuclide is a nuclide—a general term applicable to all atomic forms of the elements—that is radioactive. Nuclides are distinguished by their atomic number, atomic mass, and energy state.

Over geologic time there has been at least one instance of a natural reactor—that discovered by the French in 1972 at their Oklo uranium mine in Gabon, West Africa. About 10 million years ago a deposit so rich in uranium accumulated there that a self-sustaining chain fission reaction began, with groundwater serving to slow down the high-velocity neutrons, thus increasing the likelihood of further fission. The chain reactions continued for about a million years at an estimated power level of about 50 kilowatts.

fuel. This fuel load will be made up of as many as 190 or so fuel elements or assemblies, each 13 to 14 feet long and about 8 inches to a side and containing a bit over half a ton of uranium. The individual fuel assembly, which is not highly radioactive until irradiated, consists of an array of about 290 zirconium-clad rods containing uranium oxide pellets in which the fissionable U-235 atom is at a concentration of 3 to 4 percent.

When the reactor "goes critical," or begins sustaining a chain reaction, the fuel is affected by three kinds of reactions. First, the splitting of U-235 atoms by neutrons creates a profusion of fission fragments or products, with half-lives measured from the tiniest fraction of a second to tens of thousands of years (as in the case of technetium-99) or even millions of years (as in the case of iodine-129).[10] Strontium-90 and cesium-137, with half-lives of 28 and 30 years respectively, are important and abundant; these fission products are highly radioactive (the specific activity of strontium is 140 curies per gram), are biologically active (mocking calcium chemically, strontium tends to accumulate in the bones), and, especially in the case of cesium (which is highly soluble), are mobile in the environment.

A second important reaction results from the capture of neutrons by nonfissionable U-238 atoms, which are thereby partially converted to fissionable plutonium-239. Successive neutron captures lead to the creation of a variety of other fission by-products or transuranic elements (that is, elements having a greater atomic number than uranium). Besides being a nuclear explosive which can be chemically separated from the fuel, the plutonium is notorious as a potent and long-lived alpha-emitter; even a speck of plutonium, say 10 millionths of a gram, is sufficient to cause cancer if deposited within the lungs or admitted to the bloodstream through a cut.

The third major reaction affecting the fuel has to do with the "activation" of stable elements such as cobalt, iron, manganese, and nickel which are present in the fuel-rod cladding and in the fuel assemblies' stainless steel structural components; activation occurs when these elements capture neutrons. Cobalt-60, a potent gamma-emitter with a half-life of about five years, is a particularly abundant and important activation product.

The fuel assemblies remain in the reactor core for three years, with a third of the total load annually becoming "spent" and needing replacement. Before the cooling that comes with radioactive decay, irradiated fuel may contain as much as 180 million curies of radioactivity per ton

[10]In one half-life, half the atoms of a particular radioactive substance disintegrate, or decay, to another atomic form, which may itself be radioactive and hence in process of decay.

and generate up to 1.5 megawatts, or 1,500 kilowatts, of heat. Thus, even after the fission process is shut down, a meltdown of the core could occur from the decay heat alone. This is one reason why nuclear critics view the possibility of reactor accidents and possible breaches of reactor containment with such grave concern.

Radioactive decay proceeds quickly over the first months to a year following removal of the irradiated fuel from the reactor, then slows to a pace where the significant reductions come not in months but in years, decades, and centuries. But even after 1,000 years the amount of water necessary to dilute the radiotoxicity of 30 tons of spent fuel (or approximately the amount discharged from a large reactor each year) to drinking water standards would be 10 billion cubic meters,[11] equal to ten times the yearly flow of the Hudson River.[12] Accordingly, the radiotoxicity of spent fuel and indeed of many of its fission products and by-products is far too great, even without taking into account biological reconcentration mechanisms, for their release to water bodies or the general environment to be acceptable.

The radionuclides in commercial reactor fuel have for the most part thus far been well contained. The reactor system contains four barriers to the escape of radionuclides from containment. The first consists of the form of the fuel pellets. The uranium oxide, a ceramic, has a high melting point, at 5,080 degrees Fahrenheit (°F). This is essential for the pellets to withstand the 4,000-degree temperature at the center of the pellets during the fission process (the temperature at the exterior of the pellets normally does not exceed about 600° F). The zirconium cladding of the fuel rods that contain the pellets is the second barrier. The cladding will fail at about 3,600° F. The third barrier is the pressure vessel with its 8-inch-thick steel walls, and the fourth is the concrete containment shell whose walls are heavily reinforced with steel and may be up to 4 feet thick.[13]

The spent fuel, if it is not to be reprocessed, represents by far and away the largest part of the nuclear waste disposal problem. By the end of 1985, some 12,450 tons of spent fuel, in more than 45,250 individual fuel assemblies, had accumulated in the United States. The accumulation

[11]The amount of dilution necessary to meet the Environmental Protection Agency's drinking water standard is a common measure of radiotoxicity. As applied here it is a theoretical rather than practical measure. All the radioactivity in the spent fuel would not be present in a dissolved form.

[12]American Physical Society, Study Group on Nuclear Fuel Cycles and Waste Management, *Reviews of Modern Physics* vol. 50, no. 1, pt. 2 (January 1978) fig. 7B1, p. 2110; Ronnie D. Lipschutz, *Radioactive Waste, Politics, Technology, and Risk* (Cambridge, Mass., Ballinger, 1980) table 3-3, p. 104.

[13]Specifications vary. Those cited here apply to the Three Mile Island pressurized water reactors built by Babcock and Wilcox.

projected by the Department of Energy (DOE) in 1985 for the year 2000, if there is no further growth in nuclear capacity beyond the 115.3 gigawatts existing or under construction, was about 41,500 tons, in almost 148,700 fuel assemblies,[14] which might be packaged in as many as 89,500 canisters for geologic disposal.[15]

Nearly all spent fuel has remained in the pools at the reactor sites, with water providing both cooling and radiation shielding. The storage pools are built very near the reactor—typically in an adjacent auxiliary building—because the fuel's removal from the reactor vessel and transfer to the pool is carried out under water, by remote handling devices. The fuel assemblies are kept separated by a grid of metal racks, and boron (a neutron-absorbing material) is present to prevent fission reactions.

In recent years the spacing required between assemblies has been reduced, from 20 inches to about 12, to make room for more fuel than was originally expected. Initially, storage pools were designed and equipped to accommodate no more than one or two annual discharges, plus the discharge of an entire core in the event reactor repairs or maintenance necessitated a complete defueling. The utilities expected to send each annual load of discharged fuel to a reprocessing plant after several months of cooling. The re-racking of the pools to increase the storage density by as much as three or four times began once it became evident that reprocessing was to be long delayed.

Since 1983 the Nuclear Regulatory Commission (NRC) has been studying the risks of severe accidents and containment failure in spent fuel pools. As discussed in the NRC document *A Prioritization of Generic Safety Issues* (1983), these risks, once considered improbable, are being reexamined because of the greater storage density in the pools and because laboratory studies have indicated the possibility that fire could spread throughout the dense array of fuel assemblies in the event a pool lost its cooling water. The risks are deemed greater for boiling water reactors, which have their spent fuel pools above grade, than for pressurized water reactors, which have theirs below grade. (The NRC study is expected to be issued in 1987.) The increased attention given the risk of major releases of radioactivity from storage pools lends greater

[14]Oak Ridge National Laboratory, *Spent Fuel and Radioactive Waste Inventories, Projections, and Characteristics*, DOE/RW-0006, rev. 1 (Washington, D.C., Department of Energy, December 1985) table 1.1, p. 29, and table 1.11, p. 44.

[15]I am assuming here that the pressurized water reactor assemblies will go only one to a canister, and that the smaller boiling water reactor assemblies will go three to a canister. There are hopes of dismantling the assemblies and consolidating the rods in a much smaller number of canisters, but this would have to be done remotely, routinely, and quickly, and has yet to be demonstrated. The question is not whether rod consolidation can be done, for this has been demonstrated, but whether it can be carried out as a practical, routine, cost-effective operating procedure.

urgency to providing the means to remove spent fuel from these pools after several years of cooling.

It should be noted that radioactivity in the commercial spent fuel is already some eight times greater than that in waste from the nuclear weapons program. The disparity steadily widens, moreover, as the fuel inventory builds up.[16] This is contrary to a widespread impression that military wastes constitute the larger part of the nation's nuclear waste disposal problem. Military waste does represent a large problem, but this is principally because of the great bulk and form of the waste which resulted from the processing and storage methods used, especially in the early years. For both spent fuel and military high-level waste, geologic disposal appears to offer the best means of containment that is both technically and politically achievable.

Low-level waste.   Although less threatening than high-level waste or spent fuel, low-level waste constitutes a broad and important category which presents its own special containment problems. Low-level waste is created by reactor operators in the process of keeping the reactor clean, reducing corrosion, and preventing radiation from building up to levels that make system maintenance difficult or impossible. The waste collected as a part of this process runs to considerable volume and assumes several forms. Some of it is potently radioactive, and all of it must be disposed of. The challenge is to make sure that containment at the off-site disposal facility is secure and that the continuous cleanup which is necessary at the reactor does not create problems someplace else.

The source of the low-level waste from reactor operations is the reactor's core and cooling system. Water circulating at high temperature through the cooling system causes a slight but incessant corrosion of all exposed metal surfaces of the system's components—that is, of the reactor vessel and its internals (such as the control rod mechanism and the stainless steel structure that holds the fuel assemblies in place), the assemblies themselves, and the extensive piping that makes up the cooling circuit. The resulting corrosion products are activated within the core by the intense flux of neutrons, producing cobalt-60 and other activation products that would steadily build up in the coolant flow if not removed.

[16]Oak Ridge National Laboratory, *Spent Fuel and Radioactive Waste Inventories*, DOE/NE-0017/2, tables 1.14, 2.4. Activity in the military high-level waste at the end of 1982 totaled 1.3 billion curies, compared to about 11 billion for the commercial spent fuel. The disparity is less but still substantial if only the inventories of cesium and strontium, which account for the bulk of the longer-lived fission products, are compared. These radionuclides accounted for 0.832 billion curies in the military waste and 1.55 billion in the commercial spent fuel. See ibid., tables 1.16, 2.8, 2.9, and 2.10.

Some fission products also are picked up by the cooling water, in two ways. First, perhaps one-tenth of 1 percent of the tens of thousands of fuel rods in the core have pinhole-size leaks that allow some of the more volatile fission products, especially cesium-137, to escape. Second, fissioning takes place in the trace amounts of "tramp uranium" unavoidably left on the surface of fuel rods during their manufacture, and this too creates fission products.

If the accumulation of fission and activation products in the cooling circuit were allowed to proceed unchecked, radioactivity would quickly increase and interfere with the routine maintenance activities vital to safe operation of the reactor. For instance, because the cooling water in a pressurized water reactor is at a pressure of more than 2,200 pounds per square inch (compared to normal atmospheric pressure of about 15 psi), periodic inspection of all welds in the cooling system piping is a necessity. Yet if high radiation fields build up around the piping and steam generator tubing, inspection and repair activities either cannot be done at all or will require a profligate use of personnel. In the early 1970s Consolidated Edison in New York, at its Indian Point-1 reactor, "used up" some 1,500 welders—that is, the workers were used to the limit of the radiation exposures allowed—to repair defective welds in the steam generator system.[17] Later in the seventies, Commonwealth Edison, at its 200-megawatt Dresden-1 plant at Morris, Illinois, found that the reactor was becoming too hot to maintain because of the progressive buildup of radioactive crud inside its some five miles of piping. The Dresden-1 decontamination project involved building, at a cost of $39 million, an elaborate, remotely operated facility for flushing out the cooling system with a strong decontaminating solution and processing all the resulting wastes.

To minimize the need for major decontamination projects a continuous, day-to-day effort at purifying the cooling water is necessary. This is done chiefly by use of demineralizer filtration units in which small beadlike resins capture (by ion exchange) soluble substances such as cesium. Other types of filters capture nonsoluble substances such as the cobalt-60 that becomes attached to iron deposits. The spent filtration media can be very radioactive, the radiation sometimes exceeding 1,000 rems per hour, a lethal dose. They are transported to low-level waste burial grounds in heavy-duty, lead-lined casks, with one such cask to a large flatbed truck.

Additional low-level waste arises from other plant maintenance activities, such as cleaning up after the leaks which are inevitable in any

---

[17]Bernard J. Verna, "Radioactive Maintenance," *Nuclear News*, September 1975, pp. 55–56.

system that has miles of piping and innumerable valves. A leaking valve, for instance, may spill hundreds of gallons of radioactive water onto the reactor building floor. All of this water, and particularly the evaporator solids that result from processing it, becomes radioactive waste, together with all the materials contaminated in cleaning it up: mops, protective clothing worn by workers, detergents, the failed valve, and so on.

A single large reactor can generate over the course of a year from 600 to 1,400 cubic meters of low-level waste, or somewhere between 100 and 200 truck shipments, the number tending to vary according to the type of reactor (boiling water reactors generate more) and in proportion to the amount of waste requiring heavily shielded casks. Waste generation increases over time as a natural consequence of plant aging and wear and tear, and more leaks and equipment failures.

The total generation of low-level reactor waste is expected to approach 4 million cubic meters by the year 2000.[18] About three-fourths of this waste is still to be generated; this would be enough to fill four or five commercial burial grounds of the kind currently in use. They consist of a series of trenches, typically about 50 feet wide and 30 feet deep, in which the drums and other containers are placed and covered over. Six such commercial burial grounds have been established in the past, but today only three are operating. Of the three that have been closed, two especially presented major containment problems, as will be described in the next chapter.

Although sanctioned by the Nuclear Regulatory Commission the entire concept of shallow land burial is now in disfavor among the states and there is little likelihood of another burial facility similar to the existing ones being established again. The trend in the United States is toward surface disposal facilities such as concrete bunkers, which rely on engineered barriers. But some foreign countries have gone further In Germany the policy always has been to commit low-level as well as high-level waste to deep geologic disposal. Sweden also practices geologic disposal, though not at great depth for low-level waste. A promise of robust containment will almost surely be necessary for success in implementing the U.S. Low-Level Radioactive Waste Policy Act. This law, first enacted in 1980 and amended in late 1985, contemplates establishment of several additional regional disposal sites, with the states taking the initiative through regional compacts. Initially, the aim was to establish the new sites by the end of 1985. Now the aim is to have them by the end of 1995.

[18]Oak Ridge National Laboratory, *Spent Fuel and Radioactive Waste Inventories*, DOE/NE-0017/2, table 4.1.

Whatever the disposal methods employed, some releases to the environment are inevitable. All radioactivity cannot be contained in reactor operations. For example, there is tritium (or hydrogen-3), a beta-emitter with a 12-year half-life. Tritium is discharged as a highly dilute gaseous or aqueous waste. These releases are tolerated by scientific bodies such as the International Commission on Radiological Protection, and hence by industry regulators as well. There are several reasons for this: the amount of tritium released typically is slight; tritium is rapidly diluted in the environment by stable hydrogen; and its residence time in the human body is only a week to ten days. But the reason tritium is not captured is that it is chemically indistinguishable from the stable hydrogen in water and thus is impossible to remove at reasonable cost from reactor coolant.

Decommissioning waste. The third kind of radioactive waste generated by reactor operations is decommissioning waste. Originally it was believed that nearly all of the radioactivity in a decommissioned reactor would decay during the first 50 to 100 years. Cobalt-60 was seen as the major problem, and inasmuch as its half-life is only about 5 years, a 50-year deferral would reduce its radioactivity a thousandfold, and a 100-year deferral would reduce it a millionfold. But as more was learned about the various activation products created during the fission process, it became apparent that reactors, or at least certain parts within the reactor vessel, would remain dangerously contaminated for many millennia. In early 1976 Marvin Resnikoff, a scientist with the New York Public Interest Research Group at the State University of New York at Buffalo, together with four undergraduate engineering students, made calculations which showed that nickel-59 was going to be a problem. This activation product, with a half-life of 80,000 years, would be present in significant amounts at the end of the 40-year life normally expected of a large commercial reactor. Then in 1978 a physics professor at Cornell University, Robert Pohl, with one of his students, brought to light the still more serious problem of niobium-94, a potent gamma-emitter with a 20,000-year half-life. Like nickel-59, the niobium would be induced by the intense neutron flux within the reactor core, affecting particularly the shroud and other stainless steel internals.[19]

Present thinking of the Nuclear Regulatory Commission is that these contaminated materials should eventually go to a geologic repository. The cutting up and removal of the core shroud and other internals

[19]Marvin Resnikoff and coauthors, "The Cost of Turning It Off," *Environment*, December 1976; Colin Norman, "Isotopes the Nuclear Industry Overlooked," *Science*, January 22, 1982, p. 377.

would have to be done with a cutting torch operated remotely under water to protect workers from the intense radiation field.[20] Decommissioning may involve, at the outset, flushing out the cooling circuit (as in the Dresden decontamination) to remove radioactive crud and reduce radiation fields within the plant.[21] Washing and polishing equipment, and the scrubbing, chipping (with jackhammer or flame spaller), and sandblasting of walls and floors is expected to remove the surficial radioactivity.

A possibility once considered for reactor decommissioning was entombment. Reactors would be decontaminated to one degree or another and then entombed or sealed up. But numerous huge, partly radioactive hulks would be left for future generations to monitor and maintain to make sure that no radioactivity escaped. The other possibility, and the only one now considered an option in the United States, is to dismantle all decommissioned reactors, either immediately or after a "safe store" period of up to 50 years to permit most of the short-lived radioactivity to decay.[22]

Early dismantlement can produce enough waste to become a dominant part of the low-level waste problem and significantly increase the number of disposal sites needed. Indeed, immediate dismantlement of a large reactor would generate 18,000 cubic meters of waste (equivalent to 86,500 fifty-five-gallon drums), or as much as 1,500 truck shipments, equal in volume (though not in radioactivity) to about a third of all the low-level waste generated annually by nuclear power facilities in the United States at the start of the 1980s[23]—or about half the amount of waste the State of South Carolina allows the commercial burial facility at Barnwell to receive annually. The amount of material with high levels of induced activity, destined for geologic disposal, could require more than 150 truck shipments.[24] The cost of immediate dismantlement has

[20]Nuclear Regulatory Commission, *Technology, Safety and Costs of Decommissioning a Reference Boiling Water Reactor Power Station*, NUREG/CR-0672, vol. 2 (Washington, D.C., 1978) pp. G-5 to G-18.

[21]Ibid., appendix on decommissioning methods, pp. G-1 to G-43.

[22]Interview by the author with Carl Feldman of the Nuclear Regulatory Commission staff, March 9, 1984. Feldman is responsible for preparation of decommissioning regulations.

[23]Nuclear Regulatory Commission, *Technology, Safety and Costs of Decommissioning*, NUREG/CR-0672, vol. 2, table I.3-2, p. I-33; and NUREG/CR-0130, a companion study on the pressurized water reactor, vol. 2, table H.5-1, p. H-31. The radioactive waste volumes estimated are 19,000 cubic meters for 1,155-megawatt boiling water reactors and 17,900 cubic meters for 1,175-megawatt pressurized water reactors. Oak Ridge National Laboratory, *Spent Fuel and Radioactive Waste Inventories and Projections*, DOE/NE-0017 (Washington, D.C., 1981) tables 6.6, 6.7.

[24]Nuclear Regulatory Commission, *Technology, Safety and Costs of Decommissioning*, NUREG/CR-0672, vol. 2, table I.3-3, p. I-35.

been estimated to be on the order of $100 million per reactor,[25] and this may be far too low. Bertrand Barré, French nuclear attaché in Washington from 1980 until mid 1984, sees a "vast margin of uncertainty" in decommissioning costs and puts the "upper boundary" at a sum equivalent to the cost of building the reactors.[26] By deferring dismantlement for 50 years, the amount of radioactive material requiring burial as low-level waste would be reduced by perhaps as much as 90 percent; but the savings here would be offset in part by the need to guard and maintain the facility and carry out environmental monitoring.[27]

Three Mile Island: The Cleanup and the Wastes

The accident at Three Mile Island (TMI) illustrates dramatically how high the cost can be of even a limited failure of containment which causes no fatalities and no documentable physical health effects of any kind. Aside from this accident's political cost to the nuclear industry, the imperative to contain radioactivity would be manifest just from the massive and extraordinarily complex clean-up and waste management burdens that have been imposed. What essentially happened at Three Mile Island was a partial failure of the first several barriers in the containment system. Although not catastrophic in a public safety sense, the failure was sufficient to transform the inside of the containment building into an enormous and grossly contaminated hot cell and to drive the owner of the reactor, General Public Utilities (GPU), to the verge of bankruptcy.

This loss-of-coolant accident, resulting from a combination of equipment malfunctions and human error, led to the partial melting of the cladding of most of the fuel rods and to the melting of some of the uranium oxide pellets. Gaseous fission products such as xenon and

[25]International Atomic Energy Agency, *The Decommissioning of Nuclear Plants* (Vienna, Austria, December 1979) p. 12. See also Atomic Industrial Forum, "Analysis of Nuclear Power Reactor Decommissioning Costs," National Environmental Studies Project report, AIF/NESP 021SR (Washington, D.C., May 1981) table A, p. 1 of the summary report. Cost estimates given in Nuclear Regulatory Commission, "Decommissioning of Nuclear Facilities," NUREG-0586 (Washington, D.C., January 1981) table 0.0-2, p. 0-45, are much lower but are expressed in constant 1978 dollars.

[26]In a letter of September 9, 1985, to the author, Barré also said, "But *even were it to cost as much as the construction itself,* which I consider to be the upper boundary, it would not be enough to seriously erode the economic edge of nuclear v. coal fired electricity in the French economic context" (emphasis in the original).

[27]Nuclear Regulatory Commission, "Decommissioning of Nuclear Facilities," NUREG-0586, table 0.0-2, p. 0-45; see also ibid., para. 0.2.4.3 on p. 0-5.

krypton escaped from the failed rods, then from the reactor vessel through a stuck pressurizer relief valve. Some escaped from the containment building to the outside environment, though not in high concentrations. Pellets from the heavily damaged upper part of the fuel core, or the part left uncovered when the coolant level dropped, fell in a rubble to lower regions inside the core and to the bottom of the reactor vessel. With some of the pellets melted or cracked and crumbling, certain radionuclides, including cesium and strontium and even some of the plutonium, escaped into the coolant, either in solution or as particulates. In this way water in the primary coolant system became highly contaminated. Again, because of the malfunctioning relief valve, hundreds of thousands of gallons of water escaped from the coolant circuit and flowed into the containment building sump, carrying about a half million curies of radioactivity.

The upshot was that all surfaces and all equipment within the reactor building became contaminated by the radioactive water, steam, and gases that escaped from the fuel. An accident that began as a scary reactor safety emergency became an enormously complex and costly exercise in reactor decontamination and waste disposal which still goes on today, eight years after the accident. The cleanup is expected to go on at least until the end of 1988, with the total costs running to a billion dollars. To this must be added the other billions in costs to which the accident will have given rise, particularly from the purchase of about $3 billion in replacement power from neighboring utilities,[28] and from the loss of the TMI-2 reactor if it cannot be returned to service.

There is much about the Three Mile Island cleanup besides the huge cost to astonish the layman. For instance, although the work force at any given time is usually about 1,000, some 13,000 workers were involved altogether from the start of the cleanup through the end of 1985, and about 3,000 had experienced measurable radiation exposure by working in contaminated areas.[29] Workers in numerous different and changing skills are required as the cleanup progresses. Moreover, as was the case on the Indian Point welding job, when radiation fields are high and potential exposures large the collective dose must be spread over numerous workers to keep individual doses below the prescribed annual ceilings. A radiological control staff of 100 professionals and technicians, with an annual budget of $5 million, oversees all aspects of the cleanup.

---

[28]Not only has the stricken TMI-2 reactor been out of service, the undamaged TMI-1 reactor was not allowed to operate again until the late fall of 1985. Replacement power cost has been running $14 million a month per reactor, according to General Public Utilities spokesman Gordon W. Tomb. Interview by the author with Gordon W. Tomb, April 11, 1986.

[29]Interview with Tomb, April 11, 1986.

The highest individual doses from the beginning of the cleanup through 1985 were less than 4 rems a year; a 5-rem maximum annual dose is normally allowed for workers.[30]

Most of the solid waste from Three Mile Island will be material contaminated in the cleanup itself, and much of it will be clothing and other equipment used to protect workers from radiation. When the cleanup was getting under way in 1979, the Bechtel Power Corporation, a contractor for General Public Utilities, estimated that 1 million pairs of plastic coveralls would be needed, as would a like number of pairs of plastic boot coverings and rubber gloves, not to mention other items such as 12,000 square feet of lead sheeting and a million square feet of plastic. The cleanup has produced some very hot wastes from filtration of the sump water; some of these wastes have been shipped to a Department of Energy laboratory at Hanford, Washington, where they are to be vitrified, or made into glass logs, for eventual geologic disposal. Low-level waste from the cleanup also is going to Hanford for shallow land burial.

The most delicate phase of the cleanup is under way now: recovery of the damaged fuel, much of it lying in a rubble at the bottom of the reactor vessel. This highly radioactive debris is being loaded under water into defueling canisters, which in turn are being inserted into massively shielded casks and shipped by rail to DOE's National Engineering Laboratory in Idaho for preparation for geologic disposal when a repository becomes available. Another challenge of the cleanup is to decontaminate all surfaces within the containment building. Workers must combat problems of recontamination, as when radioactivity removed from the walls with jets of water contaminates other surfaces within the building before all of it can be captured. All told, the cleanup is expected to generate no fewer than 600 truckloads of waste, and perhaps two to three times that amount.[31]

Some of the waste from the cleanup is expected to be contaminated with enough plutonium to be considered transuranic waste that requires geologic disposal, but if all goes as expected there will not be more than about 45 cubic meters of it,[32] or enough to fill a few hundred drums. Most of this plutonium is believed to be in the sludge at the bottom of the containment building sump. The reactor vessel and primary cooling system may also be contaminated with plutonium, but a decontamination of the cooling circuit is expected to ensure that the levels will be less

[30]Ibid.

[31]Nuclear Regulatory Commission, "Environmental Impact Statement on Three Mile Island Station, Unit 2, decontamination," NUREG-0683 (July 1980) table S-1, p. 5-9.

[32]Interview by the author with Bernard Snyder, May 8, 1983. Snyder is the Nuclear Regulatory Commission official responsible for overseeing the Three Mile Island cleanup.

than the threshold of 100 nanocuries (or billionths of a curie) per gram which the Nuclear Regulatory Commission has established for transuranic waste.[33] This is fortunate, for if the steam generators and all the other bulky equipment in the primary cooling circuit had to be cut up, packaged, and disposed of as transuranic waste, the cost and difficulty of the cleanup would be increased enormously.

In sum, the Three Mile Island accident represents a major containment failure that has caused a massive waste problem and cleanup—a cleanup in which thousands of radiation workers are being meticulously rotated in and out of the reactor building's hostile environment to keep individual doses at acceptable levels. Had the uncovering of the core been more complete and prolonged, with more extensive melting of the fuel pellets, the clean-up problem could have been much worse. Three Mile Island represents a technical and political problem of a kind that is unique to nuclear power. Nothing remotely comparable could have occurred with a fossil-fuel-fired electric generating station. Recovery from the worst imaginable accident at such a station might take only a few months, and the accident would probably soon be forgotten even if there were some loss of life; recovery from Three Mile Island will not be complete in a physical sense for another several years, and in a political sense the recovery will take much longer than that.

Containment, Wastes, and Fuel Cycle Choice

Reprocessing and plutonium recycling make containment significantly more difficult, even though in principle the problem remains manageable. As illustrated by both what happened and what did not happen at Three Mile Island, the zirconium fuel rods and the ceramic fuel pellets constitute the first two barriers to the release of radioactivity. In fuel reprocessing these two barriers are deliberately eliminated and the fission products and transuranic elements are mobilized. They are, to be sure, supposed to be contained within a carefully controlled chemical process for separating the fission products and other wastes from the uranium and plutonium.

For many people in the nuclear industry the idea of choosing a once-through fuel cycle in preference to a fuel cycle providing for reprocessing and plutonium recycling represents a disturbing asymmetry. For them, not to "close the fuel cycle" by recycling is to ignore the energy value of the plutonium and uranium left in the spent fuel. But, as has become increasingly evident in recent years, the mere fact that pluto-

[33]Ibid.

nium is a fissile material does not in and of itself mean that it is worth recovering. Whether it is worth recovering turns in part on uranium supply and demand. Beyond that, it turns on how reprocessing and recycling bear on the nuclear imperatives to contain radioactivity and safeguard plutonium and on the overall complexity and political acceptability of fission energy technology. The discussion that follows focuses on how reprocessing bears on waste management and containment.

Reprocessing and Containment

In terms of the chemistry involved, fuel reprocessing is not especially complex. Yet chemical engineers who are familiar with reprocessing acknowledge that this is a far more difficult technology than conventional chemical industry processes. In a large commercial-scale plant several tons of highly radioactive fuel may be received daily. Processing of this fuel involves a mix of uneasily compatible conditions and functions: hot cells in which the environment is intensely radioactive and corrosive and in which all operations must be performed remotely; machines that have moving parts (which wear out) and must be maintained remotely as well as operated remotely; and a veritable spaghetti of piping, literally miles of it, with the ever-present possibility that a pipe will become plugged or start to leak. Despite these daunting operating conditions the process must be kept running smoothly to achieve a steady product throughput, while at the same time not allowing the workers' radiation exposure to exceed dose limits.

Spent fuel enters the plant's "front end" and is fed into a shearing device that cuts up the fuel rods into small pieces. Next, the chopped-up fuel elements, with the irradiated fuel pellets now exposed, go into an acid bath; there the fuel pellets are dissolved, but the bits and pieces of cladding are not. In the shearing and dissolution of the fuel two waste streams arise. One of these consists of the gaseous fission and activation products, most notably krypton-85, xenon, carbon-14, tritium, and radioactive iodine. At reprocessing plants in the United States and abroad most of the iodine has been captured, but the rest of the gaseous waste stream has been routinely released in dilute concentrations to the environment. The other waste stream arising at this stage consists of the chopped-up fuel cladding hulls which are radiologically hot from activation products and from the tiny bits of undissolved fuel clinging to them. These hulls eventually must go to a geologic repository as transuranic (TRU) waste.

After the fuel chopping and dissolution, the dissolved materials are run through a series of steps in which the uranium and plutonium are separated chemically from the waste products. Because this chemical separation process is not completely efficient, not less than 0.5 percent of the plutonium also is lost to the high-level waste stream, which contains about 99 percent of the activity that was present in the spent fuel. The high-level waste goes into a relatively corrosion-resistant stainless steel storage tank to await solidification and shipment to a geologic repository.

Besides the high-level waste, the cladding hulls, and the gaseous wastes, reprocessing plants generate substantial volumes of low-level aqueous waste and general process trash, with much of the latter known or suspected to be transuranic waste contaminated with plutonium and other actinides.[34] The trash, consisting of things such as spent filters, failed equipment, laboratory wastes, and used protective clothing, is packaged in drums for shipment to a shallow land burial facility and, in the case of the material regarded as transuranic waste, for storage and eventual commitment to a geologic repository.

Just how great the loss of plutonium to the transuranic waste streams will be is disputed, but according to J. O. Blomeke, a respected fuel cycle expert at the Oak Ridge National Laboratory, past losses have been put at 1.5 percent of the plutonium in the reprocessed fuel. Counting the 0.5 percent lost to the high-level waste, this brings the total reprocessing losses to 2 percent. But to this must be added the losses that occur when the plutonium that is recovered is fabricated into plutonium fuel (a mixed oxide of plutonium and uranium) for use in light water reactors or breeders. The fabrication process involves much sintering, grinding, and handling of plutonium oxide in the fashioning of the fuel pellets. Blomeke estimates that another 2 percent of the plutonium could go into the solid wastes generated at this stage of the fuel cycle, for a total of 4 percent for all reprocessing and recycling operations. Blomeke bases this reckoning of fractional losses not on industry flow sheets, which typically show total losses of less than 1.5 percent, but on actual experience in commercial reprocessing of light water reactor fuel in the United States and abroad. Commercial experience in the recycling of breeder reactor fuel is lacking, and Blomeke sees hopeful possibilities in industry efforts to greatly improve plutonium recovery.[35] In principle, major improvements are possible. But if, as a practical matter, the same fractional loss of 4 percent should continue, the term "breeder" will turn out to have been something of a misnomer, as will be explained below.

[34]The actinides, beginning with actinium, together occupy one position in the periodic table of elements from number 89 through number 105. They include uranium and all the man-made transuranic elements.

[35]Blomeke was interviewed by the author periodically throughout 1981 and 1982.

According to the classical concept, reprocessing makes waste management easier, and some people in the nuclear industry still believe this to be true. But the only incontrovertible advantage for waste management offered by reprocessing and the breeder cycle would be to reduce the need for uranium mining and milling and hence also to reduce the generation of mill tailings. The uranium "feed" for a breeder cycle could come largely from the enormous inventory of depleted uranium (that is, uranium from which most of the fissionable U-235 has been extracted) left from the many years of uranium enrichment at American and foreign gaseous diffusion plants. The greatest advantage in waste management usually claimed for reprocessing is that the high-level waste destined for geologic disposal would contain only a fraction of 1 percent of the plutonium that was in the spent fuel. But as indicated above, one must not naively assume that the other 99.5 percent will be recycled and burned up in reactors. Several things complicate the situation. The total plutonium loss in reprocessing and recycling, possibly as high as 4 percent, is compounded at each successive recycling; for instance, from a given ton of plutonium, after six recyclings about a fifth of it, or 200 kilograms, might be lost to the waste streams. Furthermore, the much greater neutron bombardment to which the plutonium is exposed in the course of repeated recyclings means that more "transplutonic" elements are created, particularly curium and americium.[36]

As a result, the radioactivity of the alpha-emitters present in the accumulated wastes from a breeder cycle is only four times less than that present in spent fuel during the period in which transuranic elements are dominant in the toxicity of the waste.[37] A reduction of this magnitude would not make a major qualitative change in the waste isolation problem. Furthermore, borosilicate glass, the matrix which the United States and other nations have chosen for solidification of high-level waste, offers little or no advantage over spent fuel in terms of leach resistance and radionuclide retention.[38] On the other hand, either glass or the uranium oxide of the spent fuel makes an adequate waste form if

[36]Remarkably, the americium-243 isotope decays with a half-life of 7,300 years back to plutonium-239. Transplutonic elements are those above plutonium in the periodic table—those having an atomic number greater than 94.

[37]Hartmut Krugman and Frank von Hippel, "Radioactive Waste: The Problem of Plutonium," *Science*, October 17, 1980, pp. 319–320.

[38]Interview by the author on March 14, 1983, with Rustum Roy of the Materials Research Laboratory, Pennsylvania State University, and formerly chairman of a National Academy of Sciences panel on high-level waste solidification. Y. B. Katayama, chemical engineer at Pacific Northwest Laboratories (where work on borosilicate glass has been carried on for many years), concurs in the view that glass offers little advantage over spent fuel in terms of resistance to leaching and of radionuclide retention. Katayama was interviewed by the author on January 30, 1986.

the waste isolation function can be left principally to the geologic formation, or, to hedge against geologic uncertainty, to the formation in combination with the kind of highly durable waste canister now contemplated in Sweden (see chapter 9).

Reprocessing's negative consequences for waste management lie principally in the multiple streams of waste that are created. In the spent fuel some 95 percent of the heavy metal is accounted for by the uranium,[39] and its removal in reprocessing makes the packaged high-level waste only about one-fifth as voluminous as the spent fuel; that is, for every gigawatt year of nuclear generation there would be some 73 canisters of spent fuel, compared to only 15 canisters of high-level waste from the reprocessing of breeder fuel. But this reduction in volume would be far more than offset by the large amounts of transuranic and other wastes that would be created in the breeder cycle, much of it sufficiently contaminated with gamma-emitters to require shielding and remote handling. The radiologically hot cladding hulls from the chopped-up fuel elements would fill some 75 canisters. In addition, there would be about 266 cubic meters of other transuranic waste, enough to fill nearly 1,280 fifty-five-gallon drums or other packages of equivalent size.[40] All of the transuranic waste, collectively containing up to three times more long-lived alpha activity than the high-level waste, must go to a geologic repository. Yet to contain the radioactivity in this voluminous and chemically heterogeneous material according to the standard required for high-level waste—no release from the engineered barrier system for the first 300 to 1,000 years, and a release of only 1 part in 100,000 per year thereafter—could be expensive.[41] Besides the streams of transuranic wastes, there would be other waste streams associated with reprocessing and recycling for breeders. These would include 427 cubic meters (or the equivalent of 2,050 drums) of low-level waste per gigawatt year and, if regulations should require, several high-activity packages of concentrated fission products to be captured from the off-gases rather than released up the stack. The latter might include the noble gases xenon

[39]In industry parlance, "heavy metal" refers to uranium, thorium, and plutonium, heavy elements high in the periodic table.

[40]Interviews by the author with J. O. Blomeke.

[41]Nuclear Regulatory Commission, 10 CFR, pt. 60, para. 60.113, in *Federal Register* vol. 48, no. 120 (June 21, 1983) p. 28224. To meet the containment standard—none has yet been promulgated for transuranic waste—might require (among other things) high-temperature incineration to reduce all of the transuranic waste to a lava-like material or slag. A slagging facility was once proposed for the military transuranic waste stored and buried at the Idaho National Engineering Laboratory; its cost was put at $555 million (in 1980 dollars). Department of Energy, "Conceptual Design Summary of the Transuranic Waste Treatment Facility," EYG-TF-5284 (prepared by T. D. Tait EG&G Idaho, Inc., November 1980).

and krypton, carbon-14, and tritium. Capture of krypton-85 would be required even under current Environmental Protection Agency (EPA) regulations.

In addition to all the above, another substantial waste problem arises from reprocessing and plutonium recycling. It is that the reprocessing and mixed oxide fuel fabrication plants must eventually be decommissioned and dismantled, and much of the resulting waste will be transuranic and require geologic disposal.[42] Less decommissioning waste would be generated than in the case of a reactor, but the cost of dismantlement might actually be greater because of the difficulties in working with facilities and equipment contaminated with alpha-emitters, including plutonium occurring as respirable dust particles—the most dangerous form plutonium can take.

Plutonium Losses and Doubling Time

Plutonium losses to the waste streams are important to what in industry parlance is known as breeder "doubling time." Doubling time has to do with the time it takes for a breeder reactor to breed enough plutonium to meet not only its own need for fissile material but to provide what is needed to start up a second breeder. The first breeders are fueled with the plutonium produced in light water reactors. But once a number of breeders have been started up they will constitute, or so goes the breeder concept, an essentially self-sustaining and self-generating "breeder economy." Once this stage were reached additional breeders could be started up without further dependence on either light water reactors or on the availability of fresh supplies of uranium. All the uranium needed would come from the huge existing stocks of depleted uranium, stocks large enough to meet even a large breeder program's requirements for fertile isotopes for centuries.

Three variables figure in the doubling time: first, the breeding ratio; second, the time out of cycle, or the time lost from productive breeding to such necessary steps as cooling, transporting, and reprocessing irradiated fuel and fabricating fresh fuel; and third, the percentage of plutonium lost to the waste streams in reprocessing and fuel fabrication. If the reactor produces only enough plutonium to meet its own needs, it is not a true breeder and might better be called a converter. A breeder, by definition, will have a compound breeding ratio of anywhere from,

---

[42]Nuclear Regulatory Commission, *Technology, Safety, and Costs of Decommissioning a Reference Nuclear Fuel Reprocessing Plant*, NUREG-0278 (Washington, D.C., October 1977) vol. 1, pp. 7-34 to 7-39, and vol. 2, pp. E-37 to E-44 and E-67 to E-73.

say, 1.12 to 1.3. But breeding and power generation are conflicting aims: if the aim is to generate power, the longer the fuel is left in the reactor, the more plutonium is fissioned and the more energy is released; but if the aim is to optimize breeding, the sooner the fuel is removed from the reactor, the more plutonium will be recovered. As for time out of cycle, it has not been many years since the value of the plutonium was assumed to be so great that rapid recycling would be an economic necessity. The shipping of irradiated breeder fuel from the reactor to the reprocessing plant after as little as thirty days' cooling was contemplated. A single one-and-a-half-ton shipment of such fuel would contain 75 million curies of radioactivity and generate 300 kilowatts of decay heat, posing a formidable challenge to those responsible for the containment of that activity and the health and safety of the public.[43] Today, the likely prospect is for several years of out-of-cycle time, to allow more time for cooling and other essential steps. As noted above, the percentage of plutonium lost to waste streams, the third variable, could be as much as 4 percent instead of the 1.5 percent postulated in industry flow sheets unless major improvements are made over past performance.

In 1983, General Electric's advanced reactor systems department at Sunnyvale, California, made computer analyses of compound doubling times for a large breeder, using various assumptions as to breeding ratio, out-of-cycle time, and plutonium losses.[44] The results were striking. Whereas in the mid 1960s the industry was thinking of doubling times of 8 or 10 years,[45] the present reality is that even to look for a doubling time of 25 to 30 years is highly optimistic.

The General Electric analyses indicated that with a breeding ratio optimized for electricity generation—GE used a ratio of 1.12 to 1—the doubling time will be 47 years even if out-of-cycle time is held to 1 year and plutonium losses do not exceed 1 percent. If these latter assumptions are increased to 2 years and 2 percent (only half what the actual plutonium losses might be), doubling time is extended to 90 years. Breeding ratios of as much as 1.35 to 1 are believed to be achievable, but at such a large sacrifice with respect to electricity generation that the only practical choice might be a lower ratio, say of 1.23 to 1 (used

[43]Alvin M. Weinberg, "Social Institutions and Nuclear Energy," *Science*, July 7, 1972, p. 37.

[44]These analyses were done under a Department of Energy contract by a core design group headed by Ron Murata, to whom the author was referred by the department's breeder program office. The type of reactor postulated in the analyses was a liquid metal fast breeder of 1,350-megawatt capacity.

[45]Atomic Energy Commission, *Civilian Nuclear Power* (Washington, D.C., February 1967) p. 32.

by General Electric in one analysis), despite the desire to give breeding greater emphasis. With this compromise ratio, a doubling time of 35 years could be expected, but only if out-of-cycle time were held to 2 years and plutonium losses to 2 percent. The doubling time is 52 years if these assumptions are changed to 4 years and 4 percent. A rapidly expanding breeder program might turn out to be technically achievable only if fissionable uranium, instead of plutonium, is used to start up the new breeders, or "converters."

Permanent, or Geologic, Isolation

The isolation or disposal of spent fuel or high-level waste in mined geologic repositories is expected to be the final step in efforts to contain radioactivity throughout the nuclear fuel cycle, whatever the particular cycle chosen might be. Technically, this should be more easily accomplished than containment of activity in reactor operations and in fuel reprocessing and recycling. Geologic disposal need not take place until the radioactivity and heat from the short-lived fission products have been much reduced by radioactive decay. Also, the disposal, or isolation, would be deep beneath the earth rather than at the surface where reactor and reprocessing operations must be carried out. Geologic disposal does require effective isolation for the long period of hazard, but this is a challenge that can be met.

No spent fuel or high-level waste from reprocessing would be committed to a repository until at least ten years beyond the fission process, and perhaps not until fifteen years or longer if the sensible rule that the older wastes go first is followed. Even with ten years' aging the radioactivity will decline by several hundredfold from what it was upon first being removed from the reactor. Similarly, the decay heat—the energy source that makes even a reactor that has been "scrammed," or shut down, susceptible to containment failures from a loss-of-coolant accident—will decline by more than a thousandfold. In geologic disposal operations, the heat energy necessary for violent disruptive events simply would not be present, although in some circumstances the heat might be sufficient to gradually compromise the geologic containment. Moreover, containment of radioactivity in a repository deep inside a geologic formation will always be more certain and secure than containment at the earth's surface. It should be noted that all of the various hypotheses as to the cause of the mysterious radiological disaster of some thirty years ago at Kyshtym, the Soviet Union's plutonium production center in the southern Urals, postulate the sudden or gradual release of radioactivity from plants or waste storage facilities on the earth's surface.[46]

The challenge of geologic isolation lies not in preventing sudden catastrophic containment failures, for failures of that kind belong in the realm of such events as a direct hit by a large meteorite, for which the probability is vanishingly small. Rather, the challenge is to delay and limit the release of radioactivity as necessary to keep radiation doses to individuals and populations from exceeding tolerable levels. The delay allows for radioactive decay and reduces the potential for releases in harmful concentrations.

The standards and regulations for geologic isolation of nuclear wastes prescribed by the Environmental Protection Agency define the limits, the periods of delay, and the dosage levels that the EPA deems tolerable.[47] The overall population dose attributable to escaping radioactivity is not supposed to cause more than 1,000 fatalities over the 10,000 years following closure of the repository, which would hold the equivalent of 100,000 tons of spent uranium fuel (or the amount discharged by 100 large reactors over more than 30 years). Individual doses during this period are not to exceed 25 millirems a year, or about a quarter of the natural background radiation and less than the deviation in natural background between mountains and seashore. To this end, even after the first 1,000 years the release rate of the radionuclide inventory from the engineered repository system is not (with some exceptions) to exceed 1 part in 100,000 per year. Groundwater travel time to the "accessible environment" (either the atmosphere or the ground and surface waters beyond a 100-square-kilometer control area) shall not as a general rule be less than 1,000 years. If an important regional aquifer or other special source of groundwater lies within or near the control area, the EPA drinking water standard of 4 millirems a year as the maximum allowable dose would apply. Thus, to reiterate, geologic isolation as presently defined has the aim of delaying releases long enough and keeping them low enough to prevent radionuclides from reaching humans by way of well or surface water in concentrations sufficient to harm more than a very few people in any one generation. In principle, even a more stringent standard, aiming for absolute containment over the period of hazard, is achievable by relying more heavily on engineered barriers.

[46]John R. Trabalka, L. Dean Eyman, and Stanley I. Auerbach, "Analysis of the 1957–1958 Soviet Nuclear Accident," *Science*, July 18, 1980; Diane M. Soran and Danny B. Stillman, "An Analysis of the Alleged Kyshtym Disaster" (Los Alamos National Laboratory, January 1982). These reports allow for the possibility of an explosion of ammonium nitrate—a chemical agent the Soviets once used in reprocessing—as one dispersal mechanism.

[47]Environmental Protection Agency, "Environmental Standards for the Management and Disposal of Spent Nuclear Fuel, High-Level and Transuranic Radioactive Wastes; Final Rule," 40 CFR, pt. 191, *Federal Register* vol. 50, no. 182 (September 19, 1985) pp. 38066–089.

The Worst Case

What a credible worst case is in the event a repository fails to perform up to expectations is a matter of considerable disagreement. But leaving aside scenarios involving direct human intrusion into a repository (such as mining salt from a salt dome that contains a repository), the worst case will have to do with the contamination of local or regional ground and surface waters beyond prescribed limits. If one asks how far beyond, the answers fall along a wide spectrum of opinion. The question may even be disputed within the same family of experts, so to speak.

Take the case of two reports from the National Academy of Sciences which came to strikingly different conclusions as to the doses to which people might be exposed. The academy's Committee on Nuclear and Alternative Energy Systems (CONAES), in its 1979 report,[48] cited a worst-case analysis done by the Department of Energy's predecessor, the Energy Research and Development Administration (ERDA). This analysis assumed that all of the shallow-buried military high-level waste tanks at the Savannah River plant in South Carolina were abandoned and left to corrode, with the waste allowed to leak out and eventually contaminate ground and surface waters within a region inhabited by 70,000 people. In these circumstances, the ERDA analysts estimated, the average individual's radiation exposure would be about twice natural background. As a consequence, whereas the normal incidence of cancer for this population would be about 10,000 cases over a 70-year life-span, the ERDA analysis indicated that there would now be an extra 100 cases— certainly an undesirable but not a catastrophic result for any one generation. (The number of cancer deaths and other health effects would be much larger if the consequences of the containment failure were projected over many generations.) CONAES took this as a generous upper bound for the hazard associated with deep geologic disposal, which certainly would offer greater waste isolation than that offered by shallow-buried tanks.[49] Although CONAES made no mention of it, the

---

[48]Committee on Nuclear and Alternative Energy Systems, National Academy of Sciences/ National Research Council, *Energy in Transition* (Washington, D.C., 1979) pp. 372–382.

[49]Two scientists with very different backgrounds and points of view also essentially accept the above as a worst case. One of them is John H. Harley, a member of the National Council on Radiation Protection and Measurement and chairman of its Scientific Committee on Technically Enhanced Radiation. The other is Thomas B. Cochran of the Natural Resources Defense Council, a nuclear physicist and long-time critic of nuclear energy. Cochran takes exception to this worst case only in that the cancer risk coefficient he would use would be somewhat higher, putting the total fatalities at about 500 instead of 100. Harley and Cochran to author, personal communications, May 1986.

One scientist who does not accept the above case as the worst that can happen is John M. Matuszek, director of the New York Department of Health's Radiological Sciences

worst consequence of failure in this scenario might come from the fear inspired within the region once contamination became known. Most cancers and birth defects might be attributed to increased radiation exposure. There might also be full or partial denial of use of the contaminated water and land resources.

Another National Academy of Sciences group, the Waste Isolation Systems Panel (WISP), did not describe a worst case as such in its 1983 report, but by implication indicated that the release of radioactivity from a repository over very long periods of time could result in a public health catastrophe. As the WISP report was being prepared, the Environmental Protection Agency was proposing only aggregate population dose limits. The panel saw the absence of a dose limit for individuals as a serious omission; it also took exception to basing repository performance on health effects projected out to only 10,000 years.[50] WISP postulated conditions under which groundwater might be heavily contaminated both during and long after 10,000 years—indeed, contaminated to the point that in some instances the doses to individuals and local populations might be lethal. Some long-lived radionuclides, such as carbon-14, would remain capable of delivering very high doses from groundwater for tens of thousands of years, the panel contended.

But worst cases, even when considered credible by technically qualified specialists, stem from circumstances and conditions which repository systems can be designed against. For instance, a critical factor in the WISP analysis—the dissolution of waste forms by groundwater flow— might be deferred for tens of thousands of years by building the repository high above the water table in a dry desert area and packaging the waste in exceptionally corrosion-resistant canisters.

Uncertainty and Geologic Discontinuities

The concept of geologic repositories goes back to the mid 1950s. Field studies of salt formations as a possible disposal medium began in the 1960s, and investigations of salt and other rock types have proceeded with gathering intensity since the mid 1970s. Yet no repository site has

---

Laboratory, who served on the National Academy of Sciences' Waste Isolation Systems Panel. Matuszek observes that the radioactivity present in a repository for commercial high-level waste or spent fuel would be far greater than that in the Savannah River tanks for military wastes. He also indicates that waste escaping from shallow-buried tanks would likely be subject to dilution by surface water, in contrast with waste moving slowly in much greater radionuclide concentrations in a deep groundwater system which might be tapped by wells. Matuszek to author, personal communication, May 1986.

[50]The latter was an objection which EPA never subsequently satisfied, although its final standards did call for a "qualitative" comparison of performance projected out to 100,000 years for candidate sites.

yet been identified as clearly suitable and licensable. Why has it taken
so long? Apart from political obstacles, the answer has had to do with
the scientific and technical problem of "characterizing" potential sites
to minimize uncertainty.

An essential quality in a site is strong evidence of predictability. In
particular, this means evidence that the repository block—the mass of
rock that would contain a maze of repository tunnels extending over as
much as 2,000 acres—is fairly homogeneous throughout, having either
no major discontinuities (such as faults) or very few, with those few
able to be well identified and judged acceptable. Also, and equally
important, it means that the groundwater regime in and around the site
be well understood in terms of where the water is coming from, where
it is going, and how fast or slowly it is moving. In addition, it is important
to know the geochemical characteristics of the groundwater and host
rock, for these characteristics can have a definite, if not easily determined, effect on radionuclide retention or mobility.

There may be many sites which could be characterized as sufficiently
predictable, but finding them has not been easy. In this connection, to
compare the geology of the continental land masses with the geology
of the subseabed in the more stable regions of the deep oceans is
instructive. The geology of the subseabed, particularly in the gently
rolling abyssal hills provinces found in the middle of the tectonic plates,
can be astonishingly homogeneous. There, a clay layer up to 100 meters
thick, which has accumulated over the eons from a gentle rain of volcanic ash and other fine-grained material from the ocean's surface, overlies the basement rock of lava or basalt which was extruded in great
flows from volcanoes nearer the edges of the plates. Core samples
obtained by deep-sea drilling have revealed a remarkable predictability,
with stratigraphic markers such as ash layers showing up where expected
in sample after sample.[51] The geology of the continental land masses, on
the other hand, generally bears the evidence of a tortured history.
Discontinuities are often more the rule than the exception, having been
caused by continental uplift and subsidence, overthrusts, earthquakes
and volcanoes, glaciation and erosion, and other phenomena.

Some of the rock formations under investigation as possible repository
sites are much more complex than others, but none are quickly and
easily characterized. Harry W. Smedes, a respected field geologist formerly with the U.S. Geological Survey and now a Department of Energy
consultant, observes that to drill a dozen boreholes at random from the
surface into a rock formation will not permit characterizing that formation dependably, just as a dozen books chosen at random from the

---

[51]Interviews by the author with Charles D. Hollister of the Woods Hole Oceanographic
Institution, May and September 1982.

shelves of a library would not be representative of the entire library. Test borehole drilling can yield valuable information, but it cannot be done all at once; the data gained from the first holes must be used to help select the places where subsequent holes are drilled.

In most of the rock types investigated the drilling must be limited, for otherwise the integrity of the formation may be compromised. Each borehole drilled into the formation creates a potential pathway for the movement of groundwater and radionuclides unless it is effectively plugged and sealed. Geophysical survey techniques, such as setting off small explosive charges along a seismic line on the surface and reading the sound reflections with sensitive geophones, are used along with the test drilling. But while geophysical surveys may provide the first indication of, say, faults, brine pockets, or the pinching out of an otherwise thick rock layer, the presence or absence of such anomalies or discontinuities generally cannot be confirmed by such surveys.

Conditions at the proposed repository horizon or level in a rock formation can be further explored by sinking shafts and going down to look and conduct tests. From test chambers at the bottom of the shaft extended rock sampling can be done by means of lateral test borings reaching up to 1,000 feet in several directions. Among the tests done at the repository horizon would be those to measure interactions between the host rock and heat-generating wastes. But exploring a rock formation from deep exploratory shafts is expensive and time consuming, with each investigation costing up to a billion dollars and taking several years to carry out. The selection of sites for these investigations must be done in a careful and deliberate manner, because to sink a shaft at great expense only to discover a major fault or other disqualifying condition that could have been detected from the surface would be politically embarrassing and economically wasteful.

Understanding of the scientific and technical problems associated with repository siting has come in two historical phases. The first phase ran from the inception of the geologic disposal concept in the 1950s until sometime in the mid to late 1970s; salt was the medium of principal interest, and the selection of sites and the design and construction of repositories were perceived much more as engineering than scientific exercises. Beginning in 1976, the waste program underwent a major expansion, with more participation by earth scientists. It was soon found that the program's scientific foundations were weak and that the traditional approach to repository siting was insufficient. The geologic and hydrologic uncertainties were deemed to be such that isolation by the immediate repository host rock could not be assured. For instance, mechanical and chemical interactions between the heat-generating wastes and the host rock might lead to the mobilization of radionuclides and open up pathways for their release.[52] A succession of engineered and natural barriers, together comprising a total waste isolation system, was now held to be necessary.[53] The groundwater regime called for special attention, because in most cases groundwater flow would be the only mechanism for radionuclide transport. But the consensus view—formally arrived at in the United States by a major interagency review, discussed in chapter 4—was that given a careful systems approach effective geologic disposal is indeed achievable.

To date, no strategy of geologic isolation has been confirmed by a practical demonstration. Nonetheless, of all the major problems of containment that nuclear power presents, the permanent geologic isolation of waste appears, in principle, to be among the more tractable ones. If the waste consists of spent fuel, and is thus without the larger volumes and variety of forms characteristic of reprocessing wastes, the problem appears to be more manageable still.

[52]J. D. Bredehoeft, A. W. England, D. B. Stewart, N. J. Trask, and I. J. Winograd, "Geologic Disposal of High-Level Radioactive Waste—Earth-Science Perspectives," U.S. Geological Survey Circular 779 (1978).

[53]This new siting philosophy and how it evolved are briefly described in U.S. Department of Energy and U.S. Geological Survey, *Earth Science Technical Plan for Disposal of Radioactive Waste in a Mined Repository* (Washington, D.C., April 1980) p. 15.

2
A Technology Ahead of Itself

How could the U.S. Atomic Energy Commission (AEC), the agency behind the start-up of nuclear power in the 1950s and 1960s, have launched this important new commercial enterprise without first knowing what was to be done with the radioactive wastes? This question is part of a larger one: How could the Atomic Energy Commission have proceeded with the development of nuclear power without first dealing satisfactorily with all of the major public health and safety issues associated with nuclear technology, in particular the waste issue, the reactor safety issue, and the safeguards dilemma inherent in the fact that plutonium is a nuclear explosive as well as a nuclear fuel? Part of the answer is that neither the AEC nor its congressional overseer, the Joint Committee on Atomic Energy, fully recognized the nuclear imperatives for what they were. Priorities thus went awry and nuclear power as an important new technology got ahead of itself.

Although the public health and safety hazards of nuclear power were not ignored, none was dealt with adequately. There were three major hazards. One was low-level radiation exposure at various points in the nuclear fuel cycle, from uranium milling to reactor operations to the recycling of uranium and plutonium and the management and disposal of wastes. Another arose from the possibility of a reactor accident or an accident involving freshly irradiated fuel elements, with the chance that high concentrations of radioactivity would be released, affecting large numbers of people and contaminating large areas. The third had to do with possible thefts, forcible seizures, or diversions of nuclear explosive materials in the course of fuel cycle operations. Plutonium would be the material of principal concern because uranium was to be used for the most part in its low-enriched, nonexplosive form.

But the demands on the nuclear enterprise were to go far beyond the need to safeguard explosive materials and to contain radioactivity inside operating reactors. Demands for containment, especially, have focused on a mix of hazards, some real, some speculative, some merely perceived, and some in the grey regions between these categories.

Early on, nuclear technology and fission energy seem to have been perceived by the public as awesome and outside its experience. First of all, fission energy was born of the bomb; as the saying goes, "It was the same as if electricity had come in with the electric chair." In this realm of public perceptions in which most judgments are beyond empirical proof, Robert Jay Lifton, professor of psychiatry at Yale University, has been one of the important investigators. Of the psychological impact of the Hiroshima bombing on the survivors of that holocaust, he observes that there is a popular fear of "invisible contamination," an "imagery of extinction," and an "extension" of the fear of nuclear weapons to nuclear energy. He has written, "I believe that the uneasiness related to both nuclear weapons and nuclear power as expressed by mass protest movements throughout the world represents the most fundamental, primal fears about the integrity of the human body, as threatened by the invisible poison of irradiation."[1]

In attempting to explain why nuclear power "appears to evoke greater feelings of dread than almost any other technological activity," social psychologist Paul Slovic and his colleagues at Decision Research also conclude that the difference may lie in nuclear power's "association with nuclear weapons."[2] As a result of its violent origins, nuclear power, they say, is perceived by people as a technology whose risks are uncontrollable, lethal, and potentially catastrophic. Slovic and his associates believe that this perception comes from the "availability heuristic," whereby an event is "judged likely if it is easy to imagine or recall instances of it." Although frequent and likely events generally are easier to imagine than rare or unlikely events, people may exaggerate the probability of events that are particularly recent, vivid, or emotionally salient.[3] The mushroom clouds from a decade and a half of atmospheric testing of nuclear weapons constituted a long series of salient events that would become etched in the public's memory and awareness.

[1]Robert Jay Lifton, "Nuclear Energy and the Wisdom of the Body," *Bulletin of the Atomic Scientists*, September 1976, pp. 16–20.

[2]Paul Slovic, Baruch Fischoff, and Sarah Lichtenstein, "Perception and Acceptability of Risk from Energy Systems," in Andrew Baum and Jerome E. Singer, eds., *Energy: Psychological Perspectives*, vol. 3 of *Advances in Environmental Psychology* (Hillsdale, N.J., Lawrence Erlbaum Associates, 1981) p. 163.

[3]When asked to judge the frequency of various causes of death, respondents in the studies conducted by Paul Slovic and others made serious misjudgments that appear to reflect "availability bias," as in greatly overestimating frequencies of death from botulism, tornadoes, and pregnancy (including childbirth and abortion) and in incorrectly judging homicides to be more frequent than suicides or even than deaths from diabetes and stomach cancer. Slovic, Fischoff, and Lichtenstein, "Perception and Acceptability of Risk," pp. 157, 163, and 165.

The mystery which radiation holds for the public as a phenomenon that works beyond the senses also has contributed to the fear of things nuclear, as has the hypothesis adopted by radiation biologists more than thirty years ago that any exposure to radiation carries a risk.[4] The red and yellow radioactivity warning sign that is posted on the fence around the uranium mill tailings pile may seem just as threatening to many people as the warning sign on the gate at the reactor station, despite the enormous difference in potential health effects each represents: day-to-day exposure to low-level radiation at the tailings pile, compared to high doses that could result from a containment failure at the reactor. The uninitiated are left to wonder where they stand along the gamut of nuclear hazards.

These hazards and perceptions would have to be effectively confronted, for otherwise nuclear power might be perceived by the public as an unacceptable risk. Standards of containment and of safeguards would have to be exceptionally rigorous and uncompromisingly applied. For instance, besides the fears and controversy which could easily be aroused by accidents involving large releases of radiation, there was the costly mess that could result from failures of containment, with large numbers of radiation workers required to repair equipment and clean up spills. Indeed, major problems of contamination at a reactor or re-processing plant could be intolerable just in terms of production delays and added costs. As for terrorist incidents involving plutonium, the potential for public panic and for true catastrophes could be even greater than in the case of reactor accidents. Finally, real or perceived failures of safeguards, leading to the fact or appearance of the spread of nuclear weapons, would pose major political trouble for the nuclear enterprise.

Early Efforts to Confront the Hazards

However evident the containment and safeguard imperatives may be today, they were much less evident in the mid 1950s. The principal challenge perceived at that time was to make good on the apparently bright promise of breeder reactors as an abundant and long-lasting source

[4]In 1949 the National Committee on Radiation Protection and Measurements took the position that any radiation exposure carries a risk. It recommended that exposures be kept "as low as practicable" and that they not be accepted without offsetting benefits, such as the diagnostic benefit from a medical X ray. Accordingly, as a practical matter, radiation has been of concern over 18 orders of magnitude, from the megacuries (10^6) in a reactor to the picocuries (10^{-12}, or trillionths of a curie) in the radon gas emanating from an uncovered uranium mill tailings pile. The committee's conclusions and recommendations, though adopted in 1949, did not formally appear until the publication in September 1954 of committee report 17.

of fission energy. The period of the 1950s and the 1960s was a time of no little technological hubris. There was talk not only of nuclear-generated electricity too cheap to meter, but of atomic airplanes and spaceships and even of using nuclear explosives for enormous earth-moving projects such as the excavation of a new Panama canal. The Atomic Energy Commission and the "atomic scientists" carried a mystique, and the public viewed them with awe.[5] While this was not the same thing as trust, for a time it would be a serviceable substitute. Furthermore, there was no tradition for pausing in the introduction of a new industry or technology, or in the expansion of an old one, to identify possible adverse effects and develop programs to prevent or control them.

It is hardly surprising, then, that the importance of the containment and safeguard imperatives was not properly recognized. Such recognition could have come only from intense public discussion and thoughtful consideration of the special origins and unusual nature of nuclear technology. Yet the nuclear industry had its beginnings under conditions that allowed no possibility of thorough public discussion. There was secrecy, promotional hype, lack of open debate, and a profound lack of knowledge and understanding on the part of the public about nuclear technology.

The evolution of the basic law and policy for the development of nuclear power began when Congress passed the Atomic Energy Act of 1946 and created the Atomic Energy Commission. The commission, an agency vested with extraordinary authority, was empowered to run, under a cloak of secrecy and national security, the elaborate establishment of laboratories and production facilities that had sprung up during the war as part of the atomic bomb project. This nuclear domain was likened to an "island of socialism" in the free enterprise system. In creating the AEC, Congress also established the Joint Committee on Atomic Energy (JCAE), to which the commission would be accountable within a "closed political system" in which these two powerful entities would function largely symbiotically, often effectively shielded from challenge by secrecy and the esoteric nature of nuclear technology.[6]

---

[5]The AEC's first chairman, David E. Lilienthal, had this to say about how the atomic scientist was regarded in the postwar era: "His views on all subjects were sought by newspapermen, by congressional committees, by organizations of all kinds; he was asked in effect to transfer his scientific mastery to the analysis of the very different questions of human affairs: peace, world government, social organization, population control, military strategy, and so forth. And his authority in these non-scientific areas was, at least at first, not strongly questioned." Lilienthal, *Change, Hope, and the Bomb* (Princeton, Princeton University Press, 1963) p. 64.

[6]Harold P. Green, "The Peculiar Politics of Nuclear Power," *Bulletin of the Atomic Scientists*, December 1982, pp. 59–60.

Passage of the Atomic Energy Act of 1954 marked what one authority has called a 180-degree turn away from the "island of socialism," but it did nothing to illuminate the risks that the public and industry itself might be assuming. The act contained no fewer than two dozen references to the "health and safety of the public," but neither the language of the act nor the voluminous record of its legislative history spoke to the nature and magnitude of the risks.[7] There was no lack of congressional debate—indeed, this legislation was the subject of the longest Senate filibuster up to that time—but the discussions were principally a renewed expression of old controversies and antagonisms between publicly owned and investor-owned utilities. To the extent that questions of public risk were considered, this was done in the closed-door sessions of the joint committee and in the off-the-record negotiations between the committee and the Atomic Energy Commission. The aim, as one longtime observer of the AEC has put it, was to prevent open discussion and "to present the principal features of a bill to Congress and the public as noncontroversial."[8]

In January 1955 the AEC started a Power Demonstration Reactor Program, which one member of the joint committee later characterized as an attempt to "force-feed atomic development" with tax dollars.[9] Various subsidies, including money for research and development, free uranium fuel for up to seven years, and in some cases money toward reactor construction costs, were to be made available to qualified companies. Under this program a number of early reactors were built, including Indian Point-1 on the Hudson River north of New York City, Fermi-1 (an experimental breeder) on the west shore of Lake Erie between Detroit and Toledo, and Dresden-1 at Morris, Illinois. These and other early reactors were in the 200-megawatt or smaller range, and most were light water reactors, either of the pressurized water type designed by Westinghouse or the boiling water type designed by General Electric.[10]

---

[7]Ibid.

[8]Ibid.

[9]Daniel Ford, *The Cult of the Atom* (New York, Simon and Schuster, 1982) p. 46.

[10]A light water reactor (LWR) is one in which ordinary water ("light water") is used as the moderator that slows down high-velocity neutrons and increases further fission. A pressurized water reactor (PWR) is a light water reactor that has both primary and secondary cooling circuits. The water in the primary circuit passes through the reactor core as moderator and coolant and is kept under pressure to permit high temperature without boiling. The heat is transferred via a heat exchanger to the secondary circuit where the steam that drives the power turbine is generated. In a boiling water reactor (BWR), which has only one circuit, the water that passes through the core is allowed to boil and generate steam.

Force-feeding with direct subsidies ceased to be necessary in the early 1960s, and orders for reactors at much higher power ratings (up to 650 megawatts) became numerous. The Atomic Energy Commission and the reactor vendors were then to claim that nuclear generation was now competitive with coal-fired generation and that the much-heralded nuclear era was at hand. By the end of the 1960s cumulative reactor orders would total nearly 74,000 megawatts, with reactor power ratings having steadily risen and finally exceeded 1,100 megawatts as vendors competed by offering larger units.[11]

Thus from the mid 1950s on important decisions were being made by the commission, the vendors, and the utilities with respect to reactor type and safety features, the size and the siting of reactors, and the management and regulation of reactor construction and operations. Moreover, this increasing commitment to nuclear power had major implications for radioactive waste management, raising the very practical questions of how the low-level waste from day-to-day reactor operations was to be disposed of, and how the storage of spent fuel pending reprocessing or disposal was to be done once reactor pools reached their capacity.

There may have been a few scientists in the national laboratories who, under different circumstances, would have spoken out boldly about the emerging problems and attendant risks—and indeed some did speak out later, as in the late 1960s when John W. Gofman and Arthur R. Tamplin of the Lawrence Livermore Laboratory in California raised an alarm about low-level radiation. But the atmosphere prevailing within the Atomic Energy Commission and the nuclear industry was anything but conducive to outspoken behavior. In 1963 David Lilienthal, who had been the first chairperson of the AEC, observed in *Change, Hope, and the Bomb* that scientists in large government-sponsored research efforts face a "fearful choice" or dilemma of conscience. "They are guests," he said, "at a tax-supported feast of plenty. If they want money to test specialities, to earn more than a university pays, or merely to stay in the main current of what is happening in their fields, they must remain at the table. To reassert their traditional functions as scholarly men they may have to cut themselves off from the big laboratories, the elaborate equipment, and the generous stipends."[12]

Peter A. Bradford, a Carter appointee to one of the AEC's successor agencies, the Nuclear Regulatory Commission (NRC), spoke to this point

[11]Atomic Industrial Forum, "Historical Profile of U.S. Nuclear Power Development (December 31, 1980)," an AIF public affairs and information program background paper (1981).

[12]Lilienthal, *Change, Hope, and the Bomb*, p. 74.

in a speech in 1982. He noted that such wondrous atomic projects of the fifties and sixties as nuclear-powered airplanes and spaceships had come to nothing; that more than half of the citizens would oppose nuclear power plants in their communities, and almost as many would oppose nuclear power anywhere; that new reactor orders had ceased; and that spent fuel was accumulating at reactor sites for lack of any other place to put it. "How on earth could all this have happened?" asked Bradford, who felt nuclear power would have been found un-questionably benign if properly developed. "What we have seen," he suggested, "has been self-delusion on a grand scale" and a "societal failure to face reality" resulting from suppression of criticism and dissent. He cited three specific episodes of suppression. First was the revocation in 1954, at the height of McCarthyism, of physicist Robert Oppenhei-mer's security clearance over questions related to his position on the hydrogen bomb and his personal loyalty. In Bradford's view, Oppenhei-mer was a critic exiled. "If so distinguished a scientist as Oppenheimer could be abused in such a manner, then clearly it was not in the interest of any [AEC] employee or any scientist at a university whose research grants came from the AEC to dissent from the party line," he said. The second episode involved the ignoring or actual suppression of AEC-sponsored studies indicating that fallout from the atmospheric bomb tests might cause fatal cancers among the soldiers and civilians exposed. Bradford's third case in point had to do with the commission's attempt in 1956 to suppress a report by its Advisory Committee on Reactor Safeguards, which said that sufficient information to certify the proposed Fermi-1 experimental breeder as safe was lacking.[13]

Atoms for Peace and the Safeguards Issue

The record of the Atomic Energy Commission's behavior toward the risks of nuclear power was highly contradictory, showing concern in some instances and heedlessness in others. But the commission was not alone in heedless or precipitous behavior. The congressional Joint Committee on Atomic Energy, it is said, was "almost always more aggressive and expansionist than the [AEC and the executive branch] and it constantly pressed for larger and more ambitious atomic energy projects."[14] In the case of President Eisenhower's Atoms for Peace proposal of December 1953, which had weighty implications for plutonium safe-guards, the prime mover was the White House. Neither the commission

[13]Speech given by Commissioner Bradford on January 15, 1982, at the Groton School, Groton, Mass.
[14]Green, "Peculiar Politics," p. 60.

nor its staff had any part in this proposal; but the commission chair, Lewis L. Strauss, by virtue of the second hat he was wearing as the president's special assistant for atomic energy, was centrally involved.[15]

How the Atoms for Peace proposal came about represents a striking early instance of a precipitous and poorly considered commitment. The three principals in the shaping of the proposal were Strauss, a Wall Street financier who had been first appointed to the AEC by President Truman; C. D. Jackson, who had come to the White House from Time–Life as the president's special assistant for psychological warfare; and President Eisenhower himself.[16] Eisenhower saw Atoms for Peace as a step toward cooperation with the Soviet Union in nuclear matters, and, he hoped, toward arms control. Jackson saw the proposal chiefly as a way for the United States to win a victory in the propaganda war by showing the world a benign image. Strauss saw it principally as a way to bring the world a source of abundant energy and other benefits, as in nuclear medicine.

By hindsight, the proposal can be seen to have been premature in almost every way. There was no power reactor yet operating anywhere. Certainly there was no early or even foreseeable prospect of providing "abundant electrical energy in the power-starved areas of the world," which was what Atoms for Peace proposed.[17] Uranium was still believed to be relatively scarce. A practical breeder reactor remained but a vision, one that would take years of research and development to fulfill, if it could be fulfilled at all. Recycling plutonium as reactor fuel was also a long way from being demonstrated commercially. If nuclear power was for anyone, it was for the more technically sophisticated countries who would have the best chance of mastering the highly challenging reactor and fuel cycle technologies, including those related to radioactive waste management and disposal.

The safeguards and nuclear proliferation problem scarcely appears to have been recognized except in the most abstract way. According to Jack Holl, chief Atomic Energy Commission historian, the only politicians taking much interest in the problem at that time were those on the Republican right wing, including Senator Joseph McCarthy.[18] The right-wingers saw Atoms for Peace, Holl says, as "a great giveaway to the commie nations" of a technology that could be used to threaten the

---

[15]Interview by the author on May 21, 1984, with Jack M. Holl of the Department of Energy, successor to Richard G. Hewlett as chief AEC historian.

[16]Ibid.

[17]Dwight D. Eisenhower, "Atomic Power for Peace," speech before the United Nations General Assembly, December 8, 1953.

[18]Interview with Holl, May 21, 1984.

United States. Just what the international atomic energy agency which Eisenhower mentioned in his Atoms for Peace address was to be, and how it was to go about safeguarding the fissile materials for a rapidly expanding nuclear enterprise, were questions to be wrestled with in the future. In a talk at an industry meeting in 1982, Gerard C. Smith, chief arms control negotiator during the Nixon administration and a staff aide at the AEC when the Atoms for Peace address was given, recalled that the commissioners, other than Strauss, had no advance notice of this address. "Later, when Molotov [the Soviet foreign minister] protested to a dubious John Foster Dulles [then secretary of state] that the proposal would result in the spread worldwide of stockpiles of weapons-grade material, I had to explain to Dulles that Molotov had been better informed than he," Smith said. "Subsequently, the Soviets asked how we proposed to stop this spread. The best we could reply was 'ways could be found.'" Smith noted that it was now a generation later, yet the search for "those ways" was still going on.[19]

The AEC Record on Reactor Safety

The Atomic Energy Commission's record with respect to reactor safety is truly replete with contradictions. In places the record shows great concern to avoid or reduce risks. For instance, in the late 1940s the commission was on the point of selecting the Argonne National Laboratory near Chicago as the place to build a series of experimental reactors; but on the advice of physicist Edward Teller the commission chose instead an isolated site in Idaho, far removed from any major population center. Although such adherents of the Argonne site as Robert Oppenheimer and Enrico Fermi felt that the dangers of a nuclear incident were being exaggerated, Teller had prevailed, contending that the "calculated risks" taken in the Manhattan Project during World War II could not be justified in peacetime.[20] It was thus that an 893-square-mile expanse of sand, sagebrush, and old lava flows on eastern Idaho's Snake River plains, a site nearly the size of Rhode Island, was chosen for the National Reactor Testing Station (known today as the Idaho National Engineering Laboratory). The consequences of a theoretically possible, if improbable,

[19]Gerard C. Smith, "Nuclear Commerce and Non-proliferation in the 1980s—Some Thoughts," speech given at a conference on April 29, 1982, at St. Charles, Ill., sponsored by McGraw-Hill, publishers of *Nucleonics Week*. Smith was director of the Arms Control and Disarmament Agency, 1969–1973, and chief U.S. delegate to the first Strategic Arms Limitation Talks, or SALT I.

[20]Robert Gillette, "Nuclear Safety (1): The Roots of Dissent," *Science*, September 1, 1972, pp. 773–774. In this and three subsequent articles on reactor safety Gillette has described in detail the AEC's record and shortcomings in the field of reactor safety research.

worst-case reactor accident were also not ignored. "WASH-740," the often-cited 1957 study done by Brookhaven National Laboratory scientists at the request of the Joint Committee on Atomic Energy, indicated that in such a case 3,400 people would die, 43,000 would be injured, and property losses would total $7 billion. From the first, the AEC required reinforced concrete containment shells for commercial power reactors, and later when the size and power rating of new reactors began to grow, emergency core cooling systems were required as a precaution against meltdowns.

Yet at the same time the Atomic Energy Commission was plunging ahead with commercial reactor development as though no hazards and no major problems of containment existed. While the commission had insisted on building the small test reactors in the isolation of the Idaho desert, the utilities were allowed to build their increasingly large power reactors near cities, as at Indian Point, only thirty miles from Manhattan. And while the commission did worst-case reactor accident studies and insisted on backup safety systems, these systems were long in being put to a practical test. The commission said in the early 1960s that its proposed Loss-of-Fluid Test (LOFT) of the adequacy of backup cooling systems addressed "perhaps the most urgent problem in water reactor safety."[21] But the first actual testing with LOFT was not to come until late 1978, after more than 80 reactors had been built and licensed and not long before Three Mile Island.

The most striking contradiction in the AEC record on reactor safety was concerned with the management and regulation of reactor construction and operation. The commission, along with the Navy, had entrusted its own reactor projects to the highly disciplined, expert, and authoritarian team of nuclear engineers under Admiral Hyman G. Rickover, who insisted on perfection in every detail from contractors. Yet in its licensing and regulation philosophy at the start of the commercial nuclear program the commission was leniency itself, as nuclear critic Daniel Ford has documented in detail from the AEC's records.[22]

From hindsight, one can argue that a fundamental error was made by proceeding with nuclear power development without first conducting an extensive research and demonstration program specifically aimed at identifying reactor types that would be safe and forgiving in commercial application. The accident at Three Mile Island would demonstrate how important it is to have a fail-safe containment system immune from even the partial failures that can bring on a scary (if nonlethal) episode.

[21]Robert Gillette, "Nuclear Safety (II): The Years of Delay," *Science*, September 8, 1972, p. 867.
[22]Ford, *Cult of the Atom*.

The AEC Record on Radioactive Waste Management

By the mid 1950s the Atomic Energy Commission knew or had reason to know that provisional arrangements for radioactive waste management of the kind improvised during the wartime atomic bomb project would not suffice in the case of large-scale commercial nuclear energy. If the nuclear power industry grew as the commission hoped, there would be some 3 billion gallons of high-level waste by the year 2000 unless waste management methods were improved.[23] AEC officials could easily envision what that would mean. Already in inventory from the reprocessing of the spent fuel from military reactors there was about 60 million gallons, stored in 145 underground tanks.[24] Most of the tanks were at the Hanford reservation in Washington state, but with the commission about to start up the Savannah River plant in South Carolina, tank farms were being built there too. Without a drastic change in waste management methods, there would be literally thousands of tanks by the end of the century, with no permanent solution to the waste problem in sight.

Although waste management at the Hanford works was clearly a secondary consideration and was hurriedly improvised, Manhattan Project scientists knew that working with radioactive materials carried health and environmental risks, and they did not take these risks lightly. Out of this concern came efforts to establish programs in health physics and radiation biology at Manhattan Project sites. The aim was to protect not only human health, but also the environment and fish and wildlife. In *Atomic Quest*, Arthur H. Compton, the distinguished physicist who directed the plutonium production program, recalled that at a meeting of the council responsible for that program Crawford Greenwalt of the DuPont company (which operated the Hanford works) took the floor. "He opened his comments with one word: 'Fish!' What would the cooling water, returning from the nuclear reactors into the Columbia River, do to the fish that were so important in the economy of the Northwest?" Experiments were undertaken immediately, and it was found that while fish were unusually resistant to the direct effects of radiation, radioactivity was absorbed within bones in surprising concentrations even from water in which fission products were present in only trace amounts. "In fact," Compton said, "a fish could thus be made so radioactive that its

[23]Richard G. Hewlett, "Federal Policy for the Disposal of Highly Radioactive Wastes from Commercial Nuclear Power Plants: An Historical Analysis," manuscript, March 1978, p. 7. Hewlett is a former AEC chief historian.

[24]Ibid.; Energy Research and Development Administration, *Waste Management Operations, Hanford Reservation*, Final Environmental Impact Statement, vol. 1, ERDA-1538 (Washington, D.C., December 1975) table II 1-4, p. II 1-36 (hereafter cited as the Hanford Environmental Statement).

skeleton would mark itself on a photographic plate. It was clear that extreme precautions would be necessary in order to prevent any radioactive materials from reaching the Columbia River water."[25]

Hanford, a site of 570 square miles at a large bend of the Columbia, was chosen for the plutonium works chiefly by virtue of its isolation and the abundant hydropower and cooling water available.[26] The climate is semi-arid and the water table is typically 200 to 300 feet below ground surface. The latter conditions made waste management easier than it would have been in a region with a wet climate and high water table. Even as it was, there were problems. One had to do with the practice of drawing water from the Columbia River, passing it through the reactors, then discharging it back to the river in a once-through cooling system.[27] The river and the Pacific Ocean waters for some miles beyond the mouth of the river were being contaminated with short-lived radioactive corrosion products. If this contamination was not at levels significant for public health, it would eventually have become significant politically had not all of the reactors using once-through cooling been phased out by early 1971.

Biological reconcentration of radionuclides present even at very low levels of contamination was always a possibility, as was borne out in the case of the Willapa Bay oysters. One of the Hanford workers was found to have detectable, though slight, amounts of zinc-65 in his tissues as a result of having eaten contaminated oysters from Willapa Bay, which is on the Washington coast some 200 miles from Hanford and 30 miles north of the Columbia River's mouth. The zinc-65, which has a half-life of about eight months, was entering the river in the effluent from the Hanford reactors, then being carried some 360 river miles to the Pacific, where ocean currents carried it northward into the bay in extremely minute concentrations.[28] As filter-feeders with a particular chemical affinity for either stable or radioactive zinc, the oysters were concentrating the zinc-65 at 200,000 times its concentration in the seawater, although even in the oysters it was present at only 40 to 50 trillionths of a curie per gram. This incident was first noted in the scientific

[25]Arthur Holly Compton, *Atomic Quest* (New York, Oxford University Press, 1956) pp. 179–180.

[26]Richard G. Hewlett and Oscar E. Anderson, Jr., *The New World, A History of the United States Atomic Energy Commission*, vol. 1, AEC document WASH-1214 (Washington, D.C., 1962) pp. 188–190.

[27]Joint Committee on Atomic Energy, "Industrial Radioactive Waste Disposal," summary analysis of hearings held on January 28, 29, 30, February 2, 3, and July 29, 1959 (August 1959) p. 14.

[28]R. W. Perkins, J. M. Nielson, W. E. Roesch, and R. C. McCall, "Zinc-65 and Chromium-51 in Foods and People," *Science*, December 23, 1960, pp. 1895–1897; interview by the author with marine biologist Edward LaRoe of the Office of Biological Sciences, U.S. Fish and Wildlife Service, March 1982.

literature more than twenty-five years ago, or well before the environmental movement gathered momentum. Had such incidents persisted and multiplied, the Atomic Energy Commission would have faced a public relations crisis whether the contamination was having certifiable health effects or not.

But the most important source of radioactive waste at Hanford was the chemical separation plants where plutonium was recovered. These plants generated the liquid high-level waste that resulted in the proliferation of underground tank farms. In the original separation process plutonium was precipitated out of the aqueous solution by adding certain "carriers," which increased the volume of high-level waste enormously. After a switch to a different process in the 1950s the output of high-level waste was reduced, but still remained high in comparison with the volume generated by the reprocessing plants that would come along later (see chapter 3). In addition, millions of gallons of intermediate-level waste were discharged to "cribs," or structures through which the waste was poured into the dry desert soil, and substantial amounts of solid waste, much of it contaminated with plutonium, were dumped into trenches and buried. In relation to the modest amount of fuel irradiated, the total amount of radioactive waste generated was very large indeed.

During the early postwar years few atomic scientists appear to have seen the nuclear waste problem as an obstacle to development of nuclear power, but there was at least one highly prominent scientist who did. This was James B. Conant, president of Harvard University, who as chair of the wartime National Defense Research Committee had been one of the shapers of the atomic bomb project. In "Our Future in the Atomic Age," an essay which appeared in 1951, Conant predicted that by the 1960s there would be "a sober appraisal of the debits and credits of the exploitation of atomic fission." Placing himself forward in time, Conant continued his prophecy: "Of course, experimental plants were producing somewhat more power from controlled atomic reactions than was consumed in the operations of the complex process, but the disposal of the waste products had presented gigantic problems—problems to be lived with for generations." He concluded that for several reasons—the waste problem, the "very great" capital investments required, and the overriding importance of atomic energy's potential military applications—people would decide "the game [is] not worth the candle" and that a "self-denying ordinance [is] but common sense."[29]

---

[29]James B. Conant, "Our Future in the Atomic Age," in *Headline Series*, Foreign Policy Association, no. 90 (November–December 1951). According to Richard G. Hewlett, Conant held to his negative views about nuclear energy to the end of his life in 1978. Hewlett was interviewed by the author on March 22, 1982.

Early Consideration of Geologic Disposal

In 1954 AEC officials had begun considering the possibilities for waste
disposal in deep geologic formations, and at the commission's request
the National Academy of Sciences took up this question at a conference
at Princeton University in September 1955. In his introductory remarks
at this conference, A. E. Gorman of the Atomic Energy Commission's
Reactor Development Division said that because of the isolated locations
of the agency's reservations it had been possible "to sweep the [waste]
problem under the rug," but that now the time was at hand to face up
to it. Joseph A. Lieberman, chief of that division's environmental and
sanitary engineering branch, was almost plaintive in his characterization
of the situation: "AEC employees who labor daily with this problem,"
he said, "are faced with the constant question, 'when can we stop
building tanks?' or 'when can we do something with these wastes other
than putting them in tanks?' " Gorman said that economical solutions to
the radioactive waste problem would have to be found before a nuclear
power industry could be expected to develop and that it was the commission's responsibility to find those solutions.[30]

Although Lieberman and Gorman recognized that there was a serious
radioactive waste problem, they also assumed there was a solution. The
conference did not disappoint them on this score. The concept of disposal of high-level waste in salt mines or in cavities expressly created
in salt for this purpose was one of the possibilities identified at this
meeting. Later, the newly established National Academy of Sciences
committee on waste disposal called this concept "the most promising"
available. It said, too, that disposal could be "greatly simplified if the
waste could be gotten into solid form of relatively insoluble character."
On disposal in salt cavities the committee, chaired by Princeton geologist
Harry H. Hess, observed: "The great advantage here is that no water can
pass through salt. Fractures are self-sealing. Abandoned salt mines or
cavities especially mined to hold waste are, in essence, long-enduring
tanks." The committee cautioned, however, that the research needed to
support its opinions and recommendations remained to be done. Salt
presents a specific weakness, it noted, in that this material flows under
pressure; research on the size and shape of openings necessary for
structural stability was said to be in order. The committee felt that the
radioactive waste problem represented a unique danger and should be
handled accordingly: "Unlike the disposal of any other type of waste,

[30]National Academy of Sciences, "The Disposal of Radioactive Waste on Land," report
of the Committee on Waste Disposal of the Division of Earth Sciences, appendix B,
"Proceedings of the Princeton Conference," NAS-NRC pub. 519 (Washington, D.C., September 1957) p. 16 (hereafter cited as the National Academy's 1957 report).

the hazard related to radioactive waste is so great that no element of doubt should be allowed to exist regarding safety. Stringent rules must be set up and a system of inspection and monitoring instituted. *Safe disposal means that the waste shall not come in contact with any living thing."* The committee also questioned the safety of existing waste disposal practices "if continued for the indefinite future," but said no criticism of the responsible officials at the AEC installations was intended, for those practices were a result of the exigencies of war.[31]

The academy's advice came at a point when the Atomic Energy Commission still had time to put together a fuel cycle and waste management policy and program before commercial reactors became numerous. The first commercial nuclear station, at Shippingport, Pennsylvania, was only then coming on line. And although reactor orders would start coming thick and fast in the 1960s, significant amounts of spent fuel would not accumulate before the 1970s. Had a concerted and high-priority effort to establish a waste management program been initiated in the late 1950s, the commission or a successor agency might have been ready sometime in the 1970s to begin receiving military wastes—plus whatever commercial waste was available—for geologic disposal.

But the Atomic Energy Commission failed to mount the necessary effort, despite some surprisingly frank, unsolicited advice from the National Academy committee. In a letter of June 21, 1960, to AEC chairman John A. McCone, Harry Hess said: "No existing AEC installation which generates either high-level or intermediate-level waste appears to have a satisfactory geological location for the safe local disposal of such waste products; neither do any of the present waste-disposal practices that have come to the attention of the committee satisfy its criterion for safe disposal of such waste."[32] His committee, Hess said, recommended that disposal facilities be established at suitable geologic sites where accumulated wastes could be processed and gotten rid of, and that chemical reprocessing activities be concentrated at sites of this kind. Moreover, in its most striking recommendation the letter urged that *"approved plans for the safe disposal of radioactive wastes be made a prerequisite for the approval of the site of any future installation of the AEC or under AEC jurisdiction"* (emphasis added).[33] Implicit here was the principle that all waste disposal policies and practices should be sound enough to remain valid when nuclear power began its expected rate of

[31]Ibid., pp. 1–7, 79.

[32]This letter was reprinted in M. King Hubbert, *Energy Resources* (Washington, D.C., National Academy of Sciences, 1982) pp. 118, 119. Hubbert was a member of the National Academy of Sciences radioactive waste committee and chair of an energy resources study prepared by the academy for President Kennedy in 1962.

[33]Hubbert, *Energy Resources*, p. 116.

growth. The committee's recommendation was a premonition of demands in the 1970s for a moratorium on further growth of nuclear power until waste disposal facilities became available.

What was needed was a comprehensive and well-integrated fuel cycle and waste management policy robust and flexible enough to accommodate the prevailing realities and uncertainties. One reality, as the AEC staff well knew, was that the proliferation of tank storage for military high-level waste was undesirable. The commission knew too, or had reason to know, that a program for permanent disposal of high-level waste must accompany development of commercial nuclear power. Yet geologic disposal was at this point only a concept, and could not be prudently implemented without further investigation and research. Another reality was that the ultimate form that the commercial nuclear program would take and the importance that it would assume were still unknown. Light water reactors were seen as precursors of breeder reactors, but the necessary research and development for the breeder and its supporting fuel cycle was far from complete (and indeed has not been completed to this day). The sensible course was to address the existing military waste problem in a way that would also help cope with whatever commercial waste problem emerged. By the same token, it was important that commercial nuclear power be developed only in ways compatible with secure and permanent waste disposal. But on both accounts AEC policies went awry.

The Atomic Energy Commission actually made a considerable effort to address the waste problem, but this effort, lacking a clear purpose, was neither concerted nor systematic. On the positive side, over the eight years from 1957 to 1965 the commission supported work on a variety of waste-processing techniques, including a fluidized-bed calcination facility at the Idaho Chemical Processing Plant that could convert liquid high-level waste into a sandy calcine for storage in stainless steel bins,[34] although the first storage bins were built without hatches for the calcine's eventual removal. At several of the national laboratories research was also being done on solidifying high-level waste in materials such as glass, ceramics, and synthetic feldspars.

Most important of all, researchers at the Oak Ridge National Laboratory were looking into the storage of radioactive waste in bedded salt deposits. Their project Salt Vault, which was to begin in the mid 1960s with the temporary commitment of spent fuel canisters to an abandoned salt mine at Lyons, Kansas, would be the first in-situ experiment with high-activity waste in a geologic formation (see the section on Embarrassment at Lyons, below). It would be useful too as a trial of engineering

---

[34]Hewlett, "Federal Policy for the Disposal of Highly Radioactive Wastes," pp. 7–10.

methods for the emplacement of waste packages requiring shielding and remote handling.

Cross-purposes within the AEC

Had the Atomic Energy Commission's waste management research and development activities been coordinated and marshalled in a determined way, they might well have given the commission important elements of a timely and effective waste disposal program. Not only was such co-ordination lacking, but some commission-sponsored waste disposal activities were at cross-purposes with those just described. There were two particularly important examples of this, one at the first commercial reprocessing plant, at West Valley, New York, the other at the Hanford and Savannah River reservations in Washington and South Carolina. (These and other reprocessing facilities are discussed further in chapter 3, especially from the point of view of the technological immaturity of the commercial reprocessing ventures attempted in the United States.)

Just as the Atomic Energy Commission had initiated, in the mid 1950s, a subsidized demonstration program to help utilities build their first power reactors, it also sought to help private industry establish a commercial fuel reprocessing facility. The upshot was that in 1963, following prolonged negotiations among the commission, the State of New York, and two private applicants, a construction permit for such a plant was granted. It was to be built by a new corporate entity, Nuclear Fuel Services (NFS), at a state-owned site on the rolling glacial terrain of Cattaraugus County to the east of Lake Erie and about 30 miles south of Buffalo. Waste management had been a major issue in the negotiations.[35] Neither NFS nor the AEC was willing to assume ultimate responsibility for the high-level waste generated. In the end, the State of New York, which was eager to get in on the start of the nuclear fuel cycle business, naively agreed to assume that responsibility, provided that Nuclear Fuel Services establish a perpetual care fund—a fund which was to prove pathetically inadequate.

While refusing to take responsibility for the waste, the Atomic Energy Commission had insisted that the price of NFS reprocessing services be set low—so low that the company could not afford to build the stainless steel tanks necessary for the waste to be stored in its acidic form, or the form that facilitates waste recovery and solidification. The commission could make demands about the price of the service because commercial reactors were then too few to give a reprocessor enough business to

[35]Gene I. Rochlin, Margery Held, Barbara G. Kaplan, and Lewis Kruger, "West Valley: Remnant of the AEC," *Bulletin of the Atomic Scientists,* January 1978, pp. 17–26.

survive. Almost two-thirds of the fuel that NFS would reprocess was to come from the AEC's N-reactor at Hanford, a dual-purpose facility built to produce both commercial power and plutonium for weapons. According to Walton A. Rodger, who became Nuclear Fuel Services' general manager late in the start-up at West Valley, the commission insisted on a price of $22 per kilogram of heavy metal reprocessed, or, he says, about half of the amount that would have been justified on the basis of the firm's capital investment.[36] The commission and its staff were not of one mind about the West Valley venture. The commissioners wanted commercial reprocessing to be a fuel cycle function assumed by private industry from the start. The staff, on the other hand, wanted the AEC itself to provide this service, at least until growth of nuclear power was more advanced and substantial amounts of spent fuel were being generated.

In any case, Nuclear Fuel Services followed the example set by the commission at Savannah River and neutralized the acidic high-level waste so that it could be stored and cooled, by suspended coils, in carbon steel tanks. The more costly alternative would have been to build corrosion-resistant stainless steel tanks. This was indeed what the British had long since elected to do at their reprocessing center at Windscale, what the French were doing at Marcoule, and what the AEC itself was doing at the Idaho Chemical Processing Plant. The problem at NFS was greatly compounded in the design of the tanks. Rodger has described the tank bottom as resembling a "giant waffle iron," which meant that the high-level waste sludge would be hard to dislodge and remove.

Before authorizing construction, the commission conducted a comprehensive review of the Nuclear Fuel Services plans for both the plant and waste management facilities. The folly of its approving those waste management plans is all too apparent today as the commission's successor, the Department of Energy, struggles to carry out a congressional mandate to recover and solidify the West Valley waste (a $700-million-dollar undertaking). But the irony is that even as the improvident licensing decisions at West Valley were being made—showing that there was no intent ever to recover the West Valley waste for geologic disposal—the national laboratories were busy investigating waste solidification technology and waste disposal in salt formations.

[36]Interviews by the author with Walton A. Rodger, September 1981 and April 1982. WASH-743, the AEC's summary report of October 1957 on a "reference fuel-reprocessing plant," did not mention high-level waste storage as a capital cost. The report estimated the operating cost charge per gallon of high-level waste at $1 to $3 for 20 years' storage. The total capital cost of the plant was estimated at only $20.5 million. Atomic Energy Commission, WASH-743 (Washington, D.C., 1957) p. 11.

The commission's efforts to deal with waste at Hanford and Savannah River were a patchwork of contradictions. At the Idaho Chemical Processing Plant calcining of the liquid high-level waste was in progress. Calcining was easier there than it would have been at Savannah River or Hanford because the Idaho liquid waste was stored in its acidic form, without bulky and thermally unstable neutralizing agents. But in principle recovery and solidification were possible for at least part of the high-level waste at Hanford and for all of it at Savannah River, where the tanks were newer and better built than at Hanford; a facility to accomplish this is in fact being built today at Savannah River, at an acceptable cost. Solidification would have put the waste into a form that was safer and suitable for geologic disposal. Yet the AEC seems to have been indifferent to these possibilities.

At a hearing held by the Joint Committee on Atomic Energy in 1959, Herbert M. Parker, manager of the Hanford works, indicated that tank storage of high-level waste could be continued indefinitely. "If the tanks turn out to have a life of 50 years," he said, "it will be very simple to be prepared at the right time with an alternative set of tanks and pump the liquid into the new tanks."[37] Parker believed that no tank had ever leaked, although he said some had undergone "suspicious oscillations."[38] It would later be discovered that the first major leak had occurred six months before, when some 35,000 gallons of waste had escaped. A U.S. Geological Survey report prepared in 1953 had warned that the "true structural life" of the tanks was uncertain.[39]

The problem of tank leaks had to be dealt with, and there was no easy solution. But the solution adopted by the Hanford managers and the Atomic Energy Commission would make it virtually impossible, as a practical matter, ever to recover the waste from Hanford's 149 older tanks. The remedial program initiated there during the 1960s involved reducing the self-boiling liquid high-level waste in these tanks to a damp "salt cake," an undertaking which would not be completed until the 1980s. An essential step was to separate as much of the cesium and strontium as possible from the waste and thus reduce generation of decay heat.

[37]Testimony of Herbert M. Parker in *Industrial Radioactive Waste Disposal*, Hearings before the Joint Committee on Atomic Energy (1959) p. 9, cited by Daniel S. Metlay, "History and Interpretation of Radioactive Waste Management in the United States," in *Essays on Issues Relevant to the Regulation of Radioactive Waste Management*, NUREG-0412 (Washington, D.C., Nuclear Regulatory Commission, May 1978) p. 11.

[38]Robert Gillette, "Radiation Spill at Hanford: The Anatomy of an Accident," *Science*, August 24, 1973, pp. 728–730.

[39]Ibid., p. 730. The report referred to remained classified until 1960 and was not published until 1973, when it appeared as U.S. Geological Survey Professional Paper 717.

But the waste remaining in the tanks, even with the greater part of the cesium and strontium removed, would remain high-level, containing tens of millions of curies of fission products and somewhere between 400 and 1,200 kilograms of plutonium.[40] As salt cake and an underlying sludge, the waste could not leak out; but neither could it be pumped out for further processing and geologic disposal. The older these tanks became, the more questionable their integrity, and the greater the risk of leaks if an attempt were made to resuspend the salt cake and sludge by sluicing with high-pressure water jets. The only way the waste could be removed from the tanks in the future would be to mine it out. But the cost of this would be so high—well up in the billions—and the collective radiation exposure for the workers so great that any cost-benefit analyst might shake his head.

The commission's waste managers regarded this situation as an unavoidable dilemma. For instance, in 1973 Frank K. Pitman, then head of the agency's waste management division, said of a colleague who was responsible for the Hanford program, "Every time he makes it more difficult [for the waste] to leak out, he makes my long-range problem more difficult, but this is necessary."[41]

Geologic disposal for military high-level waste was pursued at the Savannah River plant, but on two occasions, in 1960 and 1966, the National Academy of Sciences cautioned the Atomic Energy Commission that the geohydrologic circumstances at this reservation appeared unsuitable for on-site geologic disposal.[42] Yet the commission persisted in the hope that the liquid high-level waste in tank storage there could be disposed of by mixing it into a grout, then slurrying the grout into caverns mined deep in the reservation's underlying bedrock, where it would harden like concrete.

[40]Joseph A. Liberman, Walton A. Rodger, and Frank Baranowski, "High-Level Waste Management," statement prepared for the California Energy Resources Conservation and Development Commission, March 21, 1977. These nuclear fuel cycle specialists put the plutonium content in the Hanford tanks at 380 kilograms, but said this estimate might well be low by a factor of two to three.

[41]Hearings on the Atomic Energy Commission before a subcommittee of the House Committee on Appropriations, 93 Cong. 1 sess., pt. 4 (April 5, 1973) p. 176.

[42]In his June 21, 1960, letter to AEC chairman John A. McCone, Harry Hess, the chair of the National Academy's radioactive waste committee, said, "No existing AEC installation which generates either high-level or intermediate-level waste appears to have a satisfactory geological location for the safe local disposal of such waste products." Then, again, the academy Committee on Geologic Aspects of Radioactive Waste Disposal report of May 1966 said, "None of the major sites ... is geologically suited for safe disposal." The only "possible exception" allowed was for grout injection of intermediate-level waste into shale beds at the Oak Ridge National Laboratory. "Report to the AEC Division of Reactor Development and Technology," p. 11.

The problem, as the academy's radioactive waste committee seems to have viewed it, was not that bedrock disposal was unsafe, but rather that the presence above the bedrock of the Tuscaloosa aquifer—a major water resource for the Southeast—made such disposal technically and politically unacceptable unless it could be shown to be clearly safe. In its 1966 report the academy committee's majority put it this way: "The placement of high-level wastes 500 or 1,000 feet below a very prolific and much-used aquifer is in its essence dangerous and will certainly lead to public controversy. Any demonstration of its safety must leave no shadow of a doubt." Despite this warning the Atomic Energy Commission would choose to accept a minority recommendation to pursue a step-by-step program of geologic exploration. But the political realities would ultimately scotch that effort. In 1972 elected officials in South Carolina and Georgia strongly opposed any further consideration of bedrock disposal. Georgia's Governor Jimmy Carter wrote the AEC complaining that, in now proposing the costly step of sinking an exploratory shaft, the agency was "exhibiting a dangerously indifferent" and "fiscally irresponsible" attitude.[43]

The commission's decisions concerning waste at West Valley, Hanford, and Savannah River represented a failure to achieve even a modicum of policy consistency and discipline. Whatever else contributed to the Atomic Energy Commission's ineffective dealing with the waste problem, managerial inattention and ineptitude played a part.

Finally a Waste Policy—with Flaws

In the spring of 1968 the Atomic Energy Commission undertook for the first time to develop a coherent policy for the disposal of commercial high-level waste.[44] Before the end of the year the commission and the staff had agreed on the essentials of that policy. These were, first, that high-level waste storage at reprocessing plants should be converted to an AEC-approved solid form within five years of the time of generation; second, that the solidified waste would be delivered to a federal repository not later than ten years after the reprocessing of the irradiated fuel, with reprocessors paying a one-time fee in return for the government's assuming full responsibility for ultimate disposal. This policy was formally promulgated in November 1970.[45]

[43]Letter from Governor Carter to N. Stetson, manager of the AEC's Savannah River operations office at Aiken, S.C., March 14, 1972.

[44]Hewlett, "Federal Policy for the Disposal of Highly Radioactive Wastes," p. 13.

[45]Nuclear Regulatory Commission, "Policy Relating to the Siting of Fuel Reprocessing Plants and Related Waste Management Facilities," 10 CFR, pt. 50 (November 1970) appendix F; published in the *Federal Register*, November 14, 1970.

Milton Shaw, who had been director of the AEC reactor development and technology division since the end of 1964, was a forceful advocate of the policy, which went further than many people in the nuclear industry felt was necessary. Shaw did not believe at the time, and does not believe today, that the commission's approaches to the waste problem were laggard or inadequate. From Shaw's first days at the commission research and development work on high-level waste solidification was under way and the Oak Ridge people were busy with Salt Vault, pursuing the National Academy of Sciences' recommendation to investigate salt formations. Moreover, commercial reprocessing was only beginning and it would be years before significant amounts of high-level waste would be generated. Meanwhile, experience in the nuclear weapons program indicated that while tank storage presented problems, it could provide a satisfactory interim solution. This is how Shaw saw it, and, in a sense, that was the way it was.[46]

But before long a larger and very different reality would become manifest. By almost any objective reading, the Atomic Energy Commission's waste policy was flawed on three counts. First, the new policy applied only to the generation of commercial high-level waste in the future; initially it would not apply even to the waste at West Valley. The great bulk of high-level waste then in inventory was the military waste in tank storage at Hanford and Savannah River. If embarrassing problems should arise with respect to this military waste, the commercial nuclear power program would suffer from the political backlash. Neat bureaucratic distinctions between commercial and military waste would be lost on the public.

Second, if private industry's plans to go on with and to expand fuel reprocessing were not carried out, the entire waste management scheme would be at an impasse. Yet there was no assurance that commercial reprocessing would succeed. The Nuclear Fuel Services plant at West Valley was anything but a demonstrated success technically and economically, and would be shut down before the end of 1972. Three other commercial reprocessing ventures were in various stages of planning or construction, but all were eventually to be aborted. Reprocessing technology was immature, the costs were higher than expected, and government policy bearing on plutonium safeguards was unsettled.

Third, although the new policy contemplated geologic disposal of high-level waste within ten years, there was probably no chance that a repository could be built and licensed to receive waste in anything like that time. The technical, institutional, and political problems involved were to be much more difficult than anyone had imagined. An unfortunate early demonstration of this reality was in the offing at Lyons,

---

[46]Shaw was interviewed by the author on April 20, 1984.

Kansas: plans to convert the salt mine there from an AEC research project to a permanent repository would end in severe political embarrassment.

The Atomic Energy Commission's failure to put together a coherent and effective waste policy has been ascribed to a variety of causes. In lectures at Princeton University in 1963, David E. Lilienthal, chair of the commission from 1946 to 1950 and an early critic of nuclear energy (although he did not abandon the nuclear dream), spoke to what he felt was the failure of those engaged in the nuclear enterprise to see nuclear technology clearly and as a whole. He said: "So complex an achievement as atomic energy, with so many ramifications, coming so suddenly and under circumstances that required secrecy, required a high degree of compartmentalization.... The technicians and experts and specialists took over the atom. They took it over in pieces, not as a whole. To this day no one has ever been able—or tried very hard, I think—to put the pieces together in an overall way. The preoccupation has been with the bits and pieces."[47]

Carrol Wilson, the AEC's first general manager, made the same point in his 1979 analysis of what went wrong with nuclear power, but stressed what he saw as a preoccupation with reactor development. "No one spent much time then nor even later looking at the total system of the nuclear fuel cycle from fuel enrichment through the whole system of reactor operation, reprocessing of spent fuel, and disposal of radioactive wastes," he wrote in the *Bulletin of the Atomic Scientists*. "All attention was focused on the reactor .... It is a little surprising that no group seemed to have looked at the total fuel cycle as an interdependent system. No one appeared with enough clout to set priorities and to argue persuasively that if some critical part of the system were missing perhaps the whole system would come to a grinding halt."[48]

Lilienthal's and Wilson's views seem to have been correct, certainly as regards the situation until the late 1960s, when Milton Shaw and others pushed for a commercial waste policy. Reactor development, and especially breeder development, was getting the lion's share of attention and budget support from the Atomic Energy Commission and from Congress. The perceived imperative driving the nuclear program during the 1960s was that of preparing for the day, expected to come before the end of the century, when uranium would be in short supply.[49] It

---

[47]Lilienthal, *Change, Hope, and the Bomb*, p. 131.

[48]Carrol Wilson, "What Went Wrong," *Bulletin of the Atomic Scientists*, June 1979, pp. 13–15.

[49]See Atomic Energy Commission, *Civilian Nuclear Power* (Washington, D.C., February 1967) p. 6: "Known and estimated domestic resources of uranium are adequate to meet predicted light water reactor requirements for about the next 25 years pending the development of advanced converter and breeder reactors."

was believed that only with the breeder reactor could the use of fission energy continue to grow. In *Civilian Nuclear Power*, the AEC's 1967 report to the president, reactor safety was briefly discussed, but nothing was said about the fate of fission energy depending on strict observance, throughout the fuel cycle, of the imperatives to safeguard plutonium and contain radioactivity. Indeed, nowhere in this 56-page report was the safeguards problem even mentioned. There was but brief mention of reprocessing. Moreover, in announcing that the commission was going to leave research and development work on the reprocessing of light water reactor fuel to private industry,[50] the AEC was saying in effect that this technology was approaching maturity—a proposition being disproved almost daily at West Valley. The report said that high-level waste solidification and disposal technology had reached the pilot plant and field demonstration phase, with the results soon to be made available to industry.[51] There was no mention of trouble ahead. Glenn Seaborg, the chair of the commission at the time, believes today that the AEC was doing all that could have been expected in regard to nuclear waste management, although he concedes that viewed in hindsight the problem was "underestimated."[52]

The commission might have come more squarely to grips with the waste problem if it had not ignored its critics and in some instances suppressed their work. In May 1966 the Atomic Energy Commission received from the National Academy of Sciences' radioactive waste committee a report that viewed the waste disposal practices at AEC reservations with strong disapproval. For instance, the committee warned against the practice, at reservations such as Hanford and the National Reactor Testing Station in Idaho, of dumping transuranic waste in cribs or trenches. Reiterating its advice of 1960, the committee said that it was up to the commission to found "a discipline of waste disposal," a discipline that would become increasingly important as private operators took up waste disposal and the "profit motive [became] more prominent."[53]

Philip Boffey, in his book on the National Academy, *The Brain Bank of America*, has recounted how the Atomic Energy Commission suppressed the report and got rid of the radioactive waste committee.[54]

[50]Ibid., p. 27.

[51]Ibid., p. 43.

[52]Interview by the author with Seaborg, May 2, 1984.

[53]National Academy of Sciences, Committee on Geologic Aspects of Radioactive Waste Disposal, "Report to the Atomic Energy Commission's Division of Reactor Development and Technology," (May 1966) pp. 19–20 (hereafter cited as NAS Committee on Geologic Aspects of Radioactive Waste Disposal, "Report to the AEC," May 1966).

[54]Philip Boffey, *The Brain Bank of America* (New York, McGraw-Hill, 1975) pp. 95–98.

When the agency first saw the report in draft, there was an unsuccessful attempt to have all criticism of its operating practices deleted. Some six months after receipt of the final report, Milton Shaw sent the academy a lengthy critique disputing the committee's major conclusions, and declaring that inasmuch as personnel concerned with waste management now had the report, its further distribution or publication was not justified. The upshot was that the report was suppressed, despite efforts by the committee to have it released.[55] On top of that, the academy committee was told that its services would not be needed beyond the end of the year because the AEC's research and development program on waste disposal was far along.

The committee was ultimately reestablished at the academy's urging, but as Boffey has written, the original committee was dismissed and the new one was "established with a virtual guarantee that the AEC would have closer control over its operations."[56] In light of the commission's behavior, M. King Hubbert, chair of the academy's Division of Earth Sciences during part of the 1960s, developed a hearty sense of outrage. "There has never been an agency any more ruthless than the AEC," he told me. "They were a law unto themselves. They had [the support in Congress of] an entirely collusive Joint Committee on Atomic Energy, and there were simply no constraints."[57]

Embarrassment at Lyons

One early indication that the nuclear waste problem was still a long way from being solved came in the fall of 1971 at Lyons, Kansas, where the AEC suffered a major public embarrassment. The commission had the previous year announced prematurely, and without proper investigation, that the salt mine used earlier in its Salt Vault research project at Lyons

---

[55]The academy report was finally released five years later, but only after Senator Edmund Muskie of Maine demanded a copy.

[56]Boffey, *Brain Bank*, p. 98.

[57]Interview by the author with Hubbert, March 26, 1982. King Hubbert's early disillusionment with the AEC—and later, with nuclear energy—over the radioactive waste issue had a curious twist to it, in view of his unusual role in public discussion and debate over energy policy since the mid 1950s. A petroleum geologist for much of his career, Hubbert has received virtually every honor his peers can confer on him, including the $50,000 Vetlesen prize and gold medal, which for a geologist carries the prestige of a Nobel Prize. He is probably best known to the lay public for his early (1956) and at one time much disputed prophecy that both oil and natural gas production in the United States would peak before the mid 1970s and begin to decline. At the same time, he was also predicting that the nation's reserves of fissionable material (uranium and thorium) would allow nuclear energy to fill the breach.

was its tentative choice for the site of its first permanent repository. But the site was to prove indefensible and would have to be abandoned.

The Lyons embarrassment deserves recounting because to this day it is cited as evidence that the federal government is incapable of safely disposing of radioactive waste. The director of Salt Vault from its beginning in 1963 until its end in early 1967 was Frank L. Parker, then of the Oak Ridge National Laboratory but since then a professor of engineering at Vanderbilt University. Parker had found the Lyons mine acceptable for Salt Vault research, but this had involved only the temporary emplacement of seven canisters containing spent fuel assemblies to test the response of the salt to heat and radiation. To convert the mine, which was not in the best condition, to a repository for permanent emplacement of tens of thousands of waste canisters would be quite a different matter. Parker says, "We had not the slightest intention of ever using the mine for a repository."[58]

Nor had the commission itself, until 1970, planned to use the Lyons mine for a geologic repository. Commercial reprocessing was only beginning and AEC policy allowed high-level waste to remain with the reprocessor for up to ten years. But a fire that gutted the commission's Rocky Flats plutonium plant near Denver, Colorado, in May 1969 led to a speed-up in the agency's plans for a repository. Scrap from the plutonium metal that is machined at the plant into nuclear weapons parts is believed to have ignited spontaneously. The fire did upwards of $50 million in damages, and was one of the most costly industrial fires ever.[59] In the cleanup afterward, some 9,300 cubic meters of plutonium-contaminated debris, amounting to hundreds of railroad carloads when packaged, was shipped from Rocky Flats to the National Reactor Testing Station (NRTS) near Idaho Falls, where more than 50,000 cubic meters of transuranic waste was already buried.[60]

As the story goes, an Idaho trout farmer, alerted to the waste shipments by one of his customers back east, wrote Idaho's Governor Don W. Samuelson to air his concern about possible contamination of the Snake River plain aquifer, a prolific groundwater source that feeds the springs on which the state's trout farmers depend.[61] The NRTS waste burial site

[58]Interview by the author with Parker on March 31, 1982.

[59]Peter H. Metzger, *The Atomic Establishment* (New York, Simon and Schuster, 1972) pp. 147–150. Had the fire burned through the roof hundreds of square miles of the metropolitan Denver region might have been contaminated by plutonium aerosols. Ibid., citing testimony of Major General Edward B. Giller, director of the AEC Division of Military Applications, before a House Committee on Appropriations subcommittee hearing on the fiscal 1971 supplemental appropriations bill, October 1, 1970.

[60]Lawrence E. Davies, "Fire Cleanup Keeps Plutonium Plant Busy," *New York Times*, June 22, 1969.

[61]Metzger, *Atomic Establishment*, p. 150.

was situated directly above the aquifer, separated from it by some 600 feet of crevassed and fissured basalt rock, with several intervening layers of sediment. Further, the waste disposal practices at the Idaho station had been termed ill-advised in the 1966 report of the National Academy's radioactive waste committee, which had visited the site in 1965. The committee was left with "two unrelieved major anxieties": in some instances safe disposal for the long term was being subordinated to economy, and there was "overconfidence" that "vast quantities of radio-nuclides" would be safely contained indefinitely.[62]

The result of the alarm raised by the trout farmer was that Idaho's governor and two U.S. senators vigorously protested the shipment of the Rocky Flats waste to the NRTS. The commission promised them that all of the transuranic waste at this installation, including that which had been buried there in the past, would be recovered and committed to a federal repository in a salt mine.[63] Shortly afterward the AEC announced its tentative decision to convert the Lyons salt mine to a repository.

Former Atomic Energy Commission officials who were close to that decision emphasize that it was subject to confirmatory studies and was not final. "It was never characterized to me as anything other than tentative," recalls Owen Gormley, who became project engineer for the nuclear waste disposal effort about the time the decision was made.[64] But even a tentative decision of this kind implies a certain level of confidence and supporting information. Yet the AEC actually seems to have known little about the mine site's geologic and hydrologic setting. No inquiry was made before the decision to determine whether the integrity of the site might have been impaired by the extensive oil and gas exploration in this part of Kansas, nor had the AEC inquired whether mining practices of the American Salt Company at a mine adjacent to the AEC site had created uncertainties for the repository project. The agency thus missed the chance to learn early that by extensive solution mining—a process in which fresh water is injected into the salt and withdrawn as brine—American Salt had created an enormous cavity, later reported to be about one-half mile wide and three-quarters of a mile long.

Knowledge that such a cavity existed nearby would have posed the risk of a public relations disaster of the first magnitude for the Atomic Energy Commission. Solution mining leaves no pillars to support the overlying rock, which therefore may collapse into the cavity below,

[62]NAS Committee on Geologic Aspects of Radioactive Waste Disposal, "Report to the AEC," May 1966, p. 10.

[63]Letter from Glenn Seaborg, AEC chairman, to Senator Frank Church of Idaho, June 9, 1970.

[64]Interview by the author with Gormley on March 14, 1986.

creating a huge sinkhole. If this were to occur at the American Salt mine, the sinkhole would fill with water from aquifers above the salt, leaving a large pond several hundred feet deep about two miles from the Lyons mine, or perhaps even nearer. An AEC staff report later mildly observed that while there likely would be "no real technical significance for repository safety," so "sudden and dramatic" an event could "certainly engender unfavorable emotional and public relations problems."[65]

The governor of Kansas, Robert Docking, was not told that the Lyons mine might be converted to a repository until March 1970, or about two and a half months before the announcement of the decision. Commission officials gave the governor reassuring answers to his questions about the safety of the site, but his approval was neither sought nor obtained. According to Owen Gormley, the state raised no major doubts about site suitability until after the tentative decision was announced; indeed, as he recalls, Kansas geologists had been indicating that the salt beds were known for a predictable regularity.

But William W. Hambleton, the Kansas state geologist and a member of the National Academy of Sciences' panel on nuclear waste disposal in salt, was by no means satisfied that the Lyons site would be suitable, and was deeply concerned lest this site be chosen without proper investigation. Although aware that solution mining had been going on nearby and that unrecorded oil and gas exploration holes might be present at the site, Hambleton did not know a huge cavity had been created and he had no specific knowledge of exploration holes; his main concern was that too little was known about the regional geology and hydrology.[66]

The academy panel, which had been asked by the AEC to report on the suitability of salt as a generic rock type (but was not asked to consider the suitability of the Lyons mine), was meeting at the University of Kansas when word came of the Lyons site's tentative selection. The panel was embarrassed because, on this important siting question, its advice was going to come after the fact.[67] Its report on salt as a generic medium, already largely written, would now have to be rewritten to include an evaluation of the Lyons site. Hambleton in particular was offended by the commission's behavior. He saw it as high-handed and insensitive treatment of state officials.

[65]"AEC Staff Report on American Salt Company Operations," enclosure with a letter of September 30, 1971, by John A. Erlewine, AEC general manager, to Edward J. Bauser, executive director of the Joint Committee on Atomic Energy. See *Congressional Record* daily ed., October 15, 1971, p. 36452.

[66]Interview by the author with Hambleton on February 5, 1986.

[67]Boffey, *Brain Bank*, pp. 102–103.

Governor Docking, at Hambleton's urging, was now insisting that the Lyons site be thoroughly evaluated, and Hambleton was recommending that test boreholes be sunk at each corner of the site.[68] But according to Gormley, what he and other commission officials saw here was an attempt by the Kansas Geological Survey and certain Kansas geology professors to make AEC research grants the quid pro quo for their cooperation with a politically controversial project. To the commission, Gormley says, this smacked of "Southeast Asia," or Third World payola, and was not to be countenanced; necessary research would be funded after defensible research needs were defined through a careful "iterative process."[69] Just such a process was initiated with the Kansas Geological Survey, which later did in fact undertake two AEC-sponsored studies, one on the geology of the Lyons site, the other to identify possible alternative sites. Hambleton scoffs at Gormley's suspicions of his motivations: "The Kansas Geological Survey is a pretty well supported organization," he says. "Frankly, the job with the AEC was a pain. We didn't need any payola."[70]

Bad news for the Lyons project started coming early in 1971. The American Salt Company, which was carrying on mechanical as well as solution mining, disclosed that when small holes had recently been drilled into its mine's working face for the emplacement of explosives, water had leaked into the mine. The surmise was that an oil and gas borehole had been intercepted and was serving as a conduit for water from an aquifer either above or below the salt. The resulting leak had been grouted and sealed, but the incident raised two troubling questions.[71] First, if uncontrolled leaks of such origin should occur in the future at the American Salt mine, could the Lyons mine, only a third of a mile away, be affected? Second, did the problem at the American Salt mine not point up a potential for similar unwitting interceptions of oil and gas holes at the Lyons mine itself?

This second concern became far less speculative when twenty-nine exploratory oil and gas boreholes were later found in a survey of the Lyons mine site and a one-mile buffer surrounding it. The Oak Ridge consultant who did the survey concluded that the probability of successfully plugging three of these holes was very low.[72] This alone might

[68]William W. Hambleton, "Historical Perspectives on Radioactive Waste Management," *Kansas Water News* vol. 23, Irrigation Issue 2 (1981); interviews by the author with Hambleton on February 5, 1986, and April 2, 1986.

[69]Interview by the author with Gormley, March 14, 1986.

[70]Interview by the author with Hambleton, April 2, 1986.

[71]"AEC Staff Report on American Salt Company Operations."

[72]Ibid.

have made the Lyons project technically and politically indefensible, but more bad news was coming.

In September 1971 Hambleton learned from the head of the American Salt Company that several years earlier, when the company tried out a new solution-mining technique called hydraulic fracturing, pressure used to force water down a hole was lost after only a few hours and the water, some 175,000 gallons of it, simply disappeared. The expert consensus was that it would never be known whether the water broke through to aquifers above or below the salt, or somehow moved off laterally. "The uncertainty of the fate of this water raises the questions of whether it, and possibly other water in the solution mining cavern could be selectively migrating toward the [Lyons] salt mine," the AEC staff later observed.[73]

The upshot, as Owen Gormley now reconstructs these events, was that the American Salt official accompanied Hambleton to a meeting with AEC officials and repeated his account of the water's mysterious disappearance.[74] Coming on top of all the other disturbing news, this information had an immediate impact. On September 30, 1971, the AEC's general manager, John A. Erlewine, wrote the congressional Joint Committee on Atomic Energy that all work at the Lyons site was now "in abeyance"; the site had not been found unacceptable, but further site-oriented work would await a study of the borehole plugging problem, of the American Salt operations, and the results of a study by the Kansas Geological Survey of alternative locations.[75] The Kansas study, delivered several months later, found the Lyons area to be "the poorest candidate" of the eight areas considered.[76]

The Lyons project was not merely in abeyance, it was dead. For many months it had been in political trouble. Congressman Joe Skubitz, a conservative Kansas Republican, had become militantly opposed to the project even though Lyons lay a few counties beyond his district. His initial opposition was more on political than on technical grounds, for in his view the question of whether the repository was to be built should

[73]Ibid.

[74]Gormley interview on March 14, 1986. William Hambleton, in an interview by the author on April 2, 1986, did not have a sharp recollection of this meeting, but found no reason to contradict Gormley's account.

[75]Letter of John A. Erlewine to Edward J. Bauser, September 30, 1971.

[76]The Lyons area embracing the proposed repository site was considered the worst location largely because of numerous old oil and gas holes, a large number of producing wells, water above and below the salt, possible deep-seated structural problems, inadequate buffer zones, and high potential for development of oil and gas reserves in the future. Atomic Energy Commission Press Release no. P-20, January 21, 1972, with attached chapter 7 from the Kansas Geological Survey report, cited in Boffey, *Brain Bank*, p. 105.

ultimately be decided by the will of the people of Kansas.[77] At Skubitz's urging, the two U.S. senators from Kansas had managed to have the Atomic Energy Commission's annual authorization bill amended to prevent any work at Lyons until a distinguished technical advisory committee could certify the proposed repository site as safe.[78] Governor Docking and many members of the Kansas legislature were also opposed to the project. Indeed, with the news of the borehole plugging problem and the disappearance of the water in the hydraulic-fracturing incident, the Atomic Energy Commission's credibility had, in the words of the agency's former chief historian, "dropped to a new low."[79]

The agency had underestimated the complexities and uncertainties inherent in repository siting, which had been made all the greater in this case by virtue of the extensive oil and gas exploration and salt mining in the region. It had also failed to establish from the start a cooperative relationship with Kansas state officials. The Lyons fiasco is still remembered as a testament to the technical hubris and political naivete which the AEC brought to this first attempt at a delicate siting task.[80]

Other Embarrassments

There was other evidence in the early 1970s that the nuclear waste program was in trouble. A major tank leak at Hanford in the spring of 1973 made the lack of permanent disposal facilities all the more glaringly

[77]Hewlett, "Federal Policy for the Disposal of Highly Radioactive Wastes," pp. 17–18.

[78]Metlay, "History and Interpretation of Radioactive Waste Management," p. 6.

[79]Hewlett, "Federal Policy for the Disposal of Highly Radioactive Wastes," p. 19.

[80]A National Academy of Sciences panel report on nuclear waste disposal in salt, issued in November 1970, cautioned the AEC to look for oil and gas exploration boreholes but made no mention of American Salt Company operations. It called for further studies and confirming evidence of the site's suitability, but said the panel anticipated no intractable problems. The panel had missed a chance to persuade the AEC to give up the Lyons site at least a year before the site was abandoned. Panel member Frank Parker, who had directed Salt Vault and who today is chairman of the academy's Board on Radioactive Waste Management, is sure he learned of the hydraulic-fracturing incident from an official of the Carey Salt Company, which operated the Lyons mine, long before Salt Vault was terminated in early 1967. William Hambleton of the Kansas Geological Survey and some of Parker's former Oak Ridge colleagues suspect his memory may be playing tricks on him, and that he never actually had early knowledge of this incident. Parker himself entertains no such doubts. But it is a fact that never once did he mention the incident to Hambleton or others on the academy panel. Parker's explanation is that after Salt Vault the hydraulic-fracturing incident was forgotten until the disclosures of the fall of 1971 reminded him of it. In any case, Parker says, "the role of the panel was totally undercut by the AEC's premature decision. My actions or lack thereof had no impact." Letter from Parker to the author, March 10, 1986.

evident. Between August 1958 and June 1973, 15 of Hanford's 149 old single-wall tanks had become "confirmed leakers" and had lost several hundred thousand gallons of waste to the surrounding soil. It was to address this problem that the Hanford managers had begun reducing the liquid waste in the older tanks to a salt cake. By 1973 considerable progress to this end had been made, but further leaks were an ever-present possibility. The leak discovered on June 8 of that year in tank 106-T was Hanford's biggest single leak ever: 115,000 gallons of high-level waste, containing more than 54,000 curies of radioactivity (mostly cesium and strontium), had escaped from the tank to the surrounding soil.[81]

That leak pointed up the fallibility of Hanford's system of management as much as it did the failing integrity of the tanks. The leak had gone unnoticed for some fifty-one days, with more than 2,000 gallons escaping each day. Furthermore, monitoring data collected during this period showed that something was wrong. But the operators who took the readings of waste levels in the tanks were not expected to interpret them; instead, their charts and graphs were left with a day-shift super-visor—who let them pile up on his desk, unread, for six weeks. Another technician finally discovered the leak.[82]

For Hanford, the Atomic Energy Commission, and waste management the June 8 leak was a public relations disaster. It led to a flurry of lawsuits by environmental groups to stop the generation of reprocessing wastes at Hanford, and to some frightening headlines. A six-column banner, "Nuclear Wastes Peril Thousands," appeared in the *Los Angeles Times* twenty-two days after the leak was reported.[83] While there actually was no immediate danger to anyone—the waste was absorbed in the soil near the tank—the leak had indeed called attention to nuclear waste as a growing and unresolved problem. Newspaper and magazine writers around the country were now alerted to a new problem and a new issue.[84] Robert Gillette, writing in *Science*, pondered whether the AEC was prepared to manage the voluminous wastes from a commercial nuclear power program if it could somehow "lose the equivalent of a railroad tank car full of radioactive liquid hot enough to boil itself for years on end and knock a geiger counter off scale at a hundred paces."[85]

[81]Gillette, "Radiation Spill at Hanford."

[82]Ibid.

[83]Ibid.

[84]William L. Rankin, Stanley M. Nealey, and Daniel E. Montano, "Analysis of Print Media Coverage of Nuclear Power Issues" (Seattle, Battelle Human Affairs Research Centers, April 1978) pp. 40, 48.

[85]Gillette, "Radiation Spill at Hanford," p. 730.

Some of the commercial low-level waste burial grounds also were giving trouble. One of these was the 252-acre facility at Maxey Flats, a flat-topped ridge in northeastern Kentucky. The Commonwealth of Kentucky, back during the nuclear euphoria of the early 1960s, had entered into an agreement with the Atomic Energy Commission to establish a low-level waste burial ground in the vain expectation of attracting nuclear industries.[86] The facility was operated on state-owned land by the Nuclear Engineering Company (NECO) as a for-profit enterprise. Substantial amounts of transuranic waste continued to be buried there for several years after the commission decided that such waste should go to a deep geologic repository, and it was the state, not the AEC, which in 1974 finally stopped NECO from accepting that waste.[87]

In 1972 the state Radiation Control Branch first detected radionuclides escaping from the site, and two years later traces of plutonium were found beyond the site boundaries. Because the shale in which the waste disposal trenches were dug had very low permeability, the trenches, once filled in, tended to fill up with water—a so-called bathtub effect—and to overflow. The trenches could be pumped out and the radioactive water that was recovered could be evaporated to reduce its volume. But the evaporation process itself produced a plume that contained tritium, and there was a radioactive sludge that had to be disposed of as low-level waste.[88] Unless the trenches, some of them almost 700 feet long, could be permanently capped to keep water out, Kentucky faced a custodial problem that would never end, given the 24,000-year half-life of plutonium. In 1977 the commonwealth, badly burned by this experience, closed the Maxey Flats site. Similar problems were occurring at the low-level waste burial grounds at West Valley, and those sites were to be closed also.

The cumulative political impact of the radioactive waste issue had been growing since the mid 1960s as the public became more aware of first one, then another aspect of the overall problem. For instance, the uranium mill tailings situation, which had begun attracting wider notice in 1966 when public health officers discovered the tailings diversions

[86]Kentucky Legislative Research Commission, "Report of the Special Advisory Committee on Nuclear Waste Disposal" (Frankfort, Ky., October 1977) pp. 1–4.

[87]In 1970 the AEC had stopped shallow land burial of transuranic waste in its waste management operations at Hanford, the National Reactor Test Station in Idaho, and elsewhere. But this practice continued through most of the 1970s at certain commercial low-level waste burial facilities where host-state officials allowed it. Not until late 1982, when the Nuclear Regulatory Commission adopted regulations 10 CFR, part 61—"Licensing Requirements for Land Disposal of Radioactive Wastes"—was the shallow burial of TRU waste from commercial sources prohibited.

[88]Kentucky Legislative Research Commission, "Report of the Special Advisory Committee," p. 16.

at Grand Junction, Colorado, was the subject of widely reported hearings by the Joint Committee on Atomic Energy in 1971. At that time the committee was still worrying more about shielding the AEC from criticism than correcting the problem.[89] When the joint committee and Congress finally took remedial action the next year it was only after years of foot-dragging and denial.

The AEC Gets a New Chairman and a New Waste Policy

In 1971 Glenn Seaborg left the Atomic Energy Commission after ten years as chairman and was replaced by James R. Schlesinger. This was a change of considerable symbolic importance. Seaborg, a co-discoverer of plutonium as a brilliant young chemist at the University of California in 1941, was dedicated to creating a "new world" through nuclear technology.[90] He saw plutonium as the potentially abundant fissile material that was going to make it all possible. As chair of the commission Seaborg was, as much as anything, promoter of the technology which the commission had a responsibility to develop and regulate.[91] Schlesinger, on the other hand, was a former Rand Corporation economist who saw the commission's role as "primarily [that of] a referee serving the public interest."[92] He saw a potentially important new source of energy, yes, but he also saw problems and uncertainties. In a speech given at an important nuclear industry meeting, Schlesinger noted the industry's spectacularly rapid development, which he likened to compressing the entire history of commercial aviation "from Kitty Hawk to the Boeing 747 . . . into less than a score of years." He said that "environmentalists have raised many legitimate questions" and that "good answers" remained to be provided on several issues of concern, including reactor safety and waste management.[93]

---

[89]The behavior of the chairman, Congressman Wayne N. Aspinall, toward critical witnesses was described by the *Denver Post* as "hostile" and "abusive." *Denver Post*, "Aspinall Handling of AEC Critics on Mill Tailings Rude and Wrong," editorial, November 3, 1971, cited in Metzger, *Atomic Establishment*, p. 195.

[90]Glenn Seaborg (written with William Corliss), *Man and Atom* (E. P. Dutton, 1971).

[91]Ford, *Cult of the Atom*, p. 23, quotes Richard Hewlett, "Seaborg was the granddaddy of plutonium, and it was always extremely important to him that he discovered a new element that would be the salvation of mankind." According to Hewlett, Seaborg "saw the whole world revolving around this technology, and here he was in the center of it."

[92]Schlesinger speech before the Atomic Industrial Forum and the American Nuclear Society, Bal Harbour, Fla., October 20, 1971.

[93]Ibid.

Schlesinger would soon have the Atomic Energy Commission put forward a new answer to the question of the disposal or storage of high-level waste or spent fuel. In 1972 the agency announced plans for development of a retrievable surface storage facility, which would be a mausoleum or array of vaults to which waste or spent fuel canisters could be committed and watched over for as long as desired. "The critics of the Lyons project were asking, why take a risk on geologic disposal before more is known about it? I thought this was a perfectly sensible point. So I proposed the mausoleum approach to allow time to resolve the uncertainties," Schlesinger told me.[94] As Schlesinger viewed the proposal, the Retrievable Surface Storage Facility, or RSSF, could be either temporary, pending permanent geologic disposal if that solution were found acceptable, or it could continue indefinitely. "I wanted to keep my options open," he said. The RSSF concept, he explained, was a reflection of the "old Rand R&D approach" of seeking better information before deciding what finally to do. "This was the opposite of the AEC approach, which was to design high confidence solutions on paper, trying to convince the public without test work that the solution was acceptable. It [the AEC approach] became increasingly irrelevant when the authority of the scientific community, which was high in the 1950s and 1960s, began to erode in the 1970s." Besides the erosion of public trust in science there was to be, from 1973 on, a "prevailing distrust for government," Schlesinger observed.[95]

The commission was continuing, on a modest level, to investigate the geologic disposal option. Its most important new initiative in this field was to look into the possibility of establishing a repository in the Salado bedded salt formation near Carlsbad, New Mexico. But this new initiative, the Waste Isolation Pilot Plant (to become known by the acronym WIPP), had a long way to go (see chapter 5). The most conspicuous item by far on the AEC's waste management agenda was the Retrievable Surface Storage Facility.

Sensitive to criticism that geologic disposal was being abandoned, the commission was not presenting the RSSF as a possible alternative to deep mined repositories, however Schlesinger personally may have felt about this. (Schlesinger was not to be much longer at the AEC: in 1973 President Nixon appointed him director of the Central Intelligence Agency.) A draft environmental statement issued by the agency in September 1974 was at pains to say that "the AEC has not stated that perpetual surveillance and maintenance of surface storage is either necessary or desirable, and has kept development work toward geologic disposal as a vital part of its program."[96] But neither environmental groups nor the Environmental Protection Agency (EPA) were accepting such assurances. The EPA expressed particular concern "that economic factors could later dictate utilization of the facility as a permanent repository contrary to the stated intent to make the RSSF interim in nature."[97]

Gus Speth, at the time an attorney for the Natural Resources Defense Council, was among the critics of the RSSF proposal. "The big problem with the RSSF concept was, and is, that it implies long-term decoupling of wastes from reactors," Speth told me.[98] "That is, we could have countless reactors without any assured means of permanent disposal."

But the RSSF initiative was to go nowhere. The Atomic Energy Commission itself was to be abolished at the end of 1974, and its successor agency would withdraw funding requests for construction of this facility.

By the 1970s it was probably too late for the Atomic Energy Commission, no matter what its new attitudes and policy initiatives, to establish a societal consensus favorable to nuclear power. Indeed, it was probably even too late for the agency to save itself. Two things were to bring about its demise and replacement by the Energy Research and Development Administration (ERDA) and the Nuclear Regulatory Commission. First, the 1973–1974 oil shock following the Yom Kippur War would point up the need for a broad-based energy policy and an agency that looked at the whole spectrum of energy choices. Second, the AEC, along with its congressional patron, the Joint Committee on Atomic Energy, had lost credibility and lost trust, and there was increasing criticism of the commission's dual role as developer and regulator of nuclear power.

The Atomic Energy Commission lost credibility and trust because of its mistakes, omissions, and growing reputation for bureaucratic arrogance, particularly with respect to issues being raised about reactor safety. But also the commission had now come into a drastically altered political atmosphere. The atmosphere during the 1950s and much of the 1960s had been, for the most part, accepting, and the agency was often under pressure from Congress to be more aggressive and venturesome in pursuing peaceful applications of the atom. Making a major difference now was the environmental movement, and, within it but distinguishable

[96]Atomic Energy Commission, "Management of Commercial High-Level and Transuranium-Contaminated Waste," draft environmental statement, WASH-1539 (September 1974) pp. 1.2-8 and 1.2-9.

[97]Environmental Protection Agency comments on draft environmental impact statement WASH-1539, November 15, 1974, cited in Metlay, "History and Interpretation," p. 8.

[98]Gus Speth, personal communication to the author, July 1982.

from it, another movement made up of growing numbers of increasingly resourceful opponents and critics of nuclear power.

The Nuclear Critics and the Technology Issues

The National Environmental Policy Act (NEPA), signed into law by President Nixon on January 1, 1970, formally established the expectation that promoters of new technologies be able to show that potential environmental effects have been identified and that alternative courses of action have been considered. The act did not insist on having well in hand plans and programs to deal satisfactorily with such unavoidable effects as the generation of wastes, but that was the commonsense political implication. The Atomic Energy Commission and the nuclear industry were thus confronted with a major new political demand—a demand that could not be met because nuclear power, in full surge of development, was a technology ahead of itself in regard to reactor safety, plutonium safeguards, and waste management.

After the 1973–1974 oil shock brought a sudden rise in energy prices and a quick decline in the growth of electricity demand, an attempt would be made to justify continued major growth in nuclear capacity in the name of energy independence. Nuclear power could indeed displace part of the oil-fired electric generating capacity, but the nation's insecurity with respect to energy lay principally with potential shortages of liquid fuels for transportation. Nuclear's place in the national energy mix would have to rest on the claim of its proponents that fission energy was a benign source of electricity, ideal for baseload generation and environmentally and economically superior to coal.

Schlesinger had entertained the hope that environmentalists would recognize nuclear power as benign, and in his first appearance at an industry meeting he expressed a belief that "when all environmentalists, including ourselves, have a chance to assess the contribution of nuclear power to the reduction of sulphur and nitrogen oxides and particulates, ... [they] will appreciate the advantages of nuclear power in relation to the real alternatives."[99] Not many years past, some leading environmentalists had in fact regarded nuclear power as a benign alternative to energy projects they were opposing. For example, David Brower, as executive director of the Sierra Club in the mid 1960s, had advocated a nuclear alternative to hydropower dams in the Grand Canyon.[100]

[99]Schlesinger speech before the Atomic Industrial Forum and the American Nuclear Society, 1971.

[100]Sheldon Novick, *The Electric War: The Fight Over Nuclear Power* (San Francisco, Sierra Club Books, 1976) pp. 185–189.

Yet long before the start of the 1970s, as far back as the mid 1950s, there were beginnings of what would become a national anti-nuclear movement. The controversy in the 1950s over fallout from nuclear weapons tests may have marked the first stage in this political evolution. A number of scientists—Linus Pauling, Barry Commoner, and others— had become alarmed despite continuing AEC denials that the fallout was harmful.[101] Ultimately, the Baby Tooth Survey initiated by the St. Louis Committee for Nuclear Information,[102] as well as other investigations, revealed that for every person on earth there would be a small but not insignificant "dose commitment" of strontium-90, cesium-137, and other radionuclides from fallout. This commitment, though trivial in terms of the risk for any given individual, may have increased the lifetime incidence of cancer deaths worldwide by as much as 60,000, which compares to a total of about 2.5 million cancer deaths caused by natural background radiation.[103]

By the 1970s Barry Commoner would become a prominent and articulate opponent of nuclear power. Among others for whom the fallout debate of the 1950s was a major influence was Ruth Weiner, a doctoral candidate at Johns Hopkins University in 1959 when Edward Teller told students there not to worry about fallout. In the 1970s and 1980s she would be an environmental science professor and anti-nuclear activist in the State of Washington, and in the latter role would lead a ballot initiative campaign seeking to ban most radioactive waste shipments into

---

[101]A remarkable episode bearing on the intransigence of the AEC in regard to the fallout issue was noted by Harrison Brown in the editorial "Fallout and Falsehoods" in the April 1985 *Bulletin of the Atomic Scientists.* During the period in question Brown was a professor of geochemistry at the California Institute of Technology. According to Brown, he was one of ten Caltech scientists who during the 1956 presidential election issued a statement supporting Adlai Stevenson's proposal to ban weapons tests in the atmosphere. John A. McCone, a member of the Caltech board of trustees and chairman of its fund-raising committee, demanded that the ten be fired. Because this demand was refused, McCone resigned from the board. When Lewis Strauss vacated the AEC chairmanship two years later, President Eisenhower appointed McCone to succeed him.

[102]Baby teeth were collected by the committee as evidence of environmental contamination by strontium-90, a boneseeker.

[103]At the peak of the weapons testing the dose commitment could have been as much as 20 millirems, equal to one-fifth of a year's exposure to natural background radiation of 100 millirems, which is believed to cause 5 cancer deaths a year per million population, or 300 deaths over a 60-year life-span. According to John H. Harley, formerly director of the AEC's Environmental Measurements Laboratory in New York and chair of a National Council on Radiation Protection and Measurements committee on background radiation, on this basis the dose commitment from fallout would increase the lifetime incidence of cancer deaths by about 10 to 15 per million; for a global population of 4 billion, this is 40,000 to 60,000 deaths resulting from fallout. Harley was interviewed by the author in March 1982.

the state.[104] There were to be even more direct political effects of the fallout and weapons testing issues on attitudes toward nuclear waste disposal. In 1981, when I asked the governor of Nevada, Robert List, why many Nevadans objected strongly to radioactive waste being disposed of in their state, he replied: "It goes back to the atmospheric testing in the 1950s. The AEC misled the public, and [now] people don't trust the authorities. They are cynical. There are survivors who litigate in the courts. There is high public awareness."[105]

Also among the early challenges to nuclear power were citizen protests over some of the first reactor projects. The first such protest went back to 1956 and Fermi-1. Fermi was the ill-fated breeder that was to be built on the west shore of Lake Erie between Detroit and Toledo— a reactor which, after many delays, would finally go into commercial operation in 1966, only to suffer a partial core melt within a few months.[106] This was a project that the Advisory Committee on Reactor Safeguards (ACRS) would not certify as safe, but which the Atomic Energy Commission had allowed to proceed anyway, attempting in the process to suppress the ACRS report.[107] The United Auto Workers, with encouragement from the Joint Committee on Atomic Energy, led the protest and unsuccessfully brought suit to block it. The controversy over Fermi-1 was to have a lasting and ironical consequence. In its anger over the AEC's behavior, the joint committee, much more disposed at that time to play the role of an independent overseer than it was to be later, in 1957 pushed through amendments to the Atomic Energy Act making public release of ACRS findings mandatory, and requiring public hearings on all construction applications, whether there were issues in dispute or not.[108] The hearings requirement opened the way for numerous local intervenor actions during the years ahead.

One early action of this kind arose after the Pacific Gas and Electric Company undertook in 1961 to build a reactor on the California coast at Bodega Head, a site that would eventually be found vulnerable to active earthquake faults. This action was led by David Pesonen, a young biologist who resigned from the Sierra Club staff and undertook the struggle on his own after failing to persuade the club to oppose the plant.[109] The Bodega Head Association on one occasion released some 1,500 balloons from the site, each bearing the message that "this balloon

[104]Interview by the author with Weiner, February 1981.

[105]Interview by the author with Governor List, February 1981.

[106]William Lanouette, "Dream Machine," *The Atlantic Monthly*, April 1983, p. 44.

[107]Ford, *Cult of the Atom*, pp. 54–57.

[108]Lanouette, "Dream Machine," p. 43; Green, "Peculiar Politics," p. 63.

[109]Novick, *Electric War*, p. 241.

could represent a radioactive molecule of strontium-90 or iodine-131."[110] Plans for the Bodega Head project were cancelled in 1964. Pesonen was later to become an attorney in San Francisco and to lead the 1976 campaign for the Proposition 15 ballot initiative aimed at shutting down nuclear power in California if reactor safety and radioactive waste disposal could not be assured.[111]

Another early intervenor action followed Consolidated Edison's announcement in late 1962 of a plan to build a 1,000-megawatt reactor in the Queens borough of New York City, only a half mile from Manhattan. David Lilienthal, the former AEC chairman, touched off the citizens' protest against this proposal, which Con Ed eventually abandoned.[112] During this period Lilienthal also was warning about the radioactive waste problem. In his lectures at Princeton University in 1963 he asked, "Should not a program of large-scale atomic reactors wait at least until it has been demonstrated that this waste problem has been 'licked'?" Lilienthal added that "so long as the safe handling of radioactive waste for a large-scale national atomic power program is still in the research stage ... *the future of atomic energy as a major reliance for civilian electricity is in grave doubt*" (his emphasis).[113]

Toward the end of the 1960s intervenor actions became numerous, in part for the simple reason that nuclear power was now burgeoning. During one three-year period the number of people living within a 30-mile radius of an existing or proposed reactor suddenly had more than doubled, going from 14 million to 37 million.[114] Accordingly, there was now the potential for opposition to nuclear power to assume the character of a widespread grass-roots movement. Numerous small opposition groups sprang up in the vicinity of proposed plants, with one group often reaching out to another to establish a cooperative network. These groups have been described by sociologist Robert C. Mitchell:

Although affected by the climate of protest which prevailed in the late 1960s, [the groups] eschewed the extra-legal tactics of riots, large demonstrations, and civil disobedience .... Their tactics were based on the use of available legal remedies, technical arguments, and conventional political actions; tactics with which they, as basically conventional middle-class citizens, felt more comfortable. ...

[110]Steven Ebbin and Raphael Kasper, *Citizens Groups and the Nuclear Power Controversy* (Cambridge, Mass., MIT Press, 1974) p. 11.

[111]Luther J. Carter, "Nuclear Initiative: Impending Vote Stimulates Legislative Action," *Science*, June 4, 1976, pp. 975–977.

[112]Robert Cameron Mitchell, "From Elite Quarrel to Mass Movement," *Transaction/Society*, July–August 1981; Ebbin and Kasper, *Citizens Groups*, p. 11.

[113]Lilienthal, *Change, Hope, and the Bomb*, p. 137.

[114]Mitchell, "From Elite Quarrel to Mass Movement," p. 77.

These local protest groups became the core of the movement and its real strength. Although they developed independently of each other, the early grass-roots groups followed a similar path. In each case the announcement of a proposed nuclear power plant stimulated a small number of local residents to seek more information about its potential safety or environmental effects. Unsatisfied with the answers they received, they would begin to question the project publicly and to organize against it. These citizen-activists were typically well educated and over 30; a striking number of the early leaders were women. Although few of the early protesters had been active in the anti-war or civil rights movements, they were much more likely than their neighbors to have a liberal political orientation and a deep environmental consciousness.[115]

According to one study, these local groups were "motivated in large measure by fear—primarily fear of accident and fear of radiation exposure." Their interventions were also characterized as a "manifestation of their anger and frustration at what they interpreted as a hollow extension of procedural due process and the denial of meaningful or substantive due process." Some citizens were said to view the Atomic Energy Commission hearings as "expensive wars of attrition, designed to obfuscate rather than clarify the true risks and benefits of nuclear power."[116]

By the early to mid 1970s a national movement challenging nuclear power was taking shape. This was evident from the emergence of groups such as the Consolidated National Intervenors, the Union of Concerned Scientists, and Ralph Nader's Critical Mass; the growing involvement of national environmental groups such as the Sierra Club and the Natural Resources Defense Council in specific nuclear issues; the ballot initiatives on reactor safety and waste issues in several western states (especially California); and the nonviolent, civil disobedience campaign by counterculture groups such as the Clamshell Alliance.

A striking characteristic of this new movement was, and is, the wide diversity in the kinds of people involved and their motivations. The Union of Concerned Scientists (USC) was reformist in character, at least at its inception, and not strictly anti-nuclear; indeed its leader, physics professor Henry Kendall of the Massachusetts Institute of Technology, once asserted that there was no fundamental technical obstacle to nuclear safety.[117] The UCS developed a thorough understanding of technical issues, such as the emergency core cooling system, that won the

[115]Ibid.

[116]Ebbin and Kasper, *Citizens Groups*, pp. 247, 252.

[117]Testimony of Henry Kendall before the Joint Committee on Atomic Energy, January 1974, JCAE, pt. 2, vol. 2 (January 1974) p. 102.

respect of experts in the national laboratories.[118] Nader's Critical Mass, on the other hand, was totally and avowedly anti-nuclear; its battle cry was not "No More Nukes" or "Safer Nukes" but "No Nukes." Critical Mass was basically a political group, given to flamboyant rhetoric and calls to action; it did not engage in the kind of meticulous technical analysis used by the Union of Concerned Scientists during the Atomic Energy Commission's long-drawn-out proceedings on the cooling system issue. Nader had arrived at an unshakeable belief that nuclear technology was fundamentally unacceptable—that it was uneconomic and posed irremedial hazards from radioactive waste, reactor meltdowns, sabotage and terrorist incidents, and nuclear weapons proliferation.[119]

The Clamshell Alliance, which mounted mass demonstrations of civil disobedience at the construction site of the Seabrook power station in New Hampshire in 1976 and 1977, represented something quite different from either Critical Mass or the Union of Concerned Scientists. Inspired by the successful site occupation that German protesters staged at Wyhl in 1975, the Clams established a style of direct action that was promptly taken up by numerous other local groups—the Catfish, Palmetto, and Abalone alliances, among others—who during a single year held more than 120 demonstrations and rallies. The Clams and their imitators drew their members from the "radical, counter-culture, and peace movement subcultures." They sought to stop nuclear power not only because they regarded it as unsafe, but because many sought "to change American society fundamentally in the direction of decentralization, demilitarization, and egalitarianism," believing that a necessary condition for this change was "a shift from centralized, capital-intensive power producing systems, of which nuclear power is the epitome, to decentralized, labor-intensive, small-scale, renewable energy systems."[120]

The national environmental groups represented a still further element of diversity in the movement challenging nuclear power. The boards of directors of some of these groups, such as Friends of the Earth and the Sierra Club, either were or eventually would become anti-nuclear in the sense that they called for the phasing out, if not immediate abandonment,

---

[118]After Henry Kendall and other UCS leaders visited the Oak Ridge National Laboratory in 1971, Donald Trauger, the laboratory's associate director, wrote AEC headquarters that they had "become intimately familiar with the relevant published literature [and had] become aware of various deficiencies in the case for ECCS performance." Cited by Joel Primack and Frank von Hippel, *Advice and Dissent: Scientists in the Political Arena* (New York, New American Library, 1974) p. 217.

[119]Novick, *Electric War*, pp. 310–324.

[120]Mitchell, "From Elite Quarrel to Mass Movement," p. 82.

of nuclear energy. Others took strongly critical positions on specific issues, but without categorically calling for an end to nuclear power. For instance, the Natural Resources Defense Council (NRDC), an environmental law group, took a large and continuing interest in the radioactive waste issue. In 1971 Gus Speth, one of the Yale Law School graduates who had founded the council, argued in a suit involving the Vermont Yankee reactor that the time to consider the fate of the waste to be generated is at the time the reactor is up for licensing, not later when the wastes actually exist. This case and other litigation springing from it went all the way to the U.S. Supreme Court, with the council losing in the end but increasing the pressure on the Atomic Energy Commission and its successor agencies to develop a permanent solution to the waste problem.

In 1977 a task force made up of leaders of national environmental groups issued *The Unfinished Agenda*, a citizen's policy guide to environmental issues.[121] Citing hazards involving reactor safety, misuse of explosive nuclear materials, and radioactive waste, the report said that a review of nuclear technology "leaves one shaken with the potential for disaster." Task force leaders were speaking only for themselves and not for their groups; nonetheless, for a consensus document to be so strongly critical was clearly significant. The groups to which they belonged were all part of the American mainstream, though in somewhat disparate ways and not always broadly representative of it.

As Robert Mitchell has said, members of environmental groups tend to be very well educated, one survey showing that nearly half the members of the five groups investigated had studied one or more years beyond undergraduate college.[122] What this suggests, and what seems evident to anyone familiar with groups such as the Sierra Club and the National Audubon Society, is that these are people widely distributed in the professions and in positions of leadership in American society, and that their influence should not be underestimated. Yet the sources of

---

[121]Members of the task force, which was sponsored by the Rockefeller Brothers Fund, were from the Sierra Club, the National Audubon Society, the Natural Resources Defense Council, Friends of the Earth, the Wilderness Society, Zero Population Growth, the National Wildlife Federation, Massachusetts Audubon Society, the Nature Conservancy, the Environmental Defense Fund, the Izaak Walton League of America, and the National Parks and Conservation Association.

[122]Robert Cameron Mitchell, "How 'Soft,' 'Deep,' or 'Left'? Present Constituencies in the Environmental Movement for Certain World Views," paper in the symposium "Whither Environmentalism?" in *Natural Resources Journal* vol. 20 (University of New Mexico School of Law, April 1980) pp. 345–346. The groups surveyed were Environmental Action, the Environmental Defense Fund, the National Wildlife Federation, the Sierra Club, and the Wilderness Society.

the challenge to nuclear power have been consistently misunderstood and underrated by many people associated with the nuclear enterprise. Sociologist Robert L. Cohen, in a 1982 study, concluded that industry decision makers tend to deny "legitimacy" to their opponents by dismissing them as "special interest groups" whose "knowledge, reasoning ability, motives, integrity, and even patriotism" are suspect.[123]

Showdown in California: The Consequences of Rapid Development

Some eighteen years after the first commercial reactor went on line the political consequences of nuclear power's precipitous development caught up with the nuclear enterprise. Proposition 15, the California ballot initiative sponsored by nuclear opponents and critics, was designed to stop construction of new reactors and force a rigid phase-out of existing ones absent convincing assurances with respect to reactor safety and radioactive waste disposal. As things turned out, the ballot proposal was not destined to be approved, but the California legislature itself was to impose a moratorium on further building of reactors pending development of a demonstrated technology for the disposal of high-level waste.

The showdown came in the late spring of 1976. About eighteen months earlier attorney David Pesonen, veteran of the successful campaign to stop the reactor project at Bodega Head, had begun the ultimately successful campaign to collect the half million signatures necessary to place Proposition 15 on the ballot. California citizens were to vote on June 8, and only a few weeks remained for them to make up their minds about the proposition's fate.

The California utilities and the rest of the nuclear enterprise enjoyed major political advantages in this struggle, but they clearly were on the defensive, especially on the waste issue. No commercial reprocessing was being done, none was in prospect, and spent fuel was accumulating at reactor sites—and there was no place to put it. The Lyons repository project had been aborted, and the Retrievable Surface Storage Facility had been abandoned as a near-term alternative. The Energy Research and Development Administration, as an AEC successor agency, was attempting to launch a well-funded geologic disposal program, but this effort was proceeding in hesitation and confusion (see chapter 5). There were also major reactor safety issues to be resolved. The reassuring

[123]Robert Leslie Cohen, "The Perception and Evaluation by Decision Makers: Civilian Nuclear Power in the United States," (Ph.D. dissertation, Columbia University, 1982) pp. 410–411.

general conclusions of a recent reactor safety study led by Norman Rasmussen of the Massachusetts Institute of Technology had been hailed by the nuclear industry, but the study was receiving highly critical reviews and its reckoning of risks would ultimately be repudiated by the Nuclear Regulatory Commission.

Moreover, three middle-management engineers at General Electric's nuclear division at San Jose had grabbed the headlines by resigning and going over to the Yes-on-15 campaign, warning that grave risks of reactor core meltdowns remained. And further, earthquake risks had frustrated the finding of acceptable reactor sites in California. Coastal locations had been preferred because of the ready availability of cooling water, but it was along the coast that seismic activity was greatest. Plans for reactors at three different sites—Bodega Head, Mendocino, and Malibu—had for this reason been abandoned. Still worse, licensing of two reactors then in an advanced stage of construction at Diablo Canyon was being held up because of the discovery of a previously unsuspected earthquake fault not far off shore.

In light of all these factors, to demand that plans for nuclear expansion be given a hard second look and perhaps be put on hold was not unreasonable. In 1976 nuclear generation represented only 4 percent of California's total electricity generating capacity; but by 1995 it was expected to be the state's largest source of power, and California's dependence on this source might then be almost irreversible, whatever the risks associated with it.[124]

Proposition 15 was perhaps needlessly provocative in that it applied not only to all new reactors proposed for the future, but also to the seven reactors then existing or under construction, and might lead to their abandonment. But any ballot initiative challenging the growth of nuclear energy in California would have been powerfully opposed, as the more limited initiatives on the ballots that year in Colorado, Oregon, and other states were opposed. The sheer economic and political inertia of the multi-billion-dollar nuclear program was the main force to be reckoned with. The weight of most of California's established institutions was thrown against Proposition 15. Formally leading the opposition was a coalition of utilities, labor unions, and other business and development interests, including the chambers of commerce, the League of Cities, the water districts, the Farm Bureau, and most of the state's important newspapers. Utilities and other corporations were contributing financially to the No-on-15 campaign, which raised large sums, certainly far exceeding $2 million. In addition, a week or so before the vote, the No-

[124]Carter, "Nuclear Initiative," p. 976.

on-15 campaign brought forth a list of 5,200 scientists and engineers opposed to the initiative. The opposition campaign also was receiving substantial support from ERDA, whose San Francisco regional office ran a speakers' bureau and issued a report stating that a nuclear shut-down in California would cost every family of four in the state $7,500 in increased electricity charges over the following two decades. Taken altogether, No-on-15 represented as powerful an aggregation of interests as one could imagine.[125]

The Yes-on-15 campaign was led or energized primarily by the environmental groups, plus a politically potent Southern California group called the People's Lobby and a dedicated Palo Alto-based activist group known as Project Survival—the latter a political spin-off of the philosophical and semireligious organization Creative Initiative.[126] On university campuses there was substantial faculty support for as well as against the ballot initiative; most students were believed to support it. All told, while the identifiable sources of support for Proposition 15 were by no means negligible, they could not compare with the No-on-15 forces. Yet from the polls the undecided voters appeared numerous enough that the Yes-on-15 campaign could not necessarily be counted a loser.

It was in these circumstances that the California nuclear moratorium law was enacted, and that Governor Edward G. Brown, Jr., signed the measure only a week before the June referendum. It provided that before more reactors could be built, the California Energy Resources Conservation and Development Commission (commonly known as the energy commission) would have to determine that a demonstrated technology or means for permanent disposal of high-level waste exists and has been approved by the responsible federal agency. This legislation, together with certain companion measures pertaining to fuel reprocessing, spent fuel storage, and reactor safety,[127] was regarded at the time as a moderate alternative to Proposition 15, and it commanded wide support. There was no suggestion of a nuclear shut-down. The bills were designed, said

[125]Ibid., pp. 975–977.

[126]Ibid.

[127]One companion measure called for determining whether fuel reprocessing services would be available, but this question was subsequently rendered moot by the Carter administration's deferral of commercial reprocessing for an indefinite time for the sake of nuclear nonproliferation objectives. Another measure required the California energy commission to look to the adequacy of spent fuel storage, but this was to be done on a case-by-case basis and never became a major issue. A third measure called for a study of the feasibility of, and need for, having reactors built under ground for greater safety; the study was made, but the commission chose not to pursue the "undergrounding" option further.

the sponsoring assembly committee, to make sure before new reactors are built that the fuel cycle is not clogged at critical points.[128]

The bills' prime mover, Assemblyman Charles Warren of Los Angeles, knew that neither fuel reprocessing services nor the means of permanent disposal of high-level waste would soon be available. The previous fall his committee had held fifteen days of hearings on the Proposition 15 issues and had heard scores of witnesses, many of them experts on the nuclear fuel cycle. At the outset most members of the committee were flatly opposed to the ballot initiative, but in the end nearly all concluded that the state should indeed reassess nuclear power. The state energy commission, established by the Warren-Alquist Act of 1974 to provide energy forecasting and planning, was believed to be the appropriate mechanism for such a reassessment.

There was opposition to the Warren bills from the utilities and organized labor, but in the main even the No-on-15 campaign looked upon this legislation benignly. Newspapers that had denounced Proposition 15 embraced the legislation warmly; even certain major utilities supported it, Southern California Edison being one. The Yes-on-15 leaders had also endorsed the Warren bills, knowing full well that their passage would likely undercut whatever chance of victory there was for the ballot initiative.

As it turned out, Proposition 15 lost by a two-to-one margin, receiving about a third of the nearly 6 million votes cast. But had there been no initiative campaign to highlight the nuclear waste and fuel cycle issues, the legislature might never have acted. Ralph Nader, who had actively supported Proposition 15, dismissed the Warren legislation as diversionary and of little substance.[129] But he was wrong. In early 1978 the California energy commission, after extensive hearings the previous year, decided in the first actual test of the new moratorium law that San Diego Gas and Electric's Sundesert reactor project should not proceed. A means of waste disposal could not be certified, and an exemption for Sundesert could not be justified on a basis of need, the commission found.[130] This was an unprecedented and far-reaching action. California had stopped a major energy project, not on the grounds that the specific facility in question would cause pollution or represent a nuisance, but because the larger industry of which it would be a part was not prepared to

[128]Carter, "Nuclear Initiatives," p. 977.

[129]Ibid., pp. 317, 319.

[130]California Energy Resources Conservation and Development Commission, "Report to the Legislature: AB1852—Alternative to a Sundesert Nuclear Project," final report, February 1978.

isolate its long-lived wastes. During the next five years, five other states—Wisconsin, Maine, Montana, Oregon, and Connecticut[131]—were to follow the California example and make the availability of permanent waste disposal facilities a condition for the construction of nuclear power plants.

Following the energy commission action in the Sundesert case, Pacific Gas and Electric and Southern California Edison filed suit to have the moratorium law declared invalid. They contended that safety regulation of nuclear power was preempted by the federal government in the Atomic Energy Act of 1954. The trial court found in the utilities' favor, but the State of California appealed, with more than a score of other states filing supporting briefs as friends of the court. The definitive ruling in California's favor came in the spring of 1983. The U.S. Supreme Court accepted the state's argument that the moratorium law was economic in its rationale and was not a trespass on a regulatory function belonging to the federal government.[132]

By then, however, the nuclear moratorium issue had in a practical sense become moot, certainly for the near term. The second oil shock from the Iranian crisis of 1979 had driven energy prices to new heights, ending the nuclear industry's hopes for a recovery in the growth of electricity demand. There was little or no market for new baseload generating facilities of any kind, nuclear or fossil fuel. These circumstances alone would have led to an increase in the wave of reactor order cancellations. But there was more: the incident at Three Mile Island, which came as a partial confirmation of the critics' view that light water reactor technology was unforgiving and had been deployed without sufficient precautions. Certainly the critics gained further credibility with the public. This was evident from opinion surveys and from the fact that the critics found greater receptiveness to their message. For instance, Jess Riley of Charlotte, North Carolina, chair of the Carolina Environmental Study Group and a national Sierra Club leader on energy issues, was much concerned about the Duke Power Company's large commitment to nuclear generation, but until Three Mile Island he had found himself on the wrong side of the wall of established opinion. "A subtle change occurred after Three Mile Island," Riley told me, speaking

---

[131]These were the states listed in the amicus curiae brief filed in September 1982 by the U.S. Solicitor General in support of the utilities in their lawsuit to have the California moratorium law declared unconstitutional. (See text, below).

[132]*Pacific Gas & Electric Co. et al. v. State Energy Resources Conservation and Development Commission et al.*, 461 U.S. 190.

particularly of his relationship with the press in the region. "Now, for the first time, we were taken seriously."[133]

Nuclear power had come too far too fast. Its development and commercial application had leapt far ahead of the readiness of the enterprise to deal properly and convincingly with the nuclear imperatives.

[133]Interview by the author with Riley, spring 1981.

3
The Reprocessing Dilemma

By the time Jimmy Carter assumed the presidency in early 1977 the effort to close the nuclear fuel cycle by establishing reprocessing and using plutonium fuel on a commercial basis was already a deeply troubled undertaking. Its troubles arose in part from problems inherent in the technology and in part from the fact that here again the attempt at commercialization was premature. Success in reprocessing as a step in producing plutonium for bombs had misled the Atomic Energy Commission into believing that reprocessing as a step in the commercial fuel cycle was sure to succeed too. But this was simply wrong.

By the mid to late 1970s three attempts at commercial reprocessing had gone awry. The first, the Nuclear Fuel Services plant at West Valley, New York, had been shut down since 1972 and was not to reopen. The second, the General Electric plant at Morris, Illinois, built to an innovative and untried design, had been declared inoperable in 1974, before reprocessing the first ton of fuel. The third, the Allied-General plant at Barnwell, South Carolina, was mired in complex proceedings related to licensing and was in need of two costly high-level waste and plutonium product solidification units, which the owners were seeking to have built at government expense as new technology demonstration facilities. So when on April 7, 1977, President Carter, with the aim of reducing the risks of nuclear weapons proliferation around the world, announced the indefinite deferral of reprocessing and commercial breeder development, he was in effect applying the brakes to an undertaking that was not going anywhere without a lot of federal help.

The leaders of the nuclear industry deplored the president's action as the ultimate insult to would-be commercial reprocessors beset by federal regulatory and policy instability. There had indeed been such instability, but the fundamental question confronting the industry was whether, under even the most sympathetic federal policies, reprocessing was possible as a strictly private, unsubsidized commercial venture.

When Ronald Reagan succeeded Jimmy Carter as president he reversed the Carter policy of deferral of reprocessing. But Reagan rejected Department of Energy (DOE) overtures to have the government either buy the Barnwell plant and have a private contractor run it or subsidize its operation in some other way. With Reagan's encouragement, the department then sought diligently to have private industry take up the challenge at Barnwell, but got little response. A DOE official attempting to bring about the start-up at Barnwell appeared before a nuclear industry group in late 1982 with desperation in his voice. Beginning with the familiar litany of advantages that reprocessing and recycling are supposed to offer, he said, "Spent nuclear fuel is not nuclear waste!"[1]

But if spent fuel was not waste, neither was it a usable resource unless reprocessing and recycling could be found economically attractive, taking full account of political and regulatory demands for the safeguarding of plutonium, containment of radioactivity, and protection of radiation workers. The issue went beyond an economic test. The provisions contemplated for safeguards, containment, and radiation protection might be rejected as insufficient.

Yet back in the postwar years when development of nuclear power was getting under way, it was assumed that reprocessing and plutonium recycling would be a practical necessity sooner or later, and probably sooner. Uranium was believed to be so scarce that without breeder reactors and plutonium recycling a significant fission energy program might be impossible. Most of the uranium used in the atomic bomb project was imported from the Belgian Congo, and little was known about potential uranium resources in the United States and other countries. The first reactor to generate electricity, EBR-1 at the Nuclear Reactor Testing Station in Idaho, was a breeder. The view that fission energy would be short-lived without breeders was to persist within the nuclear industry until fairly recent years. For instance, Floyd Culler, a veteran of the Atomic Energy Commission's fission energy development effort and today president of the Electric Power Research Institute, in 1977 told the California energy commission, "There's no question. Nuclear energy is a flash in the pan, a very short flash in the pan, unless breeders come."[2]

Another early assumption was that the chemical separation of plutonium was an established technology. Such separation had been carried out successfully from the very first production runs at Hanford in December

[1]Kermit O. Laughon, "Commercial Reprocessing: What Must Be Done," speech before the Atomic Industrial Forum in Washington, D.C., November 17, 1982.

[2]California Energy Resources Conservation and Development Commission, *Hearings on Nuclear Fuel Reprocessing and High-Level Waste Management* vol. 5, *Reprocessing*, pt. 2 (Sacramento, March 8, 1977) p. 29.

1944. Moreover, the chemical separation process underwent major improvement in the early 1950s. The first plant using the now-standard process for plutonium and uranium extraction (called PUREX) came on line at Savannah River in 1954.

But the first assumption, that uranium was going to be scarce, was wrong, and the second, that reprocessing was an established technology, was of questionable relevance to commercial nuclear power. The two assumptions will be taken in turn.

The glut of uranium existing today cannot be explained by the fact that demand has been far less than predicted. In truth, uranium has turned out to be not a particularly scarce element in the earth's crust, as MIT professor Thomas L. Neff points out in his 1984 study, *The International Uranium Market*. Since the late 1960s numerous large discoveries have been made abroad, especially in northern Saskatchewan and in Australia. The average ore grade at certain Saskatchewan deposits has been very high: the Key Lake deposit averages 2.5 percent U_3O_8, and the Cigar Lake deposit (a discovery announced in late 1983) is even richer, at an astonishing 14 percent.[3] The Saskatchewan ore is to be compared to the 0.15 to 0.2 percent ore typical of the deposits mined by the American uranium industry on the Colorado plateau; in industry parlance, the 4-pound rock (ore yielding 4 pounds of uranium per short ton) of Wyoming and New Mexico contrasts with the 280-pound rock of Cigar Lake, which is seventy times richer.

George White, Jr., an expert on the uranium market and former president of the uranium brokerage firm Nuexco, has called Saskatchewan "the Saudi Arabia of the uranium business" and predicted that the discoveries at Cigar Lake and other places in the province will prove to be "but the tip of a large, yellow iceberg."[4] Thomas Neff sees little price increase in uranium resulting from ordinary market forces over the next thirty years, even if nuclear generating capacity in the noncommunist world should more than double.[5]

As to the nuclear enterprise's assumption that reprocessing was a technology ripe for commercial application, a rude awakening was inevitable. The successful chemical separation of plutonium for bombs

---

[3]George White, Jr., managing broker of the uranium brokerage firm Nuexco of Denver, Colo., interviewed by the author on February 13, 1986.

[4]George White, Jr., then president of Nuexco, in remarks to an Atomic Industrial Forum uranium seminar, Keystone, Colo., October 2, 1984.

[5]Given the possibility of such political perturbations as a ban on uranium exports by Australia in the event another non-weapons state detonates a bomb, the price of uranium may range (in constant dollars) from its 1984 level of about $20 a pound to as much as $30 or $40, Neff told me in an interview in mid 1984. "But if you could proceed with ordinary market forces, I think the price would stay at the lower end of this range," he added. His "moderate growth" demand scenario contemplates 321 gigawatts of nuclear

could proceed almost without regard to cost. But cost would be a severe constraint in commercial reprocessing. Furthermore, cost in the latter case would be influenced by societal demands for the containment of radioactivity and the safeguarding of plutonium as a nuclear explosive—demands which obviously would never be made of a secret military project in time of war or national emergency. Similarly, the experience in the military program was no test of the political acceptability of commercial reprocessing and use of plutonium fuel.

Successful Military Reprocessing: Hanford and Savannah River

To draw the distinction between military and commercial reprocessing, historical perspective is necessary. The history of reprocessing begins at Hanford, where the first chemical separation plants for recovery of plutonium were designed and built in 1943 and 1944. The circumstances and sense of urgency were extraordinary. Uranium fission had been discovered in 1938 in Berlin, and the atomic bomb project was begun in the United States in the belief that the country was in a race with Nazi Germany to develop this ultimate weapon. Cost was no object. The DuPont Company, builder and operator of the Hanford works, rushed to construct four huge chemical separation facilities, although in the end only three were built and only two actually were used before the war ended. The entire wartime Hanford works, including three plutonium production reactors, the three separation plants, and a variety of supporting facilities, were built in total secrecy by an army of workers (numbering up to 39,000) within less than two and a half years from the time the Hanford site was selected in January 1943.

Knowing that any long interruptions or delays in production would be intolerable, the DuPont project managers went to elaborate lengths to ensure dependable remote operations and maintenance. The first of the "canyon buildings," as the chemical separation plants were called, was 221-T, a structure 800 feet long, 65 feet wide, 80 feet high, and looking like a "huge aircraft carrier floating on a sagebrush sea."[6] Along its lower tier, the canyon building consisted of a row of 40 concrete cells, most of them about 15 feet square and 20 feet deep, in which the

generation by 1990 in the noncommunist world (as opposed to about 167 gigawatts at the end of 1983) and 512 gigawatts by the year 2000.

[6]Richard G. Hewlett and Oscar E. Anderson, Jr., *The New World, A History of the United States Atomic Energy Commission* vol. 1, AEC document WASH-1214 (Washington, D.C., Atomic Energy Commission, 1962) p. 220.

various steps in the separation process would be carried out. A continuous gallery ran the length of the building above these cells and their
massive concrete covers. AEC historians Richard G. Hewlett and Oscar
E. Anderson, Jr., have described the DuPont design philosophy in these
words:

The first thought in designing cell equipment was to facilitate maintenance and
replacement. Once the plant was operating, the only access to the cells would
be by means of the huge bridge crane which traveled the length of the building.
From the heavily-shielded cab behind a concrete parapet above the gallery,
operators could look into the canyon with specially designed periscopes....
They could use the seventy-ton hook to lift off cell covers and lighter equipment
to work within the cell. With special tools and impact wrenches, they could
remove connecting piping, lift out the damaged piece of equipment, and place
it in a storage cell. They would then lower a new piece of equipment into the
operating position and reconnect the process piping.[7]

All chemical separation plants built subsequently at Hanford and Savannah River followed this all-remote maintenance design philosophy,
and for the purposes of the weapons program were operated successfully. For instance, the Savannah River PUREX plant, which remains in
use today (its capacity since 1959 rated at an impressive 15 tons of fuel
reprocessed daily), has typically been available more than 80 percent of
the time, and average worker exposure has been kept to fewer than 700
millirems per year,[8] or about one-seventh the maximum allowed.

Why this success at Hanford and Savannah River has not ensured the
success of commercial reprocessing comes down in part (leaving aside
the price of uranium) to matters of capital cost and of plant maintenance,
waste management, and safeguards.

Capital cost. Commercial reprocessors were quick to persuade
themselves that they could not afford the huge, massively constructed
PUREX canyon facilities equipped for all-remote maintenance. They would
go to the alternative of much smaller plants designed for remote maintenance only in process cells having high radiation levels, while in other
cells maintenance tasks would be performed directly by radiation workers. Such a plant could be built for less than half the cost of one built

[7]Ibid., pp. 220–221.
[8]John R. White and Howard W. Harvey, "The Evolution of Maintenance in Nuclear
Processing Facilities," paper presented before the American Nuclear Society, Washington,
D.C., November 14–19, 1982. White and Harvey are both from the Remote Technology
Corporation, Oak Ridge, Tenn.

to the DuPont design.[9] The dilemma here was that such a saving on capital costs would make for maintenance delays, interruptions in production, and increased operating costs. Furthermore, the oxide fuel from commercial light water reactors would be from six to ten times more highly irradiated than the aluminum-clad uranium metal fuel normally reprocessed at Hanford and Savannah River. More irradiation meant more fission products and more radioactivity. In such an environment radiation workers would often have to be deployed in relays and be suited uncomfortably in several layers of protective clothing. It was thus all the more desirable not to have to engage workers in time-consuming tasks which could be accomplished more quickly if done remotely, and with better containment of radioactivity.

Waste management.   The waste management practices first adopted under the exigencies of war at Hanford and of cold war at Savannah River would not be defensible for the long term even in the name of national security. As early as the mid 1950s most of them were being pronounced unacceptable by the Atomic Energy Commission's advisers at the National Academy of Sciences. Commercial reprocessors, lacking the political benefit of the national security mantle, would come under strong pressure to contain radioactivity in every phase of their operations. They could expect the demands for reduced worker exposure and environmental releases to be insistent and ever more stringent. Technically, these demands perhaps could be satisfied, but only at a further and not insignificant increase in both operating and capital costs.

Safeguards.   In the context of the weapons program, the safeguards issue arose hardly at all, at least not politically. Hanford and Savannah River existed for the purpose of producing plutonium and other special nuclear materials for weapons that were believed to be urgently needed in the national defense, and that was that. Commercial reprocessing, on the other hand, might be judged unacceptable even if it added only marginally to the risks of proliferation or of terrorist incidents. Should these risks be confirmed by events—for instance, if one or more countries without nuclear weapons should use the cover of a supposedly peaceful nuclear program to build a plutonium bomb—strong political demands to discourage use of plutonium fuel might be inevitable.

---

[9]"Report to the American Physical Society by the Study Group on Nuclear Fuel Cycles and Waste Management," *Review of Modern Physics* vol. 50, no. 1, pt. 2 (January 1978) p. S51. The study group concluded that a 1,500-ton-a-year plant built to an all-remote maintenance design developed by DuPont and the Energy Research and Development Administration would cost 128 percent more than the Barnwell plant, computed on a comparable cost basis (see the section Reprocessing in Limbo: Barnwell, below).

Safeguards would have to do with a variety of situations and problems, such as security maintenance against both internal and external threats at the reprocessing and fuel fabricating plant, the protection of plutonium and plutonium fuel shipments from theft or forcible seizure,[10] and the keeping of accurate plutonium inventory accounts. Experience in the weapons program was demonstrating how difficult it was to maintain strict accountability over plutonium inventories. During the period from 1956 to 1964 the Hanford works experienced inventory "shortages" totaling some 944 kilograms,[11] or enough for more than 100 Nagasaki bombs if the shortages were real and represented thefts or diversions. As measured by the difference between the amount of plutonium in the incoming fuel and the amount recovered, the shortages ranged from as little as about 12 kilograms one year to as much as 170 kilograms another year.

The experience at Savannah River from the start of operations in 1954 through 1976 was to be similar, except that in a number of years, plant records indicated, the amount of plutonium present at the end of the inventory period was substantially *greater* than the amount supposedly present at the start of reprocessing. None of these "MUFs" (materials unaccounted for) caused alarm, for there was no lack of mundane explanations: plutonium held up in process piping, sampling mistakes, and measurement uncertainty—the latter arising from the difficulty of determining from theoretical calculations the amount of plutonium produced in the reactor, and hence the amount in the fuel going to the reprocessing plant.[12] At the same time, one could never be absolutely sure that material unaccounted for had not been stolen or deliberately diverted. Accounting techniques have since improved (an effort has been under way to perfect "real time"—or continuous—inventorying); but they have not improved enough to reduce inventory differences to levels that are of no concern.[13] In a plant receiving 15,000 kilograms of plutonium or more a year, MUF of even less than 1 percent could represent the stuff of numerous bombs.

[10]Breeder fuel, with a plutonium content of 20 to 25 percent, is treated as a potential source of explosive nuclear material. Under Nuclear Regulatory Commission regulations, breeder fuel shipments, like shipments of separated plutonium, call for the highest order of precautions in terms of the number of guards assigned, the transport equipment used, and the procedures to be followed.

[11]Energy Research and Development Administration, *Report on Strategic Special Nuclear Material Differences*, ERDA 77-68 (Washington, D.C., August 1977) (hereafter cited as ERDA *Report*, 1977).

[12]ERDA *Report*, 1977; interview by the author with Michael Smith of the Nuclear Regulatory Commission division on safeguards, January 19, 1984.

[13]Interview by the author with Robert F. Burnett, director of the NRC division on safeguards, January 4, 1984.

Commercial Reprocessing: Failure at West Valley

When Nuclear Fuel Services (NFS) at West Valley—a venture begun by the W. R. Grace Company and later taken over by the Getty Oil Company—came on the scene in the early to mid 1960s, there was no commercial market for reprocessing services because little spent fuel had yet been generated. The Atomic Energy Commission was willing to provide a baseload by letting NFS reprocess fuel irradiated at the dual-purpose N-reactor at Hanford, which was (and still is) producing electricity for the commercial grid as well as plutonium for nuclear weapons. But the commission was insisting on such a low price for this service that capital costs also had to be kept low. In a 1957 report (known as WASH-743) describing a "reference fuel-reprocessing plant," the commission estimated that a facility capable of reprocessing 300 tons of fuel a year could be built for about $20.5 million. The actual cost of the West Valley plant was to be about $35 million.[14] Even to hold capital spending to this higher figure required cutting corners, as in the choice of carbon steel storage tanks instead of corrosion-resistant stainless steel tanks that could receive the high-level waste in its more easily managed acidic form.

From its start-up in 1966 to its shut-down in 1972 the West Valley plant was beset by a variety of problems and limitations. The plant had a nominal capacity to reprocess 1 metric ton of fuel a day, or on the order of 300 tons a year; but there really was no chance that Nuclear Fuel Services would achieve even this modest rate of production because it was generally reduced to reprocessing only the N-reactor fuel from Hanford and assorted batches of fuel, many of them quite small, from some of the early commercial and experimental reactors.[15] Chemical incompatibility among the different fuel types made continuous generation of the plant impossible, and everything had to stop for a cleansing of the process lines and a material accounting between "campaigns." Actual production over the six years the plant was in operation came to about 625 tons, or roughly a third of nominal capacity. Because of maintenance problems and increased downtime, in the latter years plant availability declined.[16] Nuclear Fuel Services lost money at West Valley, and because of its small size and high unit cost of production the plant could not be expected to compete with the newer and larger facilities being planned for the 1970s.

---

[14]Robert Gillette, "Plutonium (I): Questions of Health in a New Industry," *Science*, September 1974, p. 1027.

[15]Gene I. Rochlin, Margery Held, Barbara G. Kaplan, and Lewis Kruger, "West Valley: Remnant of the AEC," *Bulletin of the Atomic Scientists*, January 1978.

[16]White and Harvey, "Evolution of Maintenance."

Nonetheless, West Valley still might have been adjudged something of a success if commercial fuels having a high burnup could have been routinely reprocessed with the radioactivity effectively contained. (Higher burn-up fuels are those left in the reactor long enough to become highly irradiated.) Although higher burn-up fuels were in fact reprocessed during the plant's last three years of operation, the record with respect to containment, radiation exposure to workers, release of radioactivity to the environment, and plutonium losses was not encouraging. Plutonium losses were not held to 1.5 percent as expected, but by Nuclear Fuel Services' own accounting totaled 2.4 percent.[17] According to one of its critics, an analysis of NFS quarterly reports shows that the loss approached 4.0 percent.[18]

As for containment, this would have meant, in an absolute sense, keeping all radioactive materials within the process equipment. To the extent that leaks could not be avoided, these materials were supposed to be confined to heavily shielded cells; ventilation within the plant was designed so that air would flow from uncontaminated to contaminated areas. Process "off-gases" would be effectively filtered for removal of radionuclides (particularly the more radiotoxic species such as cesium and iodine) before release to the environment. Similarly, any aqueous discharges from the plant would be filtered to keep the presence of radionuclides in nearby surface waters, such as Cattaraugus Creek and its tributaries, far below the maximum permissible concentration.

Compliance with these standards was expected to protect workers inside the plant and the people and the fish and wildlife outside. Judged accordingly, containment at West Valley, while never failing in a gross way, fell far short of what was desired and required. The plant had to be regarded as deficient in design and construction, even on its own terms as a plant in which some process cells would be maintained directly by radiation workers. In some places the concrete shielding separating hot cells and areas where workers had to be routinely present was not massive enough to keep radiation at safe levels. For example, in one sampling room lead sheeting had to be placed on the floor to attenuate further the radiation from the process cell below. Radiation levels in the process cells themselves often became higher than expected because of pipe leaks and spills of radioactive liquids. When pipes became plugged, as frequently happened, sometimes the plug could be

[17]Testimony of Marvin Resnikoff in *Final Generic Environmental Statement on Mixed Oxide Fuel*, Hearings before the Nuclear Regulatory Commission, NRC docket no. RM-50-5 (n.d.) p. 69 (hereafter cited as Resnikoff testimony). Resnikoff, testifying for the Sierra Club, was citing the Safety Analysis Report, NFS docket no. 50-201, p. I-2-3.

[18]Resnikoff testimony, pp. 68–77. The examples of containment failure in the following two paragraphs are also from Resnikoff's testimony.

removed only by cutting out a section of the piping. Workers in one instance had to drill through the concrete wall of the cell to get at the problem; when the pipe finally was cleared highly radioactive liquids spilled on the floor.

Equipment failures within the process cells were routine. In one particularly notorious instance, a leaking dissolver vessel was being removed by flatbed truck to the waste burial ground on the site when contaminated solution spilled from it. The dissolver, the flatbed trailer, and pieces of the asphalt roadway all had to be buried. There was trouble even in the remotely maintained parts of the plant. The remote handling devices themselves, particularly the overhead cranes and the master-slave manipulators, often failed. The cranes repeatedly knocked themselves out of action by running over the electrical cables tangled on their tracks;[19] in other cases, cables simply failed from flexing and wear and tear or from doors being shut on them.[20] The original master-slave manipulators sometimes broke down from the pressure they applied, and all eventually had to be replaced.[21]

Because of such failures, workers had to enter cells that were intended to be remotely maintained as well as those designed for contact maintenance. The entry of workers into any radioactive area meant some spreading of contamination. To reduce the radiation levels in the cell to be entered, the walls and equipment had to be washed down in advance, thus creating contaminated water that had to be processed as waste. Similarly, the radiation workers emerged from the cell with their protective clothing and tools contaminated. The clothing had either to be treated as radioactive waste and buried or sent to the laundry. At the laundry the wash water of course became contaminated. On several occasions the sump in the laundry room became clogged, causing radioactive water to back up onto the floor,[22] which then had to be scrubbed down, creating still more contaminated water. The air filters used to trap radionuclides at West Valley were themselves a significant source of worker exposure. An Atomic Energy Commission inspector reported that almost a third of all exposures involved workers who had to change filters.[23]

[19]Ibid., p. 104.

[20]Interview by the author with Walton A. Rodger, former general manager of the West Valley plant, April 21, 1982.

[21]Ibid.; Resnikoff testimony, p. 97.

[22]Resnikoff testimony, p. 67, citing AEC inspection reports, docket no. 50-201, for January 16–20 and April 17–21, 1967.

[23]Resnikoff testimony, p. 115, citing AEC inspection report for Region I, Provisional License no. CSF-1, NFS, April 14–17, 1970.

One contamination incident overwhelmed the means Nuclear Fuel Services normally applied to effect a cleanup. In early 1969 several fuel assemblies arrived from the Hanford N-reactor badly damaged, with some of the fuel rods ruptured along their entire length. The damaged condition of these assemblies was not discovered until after they had been removed from their transport cask and placed in the fuel storage pool. Gross beta radiation in the pool shot up to more than sixty times what it should have been.[24] NFS hastily removed the damaged fuel elements, put them in scrap drums, and buried them on site in a 50-foot-deep hole, encased in concrete.[25] This was done even as the Atomic Energy Commission was moving to require deep geologic disposal of all high-level waste.[26] The plant operators vacuumed the pool and installed some forty additional ion exchange demineralizer or filtration units, but failed to reduce the radiation to an acceptable level until the pool was drained and cleaned three years later.[27]

All of these incidents and problems represented failures of containment. They were never sufficient to render the plant more than temporarily inoperative or to put more than a few workers over their allowed exposure limits. But they called into question whether the plant could continue operating over the long term. The trend for worker exposure was decidedly unfavorable, with whole-body exposures for full-time workers increasing from 2.74 rems per worker in 1968 to 7.24 rems in 1971.[28] Workers were being "burned out" as they reached their exposure limit of 5 rems per year (an averaging "body bank" formula allowed low exposures in some years to compensate for higher exposures in other years). Exposures of the regular work force would have been greater still had Nuclear Fuel Services not employed large numbers of transient workers.

A showdown between Atomic Energy Commission regulatory officials and Nuclear Fuel Services took place in February 1972 over the excessive occupational exposure inside the plant. The officials charged that

[24]Resnikoff testimony, pp. 91, 115, citing NFS quarterly report for the period ending February 27, 1969.

[25]General Accounting Office, *Issues Related to the Closing of the Nuclear Fuel Services Reprocessing Plant at West Valley, N.Y.* (Washington, D.C., March 8, 1977); Resnikoff testimony, p. 91, citing NFS quarterly report for the period ending June 30, 1969.

[26]Richard G. Hewlett, "Federal Policy for the Disposal of Highly Radioactive Wastes from Commercial Nuclear Power Plants: An Historical Analysis," manuscript, March 1978; Nuclear Regulatory Commission, "Policy Relating to the Siting of Fuel Reprocessing Plants and Related Waste Management Facilities," 10 CFR, pt. 50 (November 1970) appendix F.

[27]Resnikoff testimony, p. 91, citing AEC inspection report for Region I, Provisional License no. CSF-1, NFS, September 14–17, 1970.

[28]Resnikoff testimony, pp. 48 and 61, citing AEC inspection report, docket no. 50-201, for March 16, 1972.

the company had failed "to make reasonable efforts to maintain the lowest levels of contamination and radiation" and "to adequately instruct or effectively train employees ... in the radiation hazards involved in their job assignments."[29]

Three months later the plant began what Nuclear Fuel Services described as a long-planned shut-down to expand capacity to 750 tons a year and make the plant safer and more efficient. This work was at first expected to cost $15 million, but by the time various new regulatory requirements were taken into account the estimate had soared to more than $600 million.[30] A new Atomic Energy Commission construction permit was required, and for NFS to obtain it the proposed improvements would have to include expensive new facilities to solidify high-level waste and convert the plutonium product from its nitrate (or liquid) form to a solid, probably plutonium oxide. Furthermore, a completely new safety analysis and environmental impact statement for the plant were necessary. This was to lead to demands that the plant be strengthened structurally against earthquakes and tornadoes, to an extent much greater than Nuclear Fuel Services thought necessary.[31] The upshot was that in 1976 NFS announced its withdrawal from the reprocessing business. The company said that it was exercising its right to surrender to the State of New York its responsibility for the site and all the wastes stored or buried there.

It would soon be evident that Nuclear Fuel Services had become a real burden on the public. No major release of radioactivity had occurred, and Nuclear Regulatory Commission inspectors had found no evidence that the public had been endangered,[32] but West Valley's waste management practices were nonetheless found to be a mess. There had been some difficulties with routine, day-to-day effluents and emissions of low-level aqueous and gaseous wastes, but these had stopped with the closure of the plant. The legacy of high-level and transuranic wastes stored and buried on the site, on the other hand, was going to be around essentially forever unless those wastes were recovered and removed. The greatest concentration of radioactivity by far was in high-level waste tank 8D2, the only tank actually used for the neutralized waste from the reprocessing of uranium fuel at West Valley. To empty this tank of its

[29]Gillette, "Plutonium (I): Questions of Health in a New Industry," p. 1032.

[30]General Accounting Office, *Issues Related to the Closing*, enclosure, p. 2.

[31]Walton A. Rodger, "Reprocessing of Spent Nuclear Fuel," in California Energy Resources Conservation and Development Commission, *Hearings on Nuclear Fuel Reprocessing and High-Level Waste Management*, vol. 4, *Reprocessing*, pt. 1 (1977) pp. 317–320.

[32]Ibid., pp. 313, 371, citing letter of W. Johnson, acting chairman of the AEC, to Congressman R. D. McCarthy, April 1, 1968, and letter of L. Muntzing, AEC secretary of regulation, to D. Kirkpatrick of the National Intervenors, September 9, 1967.

some 600,000 gallons of waste and 39 million curies of radioactivity was going to be an extraordinary technical challenge.

About half of the radioactivity was, and is, in a liquid form that can be easily pumped from the tank, but the other half is a stiff claylike sludge that covers the bottom of the tank to a depth of about a foot and a half. Ordinarily sludge might be sluiced out with high-pressure jets of water, but in this case there is an obstruction—a complex internal grid structure rising three and a half feet above the tank bottom to support a veritable forest of pipe columns, which in turn support the roof of the tank. The fact that there is but one unobstructed point of access, through a riser located near the center of the tank, makes the problem of waste removal all the more difficult. Yet inasmuch as the tank was designed for only a fifty-year life, failure to remove the waste might lead eventually to contamination of ground and surface water.

This problem was to become a matter of increasing public concern and controversy in western New York and metropolitan Buffalo. The consequence of the public protest, which was led by the Sierra Club and the Coalition on West Valley Nuclear Wastes, was that the State of New York ultimately persuaded Congress to have the Department of Energy take on a cleanup at West Valley as a federal responsibility. This was accomplished by the West Valley Demonstration Project Act of 1980, passed largely as the result of the tireless efforts of New York congressman Stanley Lundine. The act calls for recovery and solidification of the waste and decontamination and decommissioning of the reprocessing plant.[33] The cost? On the order of $700 million to $800 million, of which the state is obliged by the terms of the act to pay 10 percent.[34]

Although the waste in tank 8D2 represents the greatest concentration of radioactivity left on the site, it does not necessarily represent the most intractable of the West Valley wastes. The low-level wastes buried on the site—and, as a practical matter, probably buried irretrievably—pose what could be a nagging and perhaps irresolvable problem of long-term custodial care. There are two burial grounds, one originally licensed by the Atomic Energy Commission strictly for wastes generated by Nuclear Fuel Services, the other licensed by the State of New York for NFS wastes and wastes from power reactors, hospitals, and other commercial and institutional sources. Some transuranic waste was buried

[33]Public Law 96-368, October 1, 1980, 42 USC 2021a, 94 Stat. 1347.

[34]Interview by the author on September 19, 1986, with Joseph A. Coleman, director of the division of waste projects, Office of Nuclear Energy, Department of Energy. The cost estimates prepared in connection with the FY 1987 budget were $442 million for high-level waste solidification and between $250 million and $350 million for decontamination and decommissioning of the West Valley site.

at both sites, but especially at the seven-acre, AEC-licensed NFS waste facility.

The West Valley site was expected to be excellent for disposal of low-level waste because of its good surface drainage, the absence of aquifers near the surface, and the low permeability and absorption capacity of its silty, glacial-till soil.[35] But the site was in fact a poor choice. The abundant precipitation in the region (about 40 inches a year) and low soil permeability were to lead to the same bath-tubbing problem that was plaguing the Maxey Flats burial facility in Kentucky; that is, water would infiltrate the waste trenches, collect there, and eventually overflow, transporting radionuclides off site. The concentrations of radioactivity had been low, but to have any off-site migration at all mocked past assurances of secure containment. The state-licensed burial ground, containing several hundred thousand curies of radioactivity, was closed in 1975 immediately after the discovery that one of the older burial trenches was overflowing. As at Maxey Flats, one remedy for the bath-tubbing was to pump accumulated water from the trenches for treatment before release to streams on site.

On Nuclear Fuel Services' own burial ground, waste was not buried in trenches but in some 200 holes, which were up to 15 meters deep and no wider than necessary to receive the drums or other waste containers. Chopped-up fuel cladding and failed equipment, such as fuel-dissolver vessels, made up an important part of the waste and were transuranic. Total radioactivity was estimated at about 550,000 curies, but inasmuch as NFS had some difficulty dissolving the fuel in the butt end of fuel rods, questions would arise later as to whether the radioactivity in some of the holes was not substantially higher than shown on company records.[36] A report prepared for the Department of Energy by the Argonne National Laboratory a few years after the plant closed said the "waste is fully contained within the silty till and permeability tests indicate it would take about 550 years for contaminated water to travel 30 meters through the till."[37] But in November 1983 the Nuclear Regulatory Commission found, in one of its monitoring wells, some plutonium that had migrated 50 feet in not more than 15 years.[38]

A West Valley decontamination and decommissioning task force, made up of representatives of several New York state agencies, legislators, and

[35]Argonne National Laboratory, *Western New York Nuclear Service Center Study*, TIE-28905-2 (Washington, D.C., Department of Energy, n.d.) pp. 3-38 and 3-55 (hereafter cited as Argonne study).

[36]"Report of the West Valley Decontamination and Decommissioning Task Group," p. 41, incorporated as a supplement to Argonne study, p. 2-21.

[37]Argonne study, p. 3-57.

[38]Interview by the author with Arthur Thomas Clark, Jr., of the NRC staff, July 3, 1984.

citizens' groups, urged in their 1978 report that all radioactive materials be removed and that the West Valley site be eventually restored to unrestricted use. This was almost certainly not an achievable aim. The Argonne report estimated the cost of exhumation and redisposal of all of the waste from the two state- and commission-licensed burial grounds at $910 million, in 1978 dollars. This estimate, high as it was, was probably low.

A need for constant surveillance and maintenance of the two burial grounds, probably to go on indefinitely, is very much a part of the West Valley legacy.

White Elephant at Morris

The next commercial reprocessing facility to be built was General Electric's Midwest Fuel Recovery Plant at Morris, Illinois. Although its capacity was small—300 tons a year—this plant was sized to fit the reprocessing requirements of utilities in the Midwest, where its central location would eliminate the need for long-haul rail or truck shipments of highly radioactive and massively shielded fuel assemblies. Moreover, the plant was expected to be compact and efficient enough to meet competition from larger plants outside the region.

General Electric (GE) had developed a new process, dubbed aquafluor. Two major innovative procedures were involved. One was the direct fluorination of the recoverable uranium to UF_6, a compound which assumes a gaseous form at a slightly elevated temperature and can be fed into isotopic enrichment plants. The other was the direct calcining of high-level waste—without any intermediate tank storage—into a granular powder which would be canned in stainless steel containers and stored under water to await shipment to a federal repository.[39]

A General Electric brochure aimed at prospective customers pointed out that GE, as a successor to DuPont at Hanford, had long experience

[39]The discussion of the aquafluor process and Morris plant is derived principally from the following sources: General Electric, "Nuclear Fuel Recovery at General Electric," undated brochure intended for prospective customers; interview by the author with Bertram Wolfe, manager of GE's nuclear division, San Jose, Calif., January 23, 1984; and testimony and written statement of Walton A. Rodger, March 7, 1977, in California Energy Commission, *Hearings on Nuclear Fuel Reprocessing and High-Level Waste Management*, vol. 4, pt. 1, pp. 55–58 and 320–324. According to Wolfe, when GE was discussing waste form requirements with the AEC in 1969 and 1970, the agency was willing to accept delivery of high-level waste as a calcine. Actually it was not until 1983 that the AEC's successor, the NRC, formally ruled (in its regulations 10 CFR, pt. 60) that calcine was not acceptable for geologic disposal.

in reprocessing. It described the aquafluor process as a "unique coupling" of well-demonstrated techniques. "The plant itself was designed to optimize on-line availability, ease of remote maintenance, accurate accountability [of the plutonium and uranium], high product recoveries, and effective control of waste materials," the brochure said. Direct personnel access to radioactive process systems was not going to be possible, thus systems had been designed "for remote operation, repair, and replacement without direct contact."

But the system devised for the tight, compact spaces at Morris was going to prove to be part of the problem. In the early 1970s General Electric began cold testing the Morris plant with unirradiated uranium, only to find that plumbing was clogging and other equipment was failing with discouraging regularity. The remote maintenance techniques were not dependable. In March 1974, with the cost of the plant now up to $64 million and almost twice what had been expected, GE ordered a sweeping review of the problem by a high-level management committee. One fundamental problem was the lack of inter-stage surge capacity, which meant that any failure at any point in the process would shut down the process as a whole. The committee concluded "that the mean time to failure is approximately equal to the time needed to reach equilibrium," and therefore the plant would yield little or no product. Bertram Wolfe, who became manager of the GE nuclear division after the decision not to start up the Morris plant, has explained: "What we underestimated was the difficulty of going from a laboratory environment—where you have access to equipment—to a remotely operated production plant."[40]

In obtaining a construction permit for the Morris plant, General Electric had won approval for its plans from the Atomic Energy Commission staff, the Atomic Safety and Licensing Board, the Advisory Committee on Reactor Safeguards, and the AEC itself. Yet if GE had ignored the results of its own cold tests and the recommendations of its review committee and started up the plant, the wretched state of things at Morris would have been worse. The irreparable breakdown of the plant with the reprocessing of the first batches of irradiated fuel would have left the company, the State of Illinois, the public, and the Atomic Energy Commission with a monumental hulk of radioactive waste, with the process cells and miles of piping inside the ten-story facility all contaminated. As it was, General Electric and the State of Illinois were left with a white elephant, but at least it was not radioactive.

[40]Robert Gillette, "Nuclear Fuel Reprocessing: GE's Balky Plant Poses Shortfall," *Science*, August 1974, pp. 770–771.

Reprocessing in Limbo: Barnwell

Planning and construction of the Barnwell Nuclear Fuel Plant, a joint venture trading under the name of Allied-General Nuclear Fuel Services (AGNS),[41] were coming along only a few years behind the Morris project. The plant took its name from the nearby town of Barnwell, the seat of Barnwell County, South Carolina, which is also the site of the government's sprawling Savannah River plant, known locally as the "bomb plant" chiefly for the plutonium produced and separated there. The Barnwell plant was seen by the nuclear industry as only the first of perhaps a dozen or more commercial reprocessing plants to be needed before the turn of the century to allow the utilities to work off their growing inventories of spent fuel. This was a time of extraordinarily high expectations for nuclear energy. As late as 1976, some forecasters put even their "low growth" projection at more than 600 gigawatts of nuclear generating capacity in the United States by the year 2000, or the equivalent of more than six hundred 1,000-megawatt reactors.[42]

Although ultimately the demand for reprocessing services was expected to be lively, Allied-General faced what in the late 1960s and early 1970s appeared to be an intensely competitive market in the making. General Electric was scheduled soon to start up the Morris plant; Nuclear Fuel Services would be announcing plans for an expansion at West Valley; and the Atlantic Richfield Company (ARCO) was expected to build a plant at Leeds, South Carolina. In the mid 1970s Exxon would announce plans for a plant at Oak Ridge that would be the largest yet, its initial capacity to be 1,700 tons, with an option of expanding to 2,100 tons. Accordingly, Allied-General was under pressure to keep its investment costs down, while taking advantage of lessons learned at West Valley and elsewhere.

Allied Chemical (an AGNS partner) had operated the AEC's Idaho Chemical Processing Plant, a facility principally engaged in the reprocessing of high-enriched, high-burnup naval reactor fuel at very modest rates of production. This was significant, for unlike the commission's reprocessing plants at Hanford and Savannah River, the Idaho plant was designed not for remote maintenance but for direct contact mainte-

---

[41]This undertaking was initiated by Allied Chemical Company, later joined by the Gulf Oil subsidiary General Atomic and by Royal Dutch Shell. The joint venture was first called Allied-Gulf Nuclear Fuel Services, but this was later changed to Allied-General. To avoid confusion, it is referred to as Allied-General in the text.

[42]J. O. Blomeke and C. W. Kee, "Projections of Wastes to be Generated," *Proceedings of the International Symposium on the Management of Wastes from the LWR Fuel Cycle*, Denver, July 11–16, 1976, ERDA document CONF-76-0701 (Washington, D.C., Energy Research and Development Administration, 1976) p. 96.

nance. The Idaho plant's operating history had not been trouble-free, but Allied headquarters nonetheless felt fortified by the experience. Arnold Ayers, formerly manager of the Idaho plant, was assigned to head plant design at Barnwell. The Barnwell plant was to reflect the contact-maintenance philosophy, and would be expected to reprocess 5 metric tons of fuel daily, or 1,500 tons a year.

According to Ayers and others, the design strategy developed for the new plant consisted of several major elements. Remote maintenance would be provided for those parts of the process where radiation levels would be high and where equipment would have moving parts subject to failure; contact maintenance would be provided for the rest of the plant, but care would be taken not to place equipment subject to heavy wear and tear (such as motors, pumps, and valves) inside those contact-maintained cells expected to have fairly high radiation levels. Potential leaks and plugged pipes, identified as the main cause of interrupted production and downtime, would be taken into account principally in two ways. Leaks would be minimized by joining pipes by welds (instead of by bolting interfacing rims together). Pipes would be unplugged by an ingenious (some would say impractical) "freeze plug" and air-pulse method in which jackets would be placed around pipes at the points considered most subject to plugging; when plugging occurred, liquid nitrogen would be injected into the jacket to freeze the pipe and create another plug, and compressed air would then be injected into the iso-lated pipe section (through a line already attached for just such a con-tingency) to blow out the offending plug.[43]

But while a plant built to such a design strategy might represent an improvement over West Valley and some other earlier plants, it remained to be demonstrated whether Barnwell could actually achieve and sustain the high rate of production envisaged. The plant would have as much as 75 miles of piping and tubing and about 80,000 welds, and would be astonishingly complex. To achieve the quality assurance necessary to detect all bad welds and faulty pieces of equipment—and even fraudu-lent certification of equipment[44]—would be a managerial tour de force.

[43]Interview by the author with Arnold Ayers, February 22, 1983; interview by the author with Alfred Schneider, February 15–16, 1983.

[44]In November 1982, Ray Miller, Inc., a New Jersey firm that has supplied piping, tubing, and other equipment for nuclear facilities, pleaded guilty in a U.S. District Court in West Virginia to ten counts of defrauding its customers, Allied-General among them, by practices such as removing from welded and drawn tubing the identifying mark WD and misrep-resenting the material on invoices as seamless tubing. But George Stribling, Allied-General vice-president for regulation and public affairs, told me that all the material his firm has received from Ray Miller, Inc., had been accounted for and that none was "in any way fraudulent." Interview by the author with Stribling, March 14, 1983.

Moreover, in the opinion of a number of reprocessing specialists, mistakes were made in the design of the Barnwell plant quite apart from questions of maintenance philosophy. "It was built too small even for a conventional chemical plant," says Clint Bastin, a Department of Energy chemical engineer formerly in charge of reprocessing at the Savannah River plant; "in such tight quarters there is not enough room for either proper contact or remote maintenance."[45] John R. White, who as a private consultant and former Oak Ridge National Laboratory engineer has pioneered in the development of remote maintenance concepts, puts it this way: "If you want contact maintenance you would not design a plant like Barnwell."[46] For instance, he says, the five contact cells all have a number of heat exchangers which, given a design life of about ten years, will eventually wear out. Yet the cells are highly congested. "You would have to decontaminate miles of piping to replace the heat exchangers," White observes. Then, too, as critics of Barnwell have often pointed out, this plant was expected to reprocess, each year, at least fifteen times the amount of fuel typically reprocessed at West Valley, and at a higher average burnup.

The Solidification Question: High-level Waste and Plutonium

In the late 1960s managers of the Barnwell project knew that the Atomic Energy Commission would soon adopt a regulatory requirement for high-level waste to be solidified within five years of the time of generation. But their search for a suitable solidification-facility design was confronted by technical and regulatory obstacles and highly uncertain costs. A similar situation existed with respect to anticipated regulatory requirements for the solidification and safeguarding of plutonium, and was to be even more frustrating. An environmental impact analysis, of the kind the new National Environmental Policy Act (NEPA) of 1970 called for, might have brought these regulatory problems to light before large sums were invested at Barnwell. But that act, and especially the court decisions that gave weight to its requirements, came after work on the project was already under way. Eventually, after losing a court challenge on NEPA issues, the Atomic Energy Commission asked Allied-General and other construction permit holders to show cause why their permits should not be revoked pending completion of environmental statements. But the agency, in a November 1971 decision, allowed work

---

[45]Bastin was interviewed by the author on September 23, 1981.
[46]White was interviewed by the author on June 19, 1984.

on the Barnwell plant to continue. A seventeen-page staff analysis supporting the decision did not mention either high-level waste or plutonium.[47]

Alfred Schneider, one of the top-ranking technical people at Barnwell in the early years and since 1975 a professor of nuclear engineering at the Georgia Institute of Technology, was involved with both the high-level waste and the plutonium solidification questions. According to Schneider,[48] the American technology available for solidifying high-level waste by putting it into glass (vitrification) was not then ready for commercial application, but that by 1972 substantial progress on a commercially applicable vitrification process had been made in France (although the first hot-test runs of the French process would not take place until 1978). Allied-General had already decided to acquire a fuel chopping device from France and stainless steel tanks for high-level waste storage from Britain; the emerging French waste solidification technology looked good enough for Allied-General to want that too. But upon taking this up with a high-ranking AEC regulatory official, the Barnwell managers were to be disappointed: "We were told this wasn't going to fly," Schneider recalls. "I think it was a matter of national pride. We felt pretty downcast." He adds, "We were back to square one."[49]

In the early years of the Barnwell project neither the Atomic Energy Commission nor the responsible corporations gave much consideration to the special risks associated with plutonium. Schneider felt that the plutonium nitrate (a liquid) from the chemical separations process should be converted at the Barnwell site to plutonium oxide (a solid); or better yet, that the uranium and plutonium nitrates should be coprecipitated to form a mixed oxide fuel that would dissolve readily in reprocessing. An advantage of coprecipitation—at least in the light water reactor fuel cycle, where plutonium concentrations are relatively low—was that the plutonium would not become concentrated enough to be used as an explosive material, although little thought would be given to this until later.

As matters turned out, however, the corporate owners chose neither the coprecipitation nor the plutonium nitrate-to-oxide processes. The fuel manufacturers and the utilities had expected Barnwell to deliver to

---

[47]Atomic Energy Commission, Division of Materials Licensing, "Discussions and Findings ... Relating to Consideration of Suspension Pending NEPA Environmental Review of the Construction Permit for the Barnwell Nuclear Fuel Plant," AEC docket no. 50-332, November 12, 1971.

[48]Interview by the author on February 15, 1983.

[49]Schneider says that on this occasion he and other Allied-General representatives met with S. H. Smiley, deputy director for fuels and materials in the AEC directorate of licensing. This meeting, he believes, took place in 1974. Smiley is now deceased.

them, its customers, not an oxide material but plutonium nitrate. "That was the way the market looked," Schneider says, and this was the way the owners felt Barnwell should go. Another former Barnwell engineer says confidentially, "They saw reprocessing at Barnwell as just a service to launder the fuel," adding with a touch of hyperbole, "they assumed the plutonium would be put out like milk bottles for the utilities to pick up."[50]

But in 1973 and 1974 the Atomic Energy Commission, to protect the public against the risk of plutonium transport spills, issued regulations to require that, after July 1978, plutonium could not be shipped unless first converted to a solid.[51] The new regulations were seen as a green light by Schneider and others at Barnwell who wanted to include a plutonium product facility in the project plans. But before the end of 1974 this became a vanishing prospect. The commission belatedly began preparation of a generic environmental statement on recycling plutonium in mixed oxide fuel. This became the so-called GESMO proceeding, involving public hearings and extensive staff studies. Approval of plans for the plutonium product facility would now have to await completion of the statement. Persuaded that this would be no time soon, Allied-General decided in January 1975 to suspend its design work for this facility.[52]

Still other regulatory problems were descending on the Barnwell managers. On September 12, 1974, the Atomic Energy Commission proposed new regulations defining transuranic waste as any material with 10 nanocuries—or billionths of a curie—of transuranics per gram.[53] To establish this threshold, at a level almost too low to be measured, meant that large amounts of general-process trash at Barnwell, which otherwise might have been regarded as ordinary low-level waste and disposed of as such, now would have to be kept in retrievable storage and eventually shipped to a geologic repository. At the same time the

[50]Interview with Schneider, February 15, 1983; confidential interview, February 14, 1983.

[51]Atomic Energy Commission, "Packaging and Transportation of Radioactive Material," proposed rule 10 CFR, pt. 71, *Federal Register* vol. 38, no. 147 (August 1, 1973) p. 20483; the final rule is in *Federal Register* vol. 39, no. 117 (June 17, 1974) p. 20360.

[52]A. E. Schubert, president of Allied-General, in his letter of June 4, 1975, to Robert C. Seamans, administrator of ERDA, refers to this decision and the events that led up to it.

[53]Atomic Energy Commission, "Transuranic Waste Disposal, Proposed Standards for Protection Against Radiation," 10 CFR, pts. 20 and 150, published in the *Federal Register*, September 12, 1974, pp. 32921, 32922. This proposed rule never actually took effect, and its nanocurie-per-gram threshold was superseded when Nuclear Regulatory Commission regulations set the threshold at 100 nanocuries per gram in 1982. NRC, "Licensing Requirements for Land Disposal of Radioactive Waste," 10 CFR, pt. 61, in *Federal Register*, December 27, 1982, p. 57446.

Environmental Protection Agency was developing stack-gas emission standards requiring removal and storage of krypton-85, again pending commitment to a repository.[54] The agency was also looking at the possibility of limiting environmental releases of tritium and carbon-14,[55] two other radionuclides that Allied-General was going to release up the stack. EPA and the Nuclear Regulatory Commission were in fact under increasing pressure to impose ever more comprehensive and stringent standards. For instance, in a letter dated January 6, 1975, Governor Jimmy Carter of Georgia asked the NRC's Atomic Safety and Licensing Board Panel to require immediate installation of krypton-85 removal equipment at Barnwell and to initiate "serious" research on development of systems for the capture, containment, and storage of tritium at both the Barnwell and Savannah River plants, which were just across the Savannah River from Georgia.

Proposals for a Federal Demonstration Facility

In sum, Allied-General faced major regulatory changes and rising and uncertain costs with respect to high-level waste solidification, the plutonium product facility, and thresholds and standards for the management and disposal of transuranic wastes and stack gases. Furthermore, large technological uncertainties remained to be overcome. The company was to acknowledge this in a letter to the Energy Research and Development Administration in which it proposed that ERDA build the high-level waste and plutonium solidification facilities at Barnwell and let Allied-General operate them as a federally sponsored demonstration project.[56] Allied-General referred here to "ERDA-33," the report of a special task force that had looked into the status of commercial fuel cycle technology following the NFS troubles at West Valley and General Electric's decision to write off its Morris reprocessing facility as inoperable. As Allied-General noted, the report stated that the government " 'stepped out of the high level waste development program before the technologies of spent fuel reprocessing, recycle, waste disposal, and safeguards had been demonstrated,' " and that industry had not been able to demonstrate them successfully on a commercial scale. The resulting uncertainty " 'pervades the entire nuclear fuel cycle,' " the report said, adding that the task force believed " 'the government's departure

---

[54]The new standards were promulgated January 13, 1977, and were to take effect January 1, 1983. See *Federal Register* vol. 42, no. 9 (January p 13, 1977) pp. 2858–2861.

[55]Environmental Protection Agency, *Final Environmental Statement*, vol. 1, *Environmental Radiation Protection Requirements for Normal Operations of Activities in the Uranium Fuel Cycle*, 40 CFR 190, November 1, 1976 (Washington, D.C., 1976) appendix, pp. A-16 and A-17.

[56]Letter of A. E. Schubert to Robert C. Seamans, June 4, 1975.

from the scene was premature' " and that government participation in closing the fuel cycle was essential.[57]

During 1975 ERDA officials met frequently with Allied-General and a half-dozen other companies engaged in or considering ventures in reprocessing and plutonium fuel fabrication; ARCO, Exxon, and Westinghouse were among them. The agency's nuclear fuel cycle and production division wanted to mount an ambitious technology demonstration that would include high-level waste and plutonium solidification at Barnwell. ERDA was not looking at the Allied-General facility merely as a commercial entity, but as a demonstration platform: "To have some confidence that 20 to 30 chemical reprocessing plants [enough to serve from 1,000 to 1,500 large reactors] will be built in support of the light water reactor system requires a demonstration now," wrote a top official of the nuclear fuel cycle division in an ERDA memorandum.[58]

The Energy Research and Development Administration and the nuclear industry generally were still optimistic about the industry's prospects. Reactor orders already were down sharply from the peak reached in 1973, but the precipitous decline in the growth of nuclear power which was to occur was not yet expected. The price of uranium was at a record high of more than $40 a pound and was expected to continue rising. "The whole atmosphere was quite bullish, what with OPEC [the Organization of Petroleum Exporting Countries] and the high price of oil," says George B. Pleat, the ERDA official who coordinated the meetings between his agency and the fuel cycle companies.[59] Accordingly, for ERDA and these companies to have been talking of having the agency step out front and spare industry the risk of fuel cycle facility demonstrations indicates that the unresolved technical problems associated with reprocessing were not perceived as minor.

Before the end of 1975, ERDA's nuclear energy division underwent a change of leadership and priorities, and the plans for a large, costly federal demonstration program were dropped. The agency initiative did, however, lead to a conceptual study by DuPont for a 10-ton-a-day, 3,000-ton-a-year reprocessing plant whose system for remote maintenance would follow essentially the familiar Savannah River design.[60] Also, the

[57]Ibid.

[58]ERDA memorandum of May 19, 1975, prepared for F. P. Baranowski, director of the Nuclear Fuel Cycle and Production Division, by George B. Pleat, assistant division director for the fuel cycle, reviewing a visit by Baranowski and other ERDA staff to Barnwell on May 13, 1975, and describing the alternatives by which ERDA might carry out or support a demonstration there.

[59]Interviewed by the author on February 18, 1983.

[60]Interview by the author with Clint Bastin of the Department of Energy, February 17, 1983, with reference to DuPont's "Spent Fuel LWR Recycle Complex, Conceptual Design," issued in November 1978.

Oak Ridge National Laboratory's fuel recycling division was developing a new "remotex" concept that contemplated putting the process equipment into two large, barnlike cells where repairs could be made in place from a center isle by mobile manipulators directed by operators at a television console; as a backup, an overhead crane would be available (as at Savannah River) to lift out entire equipment modules for maintenance at a repair station.[61] The remotex plant was expected to be cheaper than one of the DuPont design because of the reduction in the size and mass of the building. But neither the Oak Ridge nor the DuPont studies amounted to more than interesting concepts. There was no money to carry either one forward, and even had this not been the case Allied-General's plight at Barnwell would have been in no way relieved.

The abandonment of federal plans to demonstrate high-level waste and plutonium solidification at Barnwell could have itself meant the beginning of the end for the Allied-General project. The plant's corporate owners were already taken aback by the enormous increase in cost of their reprocessing venture. Originally estimated at $80 million, their investment exceeded a quarter of a billion dollars. Unless the whole venture were to be abandoned, they would have to commit perhaps another half billion. But as matters turned out, the willingness of the owners to see the project through despite the soaring costs and technical uncertainties was never tested. Dramatic changes in federal policy arising from concerns about plutonium safeguards and nuclear proliferation were now in the offing.

Safeguards, Proliferation, and Presidential Policies

When the Atomic Energy Commission, in early 1974, began the GESMO proceeding on the generic question of whether to allow plutonium recycling, the focus of concern was on the possible radiological and environmental effects, not on the fact that plutonium is an explosive nuclear material that might be stolen or diverted. In the draft GESMO document issued in August 1974 the commission staff concluded that the widespread use of plutonium in light water reactors should be approved as being desirable in itself and as a step toward plutonium recycling in breeders. The document provided no detailed analysis of

[61]J[ohn] R. White, B. F. Bottenfield, and E. D. North, "Facility and Equipment Concepts for the Hot Experimental Facility," paper presented at the American Nuclear Society Topical Meeting, "Fuel Cycles for the '80s," September 1980; and M. J. Feldman and John R. White, "Remotex—A New Concept for Efficient Remote Operations and Maintenance in Nuclear Fuel Reprocessing," paper presented at the 28th Conference on Remote Systems Technology, American Nuclear Society winter meeting, 1980.

the problem of safeguarding plutonium, but concluded that the problem would not be unmanageable.

At the end of 1974 the responsibility for GESMO passed to the Nuclear Regulatory Commission, one of AEC's successor agencies. On January 20, 1975, Russell W. Peterson, chair of the President's Council on Environmental Quality, wrote to William A. Anders, the first chairman of the Nuclear Regulatory Commission, to say that no decision on the recycling of plutonium should be made in the absence of a systematic analysis of alternative safeguard systems. Shortly thereafter, the NRC agreed that a supplementary GESMO document on safeguards would be prepared. This meant that another year or more would pass before the commission could decide whether the Barnwell plant should be licensed. The commission would first consider what policies should apply to the design, product specifications, and security arrangements for facilities such as the Barnwell plutonium product facility and the plutonium fuel fabrication plant that Westinghouse was proposing to build at Anderson, South Carolina. But as things turned out, the GESMO proceeding was to be overtaken by presidential policy decisions and aborted.

Proliferation Traumas in 1974

To the extent that GESMO was coming to be concerned with safeguards, the focus was on theft or seizure of explosive nuclear materials by individuals or subnational groups for terrorist purposes. But it was the other side of the safeguards problem—the possibility of non-nuclear weapons states using ostensibly peaceful nuclear facilities or materials to make nuclear weapons—that gave rise to major changes in U.S. policy by Presidents Gerald Ford and Jimmy Carter in 1976 and 1977.

Leading up to those changes were several traumatic happenings in 1974. The first of the traumas was India's detonation of a plutonium bomb, referred to as a "peaceful" explosive, on May 18 of that year. The Indian bomb stood out as a classic illustration of the proliferation problem.[62] The Atoms for Peace program had helped the Indians obtain, at places such as Oak Ridge and Canada's Chalk River laboratories, the expertise necessary for their own broad-based program of fission energy research and development. The Indians' research facilities included the Canadian-designed heavy water Cirus research reactor, which runs on natural uranium and thus makes enrichment services unnecessary, and a small reprocessing plant to separate the plutonium from the Cirus fuel.

[62]Interview by the author in 1974 with Charles Van Doren, then deputy assistant director for nonproliferation at the U.S. Arms Control and Disarmament Agency.

With India's nuclear test accomplished, the Pakistanis were protesting and would be covertly pursuing a bomb of their own. In many quarters there was now concern that the taboo against nuclear weapons spread was weakening.

This trauma was quickly followed by two others. In June of 1974 President Nixon, on the point of being driven from office by Watergate and seeking new foreign policy achievements, visited Egypt and Israel. Nuclear cooperation agreements were announced under which these countries, only recently at war, would buy power reactors from U.S. vendors, but without the "fullscope" safeguards necessary to make all their nuclear facilities subject to inspection by the International Atomic Energy Agency.[63] In July, Dixy Lee Ray, then chair of the Atomic Energy Commission, announced that the United States, long the noncommunist world's supplier of uranium enrichment services, could accept no further foreign applications for enrichment contracts because its capacity was fully booked. Moreover, approval of some of those applications already received would depend upon the countries concerned authorizing the recycling of plutonium. According to Charles Van Doren, who was with the U.S. Arms Control and Disarmament Agency at the time, the agency viewed these developments with a growing sense of dismay.

Fred C. Iklé, ACDA's director under Presidents Nixon and Ford, would later speak of an "egregious" train of errors going back to the start of President Eisenhower's Atoms for Peace program. Atoms for Peace had led to the declassification and release of an "avalanche" of more than 11,000 technical papers on plutonium recycling; the drafting of bilateral agreements under the program had often been "sloppy" (so-called peaceful explosions were not clearly prohibited, for example). Iklé saw a tendency to "dump" the problems of safeguarding nuclear technology into the "closed magic box" of the International Atomic Energy Agency (IAEA), by which he meant that with the agency treating much relevant data as confidential, the effectiveness of the safeguards regime could not be adequately monitored by outsiders.[64]

Following the traumas of 1974 the existing safeguards regime—made up of the IAEA, the Nuclear Nonproliferation Treaty of 1968,[65] and the web of bilateral agreements between nuclear supplier and recipient states—was increasingly coming to be seen as flawed, insufficient, and

[63]The new reactors would be under the safeguards regime, but facilities built earlier, such as Israel's Dimona research reactor, would not. The Dimona reactor was seen as the source of plutonium for clandestine weapons development.

[64]Fred C. Iklé in his foreword to Albert Wohlstetter, *Swords from Plowshares* (Chicago, University of Chicago Press, 1977) pp. vii–xiv.

[65]A number of key countries, such as India, Pakistan, and Israel, refused to sign this treaty.

too much influenced by commercial and economic interests. Several congressional committees were determined to check what they saw as an alarming and almost promiscuous spread of dangerous fuel cycle technologies. This led, most notably, first to an amendment to the Atomic Energy Act that made all new bilateral agreements subject to congressional veto, and ultimately to the Nuclear Nonproliferation Act of 1978, which put major new restrictions and clearance requirements on the transfer abroad of nuclear fuels and technology of U.S. origin.

A new study, *Moving Toward Life in a Nuclear Armed Crowd?*, released by the ACDA in late 1975, sharpened the perception in Congress and elsewhere that fuel reprocessing was going to give more and more countries enough separated plutonium to build nuclear weapons if they chose. This study by University of Chicago political scientist Albert Wohlstetter and his associates at Pan Heuristics concluded that use of plutonium fuel not only represented a grave proliferation hazard, but was unnecessary. It argued that projections as to the rate of growth of nuclear power around the world had been much exaggerated; that future supplies of uranium had been grossly underestimated; and that plutonium fuel, when all costs were counted, was going to come at a relatively high and unattractive price.

The Ford and Carter Policies and the INFCE

Five days before the 1976 election, but following an intensive White House policy review, President Ford issued a strong nuclear policy statement bearing on the proliferation risks inherent in plutonium recycling. The president said that while use of nuclear energy should be increased, "nonproliferation objectives must take precedence over economic and energy benefits if a choice must be made." Calling for a deferral of commercial reprocessing in the United States, Ford was seeking to persuade France, Great Britain, Germany, Japan, and all other nations not to proceed with their reprocessing programs unless the proliferation risks could be overcome. He also called for a "new international regime . . . for storage of civil plutonium and spent reactor fuel," and for "centrally located, multinationally controlled nuclear waste repositories" for permanent disposal of either spent fuel or high-level waste.[66]

Jimmy Carter, as a presidential candidate, had spoken out on this issue several months before Ford, saying that if commercial reprocessing was

[66]"Nuclear Power," statement by the president, October 28, 1976, in *Presidential Documents: Gerald R. Ford, 1976*, vol. 12, no. 14, pp. 1624–1631.

to proceed at all it should be only under international auspices. As president, Carter issued his policy statement a few months after taking office, on April 7, 1977.[67] He said reprocessing should be indefinitely deferred. Barnwell would receive "neither federal encouragement nor funding from us for its completion as a reprocessing facility," the president added. The GESMO and Barnwell licensing proceedings were not mentioned, but the Nuclear Regulatory Commission eventually suspended both proceedings, with White House encouragement.

A key element in the announced Carter policy was the International Nuclear Fuel Cycle Evaluation (INFCE), a study to be conducted under IAEA auspices by working groups on every major aspect of the fuel cycle, from uranium supply to reprocessing, breeders, and waste management. All of the nuclear supplier and customer nations were invited to participate. President Carter's purpose was to seek more proliferation-resistant fuel cycles and to encourage national and international measures that would assure nuclear fuel supplies and a means of spent fuel storage to all nations "sharing common nonproliferation objectives." The Carter policy also called for commercial use of breeders to be deferred. Subsequently the administration sought to terminate the Clinch River commercial breeder demonstration project, but that project's supporters in Congress would manage to keep it alive for another six years.

The International Fuel Cycle Evaluation gave President Carter some successes and some failures. On the plus side, INFCE participants arrived at several useful statements of consensus. One was that even without fuel reprocessing and recycling enough uranium would be available through the year 2000 to support substantial growth in nuclear generating capacity—up to 1,200 gigawatts (which compares to the operable capacity in the world at the end of 1985 of about 250 gigawatts). Another was that spent fuel itself can safely be disposed of directly as high-level waste without reprocessing.[68]

Probably the most important gain from INFCE and the Carter policy was to raise awareness of proliferation risks. For instance, the French and the British announced that they would not export fuel reprocessing facilities, and for the French this meant, in particular, cancelling delivery of reprocessing equipment to Pakistan.[69] The Germans also said they

<hr>

[67]"Nuclear Power Policy," statement by the president, April 7, 1977, in *Presidential Documents: Jimmy Carter, 1977*, vol. 13, no. 15, pp. 502–507.

[68]International Nuclear Fuel Cycle Evaluation, *INFCE Summary Volume*, INFCE/PC/2/9 (Vienna, International Atomic Energy Agency, 1980) pp. 4, 8–13, 21 (hereafter cited as *INFCE Summary*).

[69]Brian G. Chow, "Reprocessing: An Uneconomic and Risky Venture," statement prepared for the House Committee on Foreign Affairs Subcommittee on Economic Policy and Trade and the Subcommittee on International Security and Scientific Affairs, August 10, 1982.

would not export such equipment, but they did not rescind a deal to set up a pilot reprocessing plant in Brazil.

INFCE's shortcomings and failures were more notable than its successes. INFCE working groups discussed various technical and institutional approaches to making all fuel cycles more proliferation-resistant, but their reports were not specific enough to provide a basis for negotiations that would have concrete results. Similarly, discussion of such topics as the need for international arrangements and controls for the storage, transfer, or disposal of plutonium and spent fuel failed to go beyond generalities. Subsequently, groups set up under the IAEA to look at concepts of international plutonium storage and international spent fuel management also finished their work with little to show for it.[70]

But INFCE was perhaps most disappointing to the Carter administration in its refusal to agree that the separation and recycling of plutonium is a greater proliferation hazard than are alternative fuel cycles. In its final report, issued in Vienna in February 1980, INFCE said that it is not easy to judge whether a fuel cycle involving the separation of plutonium is more proliferation-prone than the once-through cycle in which the unreprocessed and highly radioactive spent fuel is eventually disposed of as waste.[71]

By the time Carter left office in January 1981 leaders of the nuclear enterprise, with some support from independent observers, were calling the Carter nuclear policy a failure.[72] That policy, it was argued, had not discouraged reprocessing abroad, but was creating doubts about the reliability of the United States as a nuclear supplier, which not only hurt American commercial interests but weakened U.S. influence on proliferation problems. If this judgment was overly harsh, it was beyond dispute that the Carter policy had fallen short of its aims.

---

[70]According to Charles Van Doren, former Arms Control and Disarmament Agency official and head of the U.S. delegation to the International Atomic Energy Agency's expert group on plutonium storage from 1978 through 1980, the International Plutonium Storage study came close to producing an important analysis of policy options, but in the end "fell apart" and yielded no substantial results. Van Doren was interviewed by the author on June 27, 1982, and February 2, 1983.

[71]*INFCE Summary*, pp. 23–24. Some observers were disdainful of how INFCE turned out. Victor Gilinsky, a member of the Nuclear Regulatory Commission who had collaborated with White House and State Department officials in bringing about the new Carter policy, has expressed his disenchantment this way: "We called together the nuclear bureaucrats of the world and said, 'Have you been wrong for the last twenty years in what you've been doing?' They thought about this for about a microsecond and said, 'No, we haven't.' So our people went rushing home to report this to the President. The whole thing was in the hands of children." Gilinsky was interviewed by the author on October 29, 1982.

[72]See especially *ENERGY: The Next Twenty Years*, a report sponsored by the Ford Foundation (Cambridge, Mass., Ballinger, 1979) pp. 59–61, 442–465.

Reagan's Uncertain Reprieve for Reprocessing

With a new president came a new policy. President Reagan, in his statement on nuclear policy of October 8, 1981, formally lifted the previous administration's ban on commercial reprocessing and promised to eliminate "regulatory impediments." But Reagan also said that responsibility for commercial reprocessing services belonged to the private sector. The closest he came to offering financial help was to indicate that the government might buy the private reprocessors' plutonium for the breeder program. This latter suggestion had found its way into the president's statement at the last minute through the urging of George A. Keyworth II and John Marcum, then director and assistant director, respectively, of the White House Office of Science and Technology Policy (OSTP). Keyworth, formerly a scientist at the Los Alamos Laboratory, and Marcum, formerly a National Security Council staffer, believed that reprocessing was the only proper way to close the commercial fuel cycle.

That Ronald Reagan came to the White House wholeheartedly in support of nuclear energy was evident from those he chose to head the Department of Energy. His choice for secretary was former governor James B. Edwards of South Carolina. Edwards had no professional background in energy policy matters, but as governor he had enthusiastically supported the Barnwell project and other nuclear developments. W. Kenneth Davis, the president's choice for deputy secretary, had at one time headed reactor development for the Atomic Energy Commission and more recently had been vice-president of the Bechtel Corporation, which designed the West Valley and Barnwell plants.

Secretary Edwards soon proposed that the government buy Barnwell and let a private contractor operate the plant in the same manner that government-owned uranium enrichment facilities at Portsmouth, Ohio, and elsewhere are operated. But in a new Republican administration pledged to curb federal spending and promote free enterprise, this proposition was not salable. In a memorandum to Edwards, President Reagan said that for the federal government to acquire Barnwell would not be appropriate. He proposed instead that the Department of Energy determine which regulatory barriers were of greatest concern to industry and submit for his review recommendations "on how to create a more favorable climate for private reprocessing efforts."[73]

The immediate upshot was that the department held a meeting on April 23, 1981, to which about a dozen top-level people from the nuclear

[73]Reagan decision memorandum on Department of Energy budget appeals, March 20, 1981.

supply business, the utilities, and Wall Street were invited. The tenor of the meeting was decidedly pessimistic. Two officials present for the owners of AGNS indicated that reprocessing had become "intertwined with state, national, and international jurisdictions and environmental, energy, and security policies" in such a way as to make it unattractive for private investors. "No foreseeable combination of regulatory reform and federal policy initiatives" would suffice to make reprocessing practicable, they said.[74] A utility official observed that American utilities had committed about $150 billion to nuclear energy; his own company had committed the equivalent of twice its book value to nuclear plants and could not invest in reprocessing ventures. Other utility representatives, noting that utility stocks were selling under book value, said that inasmuch as nuclear enterprises were considered risky, stock in a reprocessing venture might find no market at all. A nuclear supply company representative expressed the view that the transfer of reprocessing technology from the government to industry had never been completed. Another speaker observed that for his company to invest in nuclear projects there must be known costs, schedules, performance, and risks—factors which for reprocessing were all unknown. Several speakers referred to reprocessing as a prerequisite for the breeder, and some called it important to radioactive management, but none seemed to regard reprocessing and Barnwell as a prudent private investment opportunity.[75]

In late 1981 and on through 1982 DOE and OSTP officials tried unsuccessfully to put together, within the strictures laid down by the president, a plan and set of policies to overcome the private sector's skepticism about investing in reprocessing. But while this was going on two illuminating episodes occurred. The first involved what for the nuclear industry was an ill-timed and embarrassing suggestion by Secretary Edwards that part of the plutonium created in commercial power reactors should be recovered for use in nuclear weapons. The second had to do with the appearance of a strongly critical article about Barnwell in a Department of Energy periodical published by Assistant Secretary Brewer's own office, and with Brewer's subsequent attempt to pass off the criticism as coming from nameless phantoms.

Edwards's gaffe came at a September 1981 meeting of the Energy Research Advisory Board. Speaking as though totally unaware that for

[74]Brian D. Forrow (Allied Chemical Corporation) and W. Creighton Gallaway (General Atomic Company), "Prospects for Private Investment in Nuclear Fuel Reprocessing," paper circulated at an April 23, 1981, meeting at Department of Energy headquarters.

[75]From minutes of the April 23, 1981, DOE meeting cited in note 74. The meeting was private, and the identity of individual speakers was not given in the minutes.

more than two decades the United States had been urging on the world a policy of not diverting civilian reactor plutonium to weapons, the secretary said, "We are going to be needing some more plutonium for our weapons programs, and the best way I can see to get that plutonium is to solve your waste problem. Reprocess it, pull out the plutonium. . . . We could also use the plutonium in the breeder reactor and get that technology proven. . . . We would solve two problems at one time. It just makes a lot of sense to me to go that route."[76]

Edwards and other DOE officials later said that the administration favored maintaining a clear distinction between commercial and military plutonium and merely wanted the option to divert plutonium from commercial fuel to weapons in case of a clear national security need. But Congress found this episode disturbing enough to decide, by an amendment to the Atomic Energy Act adopted overwhelmingly by both Senate and House, that transfers of commercial fissionable material to military uses should be flatly prohibited.

Neither Secretary Edwards nor Deputy Secretary Davis seemed to have recognized that to pursue the Edwards proposal would not only have undermined U.S. nonproliferation policy, but would very likely have created enormous new political difficulties for the nuclear industry, especially abroad. When Davis visited the International Atomic Energy Agency in Vienna shortly after the first news of the Edwards proposal, Sigvard Eklund, the Swede then serving as the agency's director-general, is reported to have warned him that should such a course be pursued new waves of hostility to nuclear power could be unleashed in Sweden and in other countries of Europe.[77]

Assistant Secretary Brewer's embarrassment first arose as the result of what Brewer later called an "incomplete staff review" of the January 1982 issue of "Worldwide Nuclear Power," an occasional publication prepared in his office. This issue carried an article by an Argonne National Laboratory technical writer which cited numerous shortcomings of the Barnwell plant and its design. It said that the potential maintenance and licensing problems at Barnwell were such that "independent groups" which had reviewed the plant feared that reprocessing there could suffer the same fate as at West Valley and "give the industry a further black eye." Brewer later sent all recipients of "Worldwide Nuclear Power" a letter saying that the "views expressed by [the author] are strictly his own."[78] But in fact the article's characterization of the Barnwell plant

---

[76]Excerpt from the transcript of the Energy Research Advisory Board meeting, reprinted in the *Congressional Record* daily ed., March 30, 1982, p. S2979.

[77]David Fishlock, "IAEA Aghast that America May Turn to Civilian Reactor Fuel for Weapons Plutonium," *The Energy Daily*, September 28, 1981.

[78]The Brewer letter was dated March 4, 1982. The article in question, "Status of World-Wide Reprocessing," was by F. Robert Lesch.

had been extracted verbatim from a 1980 Argonne report on Barnwell, which had come with a covering letter indicating that the report's "negative attitude" reflected the views of fuel cycle specialists consulted at Oak Ridge National Laboratory, Exxon, and Argonne.

The above episode suggested that the bureaucratic commitment to commercial reprocessing and to Barnwell was now all but blind, with top Department of Energy officials not only rejecting the views of responsible and knowledgeable critics, but seeking to deny that such critics even existed. But before the end of 1982 it was to become clear that Brewer and his associates were engaged in a quixotic endeavor. By the spring of that year Brewer and his staff had set forth the general outlines of a plan to start up Barnwell. Once cleared by Secretary Edwards, the plan could be taken to the White House for discussions in the cabinet council on natural resources, then headed by Secretary of the Interior James Watt. In this plan, developed after consultations with industry people, the Department of Energy was seeking to identify the conditions and policies needed to bolster investor interest and confidence in a start-up of Barnwell. Among its prescriptions were: a negotiated agreement whereby the government would promise to buy plutonium from Barnwell to sustain breeder development (in the fall of 1982 the Office of Management and Budget rejected this idea); an "improved regulatory and licensing environment"; a contract or agreement to compensate investors fully in the event future policy changes again brought reprocessing ventures to a stop; and an "unambiguous" agreement whereby the government would accept wastes for final disposal.[79]

Foreign participation in Barnwell would be "highly desirable," according to the plan, and should be encouraged to the extent desired by potential U.S. investors and permitted by U.S. nonproliferation policy. Brewer's office had in fact found that the most concrete expression of interest from any quarter in investing in Barnwell had come from Germany. More specifically, it came from the nuclear fuel cycle company, DWK, owned by a group of German utilities which under German law were required to provide for the reprocessing of their spent fuel.

Time Runs Out for Barnwell

By November 1982 the Department of Energy officials struggling to bring about the start-up of Barnwell had become plainly discouraged. The official immediately in charge of this undertaking, Kermit O. Laughon,

---

[79]Department of Energy, "Cabinet Council Paper on Nuclear fuel Reprocessing and the Barnwell Plant," draft, May 18, 1982.

in his appearance before the Atomic Industrial Forum that month, said frankly it was "frustrating" that on this matter of reviving reprocessing the industry should be "conspicuously silent." "Time is running out for the Barnwell plant, and industry must step forward in the near future if this plant is to be completed and operated as a commercial venture," Laughon said.[80]

About a year later the corporate owners of Allied-General formally abandoned the Barnwell venture and claimed a tax write-off. The plant was officially closed on December 31, 1983; at this point nearly all of Allied-General's technical staff had been dismissed, and some of the plant equipment had been sold.[81] The corporate owners had pending a $500 million lawsuit against the government, claiming that policy changes and regulatory failures had made operation of the Barnwell plant impossible. Later there were some further efforts, initiated by industry people, to get commercial reprocessing started again, but they got nowhere. The last such effort, put forward by Bechtel National, would have depended on contractual commitments by DOE to pay for "toll reprocessing" services to recover uranium, plutonium, and high-level waste. Secretary of Energy Donald Hodel replied to the Bechtel proposal that the administration could not commit itself to this venture, of which the government would (as Hodel noted) be the principal customer. But he hoped that his negative reply would "not discourage the further pursuit of efforts to establish a private reprocessing venture."[82] So the Reagan administration and the nuclear industry were again exhorting each other to support reprocessing, yet neither was willing to step out front and promise the necessary investment.

Even if the Reagan administration had found the Bechtel proposal attractive, a start-up of Barnwell would probably have met strong resistance in Congress. The demise of the long-controversial Clinch River breeder reactor project in the fall of 1983 would have been part of the problem. Reprocessing had its rationale in the breeder, but Clinch River, the only U.S. commercial demonstration breeder in the making, was killed by a Congress exasperated at the apparent unwillingness of private industry to put any money at risk to save the project, which altogether might have cost up to $4 billion.[83]

Much of the congressional opposition to Barnwell would have come from members worried about safeguards and proliferation. Despite the

[80]Laughon, "Commercial Reprocessing: What Must Be Done."

[81]General Accounting Office, *Status and Commercial Potential of the Barnwell Nuclear Fuel Plant*, GAO/RCED-84-21 (Washington, D.C., March 28, 1984) p. 12.

[82]Letter of Secretary of Energy Donald P. Hodel to R. P. Godwin of Bechtel National, Inc., February 21, 1984.

[83]Eliot Marshall, "Clinch River Dies," *Science*, November 11, 1983.

Carter policy initiatives and INFCE, the proliferation hazard was not perceived, on balance, to have been much reduced. On June 8, 1981, Israeli aircraft destroyed the French-built Osirak Research Reactor at Baghdad, even though Iraq was a party to the Nuclear Nonproliferation Treaty and the reactor was subject to IAEA inspection. Also, there was new evidence of political terrorist activity in the United States, including the detonation of a bomb just outside the Senate chamber in December 1983. All this, together with continuing violence in Lebanon and the rest of the Middle East, would have given a sharper edge to arguments against the separation of plutonium at Barnwell.

The Barnwell project had even lost much of its support in South Carolina, its host state. Those opposing a start-up of Barnwell included South Carolina's governor Richard Riley and one of the state's most influential congressmen, Butler C. Derrick, Jr., a member of the Committee on Rules which controls the flow of legislation to the House floor. As a state legislator in the early 1970s, Derrick had been an early critic of Barnwell; yet later as a congressman he had joined with Senator Ernest Hollings in an effort to keep the project alive. Federal research and development funds they obtained for work on safeguards at Barnwell had allowed Allied-General to maintain the plant and hold the technical staff together after President Carter called for commercial reprocessing to be deferred. But in the end both Derrick and Hollings opposed further funding.

A major concern of Derrick's, and of Governor Riley's, was that for Barnwell to be used either for interim storage of spent fuel—which DOE was considering as a possible stopgap for utilities—or for reprocessing could leave South Carolina stuck with a new accumulation of high-level radioactive waste. In a speech before an American Nuclear Society meeting in Savannah, Georgia, on September 27, 1982, Governor Riley said, "I do not oppose reprocessing in principle, if it can be shown to be safe, but I do oppose the start-up of the Barnwell plant until we see demonstrated progress on a permanent solution for our high-level waste. We are not going to create the second [waste] tank farm in my state until we have cleaned up the first one." The governor was alluding to the existing tanks at the Savannah River plant; the Department of Energy was moving to recover and solidify the Savannah River waste, but still had no place for that waste's terminal disposal.

Preventing commercial spent fuel from being sent to Barnwell had become a major objective of nuclear critics and anti-nuclear activists in South Carolina. In 1982 and 1983 the Energy Research Foundation, an activist group based in Columbia, the state capital, campaigned hard to persuade county and municipal governing councils to oppose the sending of spent fuel from other states or foreign countries into South Carolina. Focusing on counties situated on major interstate highways or at ports of entry, the campaign had considerable success, generating headlines such as "South Carolina: Nuclear Graveyard of the Nation and the World?" and getting several local government endorsements. The South Carolina press had given top play to the news that the U.S. secretaries of state, commerce, and energy, in an effort to help American vendors sell reactors to Mexico, had in January 1982 sent the Mexican government a cable to encourage "Mexico's interest in acquiring [reprocessing] services or in becoming a close partner in U.S. [reprocessing] services."[84] In 1983 the state legislature, responding to these political stimuli, enacted a law prohibiting shipment of commercial spent fuel into South Carolina from abroad.

In sum, by the mid 1980s prospects for commercial reprocessing had further deteriorated, whether viewed from a technical, economic, regulatory, or political point of view.

[84]On January 23, 1982, a letter to that effect was dispatched by diplomatic cable to Secretary of Foreign Relations Jorge Castaneda de la Rosa of Mexico by Secretary of State Alexander M. Haig, Jr., Secretary of Commerce Malcolm Baldridge, and Secretary of Energy James B. Edwards.

Part 2
Searching for a Waste Policy

4
Policy Struggles in the Bureaucracy

Despite the differences in principle proclaimed, the Reagan administration policies on domestic reprocessing turned out in practice not to be all that different from President Carter's. Similarly, the Reagan policies on management and disposal of spent fuel and high-level waste have not been markedly different from those that evolved during the Carter years. In fact, the Reagan policies had their origins in those of the Carter administration, discussed in this chapter, as well as in the Nuclear Waste Policy Act of 1982, discussed in chapter 6. But the evolution of these policies actually begins in the mid 1970s during the Ford administration.

The Energy Research and Development Administration, as successor to the Atomic Energy Commission, proposed during 1975 and 1976 an exceedingly ambitious plan for the disposal of high-level and long-lived wastes from the commercial nuclear fuel cycle. It called for the siting and development of as many as six geologic repositories. The first two repositories were to be built in salt and to start operating at pilot-plant scale in 1985. The next four would be built in other kinds of rock, such as granite and shale. All were proposed to be operating by sometime in the 1990s. Six was the number believed to be needed to accommodate the waste generated over the thirty-to-forty-year life of the several hundred reactors expected to be on line by the year 2000. But so ambitious a plan was neither needed nor achievable. Electricity demand already was falling far short of past projections. With orders for nuclear reactors off sharply, the need for repositories to receive nuclear waste was correspondingly less.

By the late spring of 1977, President Carter's decision that reprocessing of commercial nuclear fuel should be indefinitely deferred had led the Energy Research and Development Administration to start rethinking waste program plans.[1] A key figure in the rethinking was Colin A. Heath,

[1]This rethinking of program requirements was evident, for example, in the paper "ERDA Waste Program Management" presented by Carl W. Kuhlman of the Energy Research and Development Administration at the Atomic Industrial Forum Fuel Cycle Conference in Kansas City, Mo., April 26, 1977.

the nuclear engineer who that spring took over as ERDA director of geologic disposal, a position he was to hold until late 1981. Before joining ERDA in 1976 Heath had held jobs with the National Aeronautics and Space Administration (NASA) and the General Atomic Company. At the Energy Research and Development Administration and the Department of Energy the frustration over geologic disposal made him wonder whether any nuclear waste would ever get below ground. Heath's politically beset job as director of geologic disposal was far harder than the normally quiet job of the aerospace engineer. He and his superiors at ERDA would be constantly at the beck and call of governors, congressional committees, and citizens' groups, all demanding that DOE explain or justify this or that program action or decision.

It became apparent by mid 1977 that a graceful way had to be found to get the waste program off the limb where it had been put late the previous year when ERDA informed thirty-six of the fifty governors of its plans to search for repository sites in their states. Field investigations were supposed to start in thirteen of the states before the end of 1977, with sites for the first two repositories to be selected before the end of 1978. But to carry on such a nationwide search for sites meant that money and resources would be spread unacceptably thin. The new budget for these investigations was $34 million, seven times greater than the budget for the preceding year,[2] but even so this was far too little to support the far-flung siting effort envisioned. In view of the cost of drilling deep boreholes, running seismic lines, and analyzing data, no real progress could be made without a more concentrated effort.

There were major political problems, too. In theory, the plan's broad national scope and the expectation that a half-dozen repositories would be built in various regions across the country showed a commitment to objectivity and fairness. Selection of sites in one region and one kind of rock would be accompanied or followed by selection of sites in other regions and other kinds of rock. Furthermore, ERDA had promised the governors that fieldwork in a state would be terminated if differences related to site selection criteria could not be resolved to the agency's and the state's mutual satisfaction. But rather than avoiding political trouble, the ERDA plan was fomenting it, in some cases gratuitously because protests were coming from states where there was no prospect of early site investigations being mounted. Knowing that the plan was politically sensitive, the agency had seen to it that all of the governors and U.S. senators and congressmen from the thirty-six states of interest

---

[2]W. C. McClain and C. D. Zerby, "Waste Isolation in Geologic Formations in the United States," in *Proceedings of the International Symposium on the Management of Wastes from the LWR Fuel Cycle* (Washington, D.C., 1976) pp. 567–579. The symposium was held in Denver, July 11–16, 1976.

were informed of the plan almost simultaneously. Letters had gone out on November 26, 1976, and were hand-delivered by scores of messengers. Yet the reaction had been far more negative than expected.[3] Don Vieth, a Department of Energy staff aide who was monitoring the response, recalls that a number of governors and members of Congress were alarmed and peremptorily asked, " 'Why are you coming into my state?' "[4] A former governor of New Mexico, Jack M. Campbell, an ERDA consultant, thought the plan politically ill-advised. "To start exploring in thirty-six states and provoke thirty-six emotional outbursts was simply going to prolong the agony," Campbell says, looking back on that soon-to-be-aborted initiative.[5]

In an early step away from the 1976 policy, Heath received an important new policy direction from G. W. Cunningham, director of ERDA's division of waste management, and from one of Cunningham's deputies, Carl Kuhlman. Heath was told that the siting effort was "too widespread" and "should be focused on only those states [offering] a reasonable near-term probability of finding a site for the first repository." Heath says, "I was told that this direction came from James Schlesinger who [as the White House energy adviser] was receiving complaints from politicians who felt that their states, which had not previously been considered, should not be so now, since obviously they would not be candidates for the first repository."[6]

The outcome was that Heath prepared for his superiors a policy memorandum which proposed limiting the siting search to six "salt states" having formations of either domed or bedded salt, plus the states of Washington and Nevada—Washington, because the Hanford site with its underlying basalt flows was there; Nevada, because of the presence of the Nevada Test Site (NTS) and its variety of rock formations.[7] Over

[3]Daniel S. Metlay, political scientist at Indiana University, reported in a 1981 study for the Office of Technology Assessment (OTA) that the response of state officials was such that the ERDA initiative "soon got mired down in the reluctance of state officials even to contemplate a facility on their soil." Metlay, "Radioactive Waste Management Policy Making," appendix A-1 to OTA, *Managing the Nation's Commercial High-Level Radioactive Waste* (Washington, D.C., March 1985) p. 223. The Department of Energy's *Report of the Task Force for Review of Nuclear Waste Management* (1978) said the ERDA initiative was "too ambitious" and was poorly designed for effective federal-state-local interaction, with the result that "public concerns were aggravated rather than resolved." P. 12.

[4]Interview by the author with Don Vieth, April 1, 1983.

[5]Interview by the author with Jack M. Campbell, April 12, 1983.

[6]Letter of Colin A. Heath to the author, December 24, 1984.

[7]This memorandum, dated August 12, 1977, was sent by Robert D. Thorne, acting assistant administrator for nuclear energy, to Robert W. Fri, acting administrator of the Energy Research and Development Administration (hereafter cited as ERDA memorandum, August 12, 1977).

the previous half year a number of governors and members of Congress had strongly advised ERDA to look first to such government-owned reservations as Hanford and the Nevada Test Site.[8] In keeping with such advice, ERDA's assistant administrator for nuclear energy, Robert Thorne, was already pushing to make Hanford and the Nevada Test Site prime candidate areas for a repository. Thorne's special assistant, Roger Richter, dispatched to the two reservations, had returned recommending that the siting effort be made there.[9]

That these federal reservations offered some significant advantages for repository siting was indeed apparent. They were located in remote desert regions; as special national security areas they were well policed; and they were already contaminated, at least in part, from the production of plutonium or the testing of nuclear weapons. Moreover, the U.S. Geological Survey (USGS), quite independently of the governors and other politicians who were advising ERDA, had told the agency that the Nevada Test Site deserved a high priority.[10] As the USGS pointed out, a wealth of geologic, hydrologic, and geophysical data on the reservations was available from studies done in connection with underground weapons tests. Also, potential disposal sites were thought to be available in a variety of geologic media, including shale, granite, and tuff.

In Heath's view, the hard-rock sites at Hanford and the Nevada Test Site and the salt sites available elsewhere were distinctly different in their potential. He felt that ERDA's best chance to meet its goal of developing two repositories by 1985 was in salt, because of the previous work done in that medium. On the other hand, the basalt, granite, shale, and tuff at Hanford and Nevada offered, in his opinion, a better prospect for repositories in which unreprocessed fuel elements could be stored subject to possible later retrieval. His memorandum pointed out that salt appeared less suitable than hard rock for such storage. He was alluding to salt's tendency to "creep" or flow plastically from the pressure of the overlying rock mass, and eventually to close up any tunnels and rooms created by mining. Retrievability had assumed a new importance in light of the Carter decision to defer commercial reprocessing, Heath noted.

Heath did not favor relying on surface facilities as the sole or principal means of retrievable storage. In his view, retrievable storage in deep geologic repositories would be superior, both economically and politically—economically, because the repository ultimately could offer permanent storage and thus would not present problems of long-term

[8]Interview with Don Vieth, April 1, 1983

[9]Colin A. Heath, personal communication to the author, December 24, 1984.

[10]Letter of July 9, 1976, from Vincent E. McKelvey, director of the U.S. Geological Survey, to Richard W. Roberts, assistant administrator for nuclear energy, ERDA.

maintenance and custodial care; politically, because to initiate retrievable surface storage before repositories were sited and built could provide competition for or divert attention from permanent geologic disposal as the national goal. And Heath doubted that siting surface storage facilities would be any easier politically than siting geologic repositories.

In sum, Heath had settled on what he saw as a promising redirection for the overall geologic disposal program: press ahead in the salt states for the first two disposal sites, while at the same time accelerating work at Hanford and the Nevada Test Site to identify acceptable hard-rock sites for repositories to accommodate retrievable storage of unreprocessed fuel. The thirty-six governors were now to be sent a new letter (or so Heath and his superiors were intending) to explain that, in light of President Carter's deferral of reprocessing, there no longer was a requirement for six geologic repositories, although retrievable storage of unreprocessed fuel might now be needed more than before.[11] Accordingly, field investigations would be limited to only eight states. Their governors would be asked for early meetings at which the new plan could be discussed. The governors of other states could now rest easy, knowing that their states were no longer targeted for early investigation.

But the new letter was never sent. The Energy Research and Development Administration was about to be absorbed into the new Department of Energy, of which James Schlesinger was to be secretary. According to John F. Ahearne, then Schlesinger's aide on nuclear issues and later chairman of the Nuclear Regulatory Commission, "We had reached the conclusion that there wasn't a [nuclear waste] program we could defend. We thought the entire program needed reexamining."[12] Nonetheless, the Heath memorandum had fairly well described the strategy that DOE would attempt to follow in the potential host states over the next several years, even as new policy studies and deliberations were going on in Washington. But the governors of several states were to deny the department the access needed for field studies, and in most of the other states major conflicts were also to arise (see chapter 5).

Policy Initiatives under Carter

The Spent Fuel Policy

In October 1977 the new Department of Energy, in one of its first major policy actions, announced a spent fuel policy which President Carter

---

[11]Letter drafted by Robert W. Fri, enclosed with ERDA memorandum, August 12, 1977.

[12]Interview by the author with Ahearne, April 11, 1983.

had himself approved. This was a follow-up to the deferral of commercial reprocessing. The department was now proposing to accept and take title to spent fuel that the utilities would deliver to government-approved sites. A one-time fee would cover the cost of both interim storage and ultimate disposal. The department would also accept for storage a limited amount of foreign fuel when this would "contribute to meeting nonproliferation goals."[13]

Although the new policy embraced the concept of retrievable storage of spent fuel in geologic repositories, the DOE officials responsible for it evidently assumed that repositories in salt as well as in hard rock could be designed for this purpose. "Maximum use" was to be made of the first repository for retrievable spent fuel storage, and this facility was still expected to be built in salt and become operational in 1985. Such storage "would go further toward demonstration of a waste management capability than would above ground storage," said the department. But surface facilities would meet storage needs arising before a geologic repository became available. Private industry would be encouraged to provide that interim surface storage under government contract, but if necessary DOE itself would provide the storage. Altogether, the spent fuel policy represented an exceedingly ambitious technical and political commitment. The proposed opening of the first repository was only seven or eight years away, and no site had been identified, much less approved. One or more away-from-reactor (AFR) surface storage facilities were expected to be ready within five years, but the department could not count on being able to obtain sites for these potentially controversial facilities.

The Deutch Report

Compared to its spent fuel policy, the Department of Energy's next policy initiative was in some respects more cautious and realistic, but not in all. In announcing that reprocessing would be deferred, President Carter also had called for a review of radioactive waste management policy. To that end, the department later established a task force under its director of research, John Deutch, a chemistry professor on leave from the Massachusetts Institute of Technology. The Deutch report, made public in March 1978, concluded that a repository could not be opened by 1985 because there would be too little time for site selection and licensing. The task force believed that there was an "independent technical consensus" in support of the geologic disposal concept, but

[13]The spent fuel policy is described in the February 1978 draft of the DOE *Report of the Task Force for Review of Nuclear Waste Management*, appendix F, pp. 60–72.

that the necessary data and predictive models relevant to effective waste isolation would not be obtained without substantial research. Salt was recommended as the geologic medium for the first repository site. The alternative of comparing "the best salt design with the best design in other media" was rejected as "unnecessarily conservative." The report said that from the standpoint of safe disposal it made no significant difference whether the waste committed to a geologic repository was in the form of spent fuel from reactors or of waste from a reprocessing plant.[14]

The Deutch task force also proposed that the Waste Isolation Pilot Plant (WIPP) project in New Mexico, which had been planned for military transuranic waste only, be adapted for an early demonstration of geologic isolation of spent fuel—a recommendation that turned out to be particularly provocative politically. What happened in response to this incautious proposal will be described in the next chapter.

Interagency Review and National Academy Appraisal

The Deutch report was meant as only the first step in defining a national nuclear waste policy. Top Department of Energy officials knew that DOE, as successor agency to the AEC and ERDA, lacked the credibility to carry off this policy-making task by itself.[15] Moreover, President Carter had been elected with the help of environmentalists, who had greater representation in his administration than in any previous one. A broadly constituted interagency review group could bring the concerns of this important environmental constituency directly to bear on the waste issue. Accordingly, at the time the Deutch report was issued the White House, acting on Secretary of Energy Schlesinger's recommendation, announced the appointment of a top-level Interagency Review Group on Nuclear Waste Management. Represented in this group, known as the IRG, were the Department of Energy and eight other agencies, together with several entities within the Executive Office of the President, including the Council on Environmental Quality (CEQ).

The agencies and executive office entities that were to participate actively and continuously in the IRG process were DOE, as the agency responsible for nuclear waste management; the Department of the Interior, as custodian of the federal public lands and parent organization of the U.S. Geological Survey, the agency with the relevant expertise in

[14]Department of Energy, *Report of the Task Force for Review of Nuclear Waste Management* (Washington, D.C., February 1978) (hereafter cited as the Deutch report).

[15]Ted Greenwood, "Nuclear Waste Management in the United States," in E. William Colglazier, Jr., ed., *The Politics of Nuclear Waste* (Pergamon Press, 1982) p. 20.

the earth sciences; the White House Office of Science and Technology
Policy, as the agency to lead a review of the state of the art in geologic
disposal; and CEQ, as the agency to oversee compliance with the National Environmental Policy Act and serve as the chief link between the
White House and the environmentalists. The secretary of energy—or
his delegate, John Deutch—was to chair the IRG. But while the Department of Energy's influence within the Interagency Review Group was
to be strong, it was not to be dominant. In fact, on those few major
issues on which agreement could not be reached, DOE was to be in the
minority.

President Carter's memorandum establishing the review group mentioned support for U.S. nonproliferation policy as one of its objectives,
and the group's participants included Joseph Nye of the Department of
State, Jessica Tuchman Mathews of the National Security Council, and
Charles Van Doren of the Arms Control and Disarmament Agency, all of
whom had helped shape that policy. But nonproliferation and support
for the president's policy on deferral of commercial reprocessing were
not to figure as central concerns of the Interagency Review Group. Its
work was regarded throughout as essentially a domestic policy matter.
Indeed, the national security agency officials were not present at most
of the review group meetings, although some of them were usually on
hand to support the Department of Energy's ultimately losing efforts to
expedite repository siting and development.[16]

Early in the Interagency Review Group exercise the scientific foundations of geologic disposal were found to be much weaker than had
previously been assumed. This scientific reappraisal took place both as
a part of the IRG effort and independently of it.[17] Yet all concerned
concluded that however promising geologic disposal was as a concept,
some major technical issues remained to be resolved.

The authors of a U.S. Geological Survey paper observed that there are
"a variety of possible interactions among the mined opening of the
repository, the [heat-generating] waste, the host rock, and any water
that the rock may contain," and that many of them are not well understood. "This lack of understanding contributes considerable uncertainty
to evaluations of the risk of geologic disposal," the USGS scientists said.[18]

[16]Interview by the author with Charles Van Doren, September 7, 1984.

[17]Professional papers from the U.S. Geological Survey, the Environmental Protection
Agency, and the Office of Science and Technology Policy (OSTP) all bore importantly on
this question, but only the one from OSTP was prepared specifically for the IRG.

[18]J. D. Bredehoeft, A. W. England, D. B. Stewart, N. J. Trask, and I. J. Winograd, "Geologic
Disposal of High-Level Radioactive Wastes—Earth Science Perspectives," U.S. Geological
Survey Circular 779 (1978) p. 3.

An Environmental Protection Agency panel of earth scientists was "surprised and dismayed to discover how few relevant data are available on most of the candidate rock types," and commented that "except for the modest effort on salt, the geological aspect of the repository problem had been largely neglected by our generation until a year or so ago. It will not be solved without a strong commitment of money and manpower, lasting beyond 1985."[19] An Office of Science and Technology Policy paper found a "rather general consensus" that a knowledge and technology base "sufficient to permit complete confidence in the safety of any particular repository design or the suitability of any particular site" was still lacking.[20]

Besides the reviews being done by agency groups, there was an extraordinary behind-the-scenes reappraisal of salt siting going on at the National Academy of Sciences. The academy had continued to favor salt since 1957, when it had first proposed this rock type as a promising disposal medium. In March 1977, Konrad Krauskopf, a Stanford University geologist and longtime member of the academy's radioactive waste committee, had told the California energy commission that the "properties of salt have been thoroughly investigated" and that "most geologists would agree that salt was an excellent choice for disposal."[21] But now an academy member who had no connection with its waste committee was running up a warning flag. This was E-an Zen, a distinguished Chinese-born research scientist at the Geological Survey's headquarters at Reston, Virginia. Zen's own specialty was petrology, the study of the origin of rocks, but through discussions with his colleagues David Stewart and Edwin Roedder in the survey's experimental geochemistry and mineralogy branch he had become convinced that past AEC-sponsored studies of the potential interaction between high-level waste and rock salt were inadequate and misleading. He had decided, moreover, that his duty as an academy member was to alert others in the academy to the potential problems with salt as a repository siting medium.

A "Dear Colleagues" letter was prepared by Zen in March 1978 and subsequently endorsed by twenty-four of his colleagues among the academy's geologists and geophysicists. Referring to potential brine migration

[19]Environmental Protection Agency, Ad Hoc Panel of Earth Scientists, "State of Geological Knowledge Regarding Potential Transport of High-Level Radioactive Waste from Deep Continental Repositories" (1978) pp. 44–45. The panel was co-chaired by Bruno Giletti of Brown University and Raymond Siever of Harvard University.

[20]Office of Science and Technology Policy, working paper for the Interagency Nuclear Waste Management Task Force, May 9, 1979, draft.

[21]Testimony of Konrad Krauskopf, *Hearings before the California Energy Resources Conservation and Development Commission on Nuclear Fuel Reprocessing and High-Level Waste Management*, March 21, 1977, vol. 7, pt. 1, pp. 98–103.

toward the waste canisters, the letter said that the canisters would be corroded and broached in a relatively short time and that "the corrosive brine could interact with the waste material directly to become radio-active" and weaken the rock mechanically. "Abnormal hot spots" beyond design safety might be created as the contents of canisters interacted, Zen and his colleagues observed. The letter said that inasmuch as the academy's past reports constituted "the intellectual basis" for salt's status as the Department of Energy's primary disposal medium, the academy had a "considerable moral responsibility" to see that any decision to build a repository in salt was scientifically well founded. Even though the need for a repository was urgent, it said, decisions should be made on the basis of adequate facts "and not be in response only to political considerations."[22]

According to Zen, these views were sympathetically received by the academy's president at the time, Philip Handler; by its Earth Sciences Advisory Board; and by some key members of its radioactive waste committee (including Krauskopf, who in 1977 became chair of this body, now elevated to the status of an academy board).[23] Moreover, Zen later saw evidence in academy reports of a more cautious and reserved attitude about salt. The letter also probably had an influence on the Interagency Review Group. Although the letter was never publicly dis-closed by the academy, word of it circulated via the Washington grape-vine.

In its final report, issued in March 1979 after extensive public review of an earlier draft, the Interagency Review Group found no scientific or technical reason why a site suitable for a repository could not be iden-tified, but insisted that a "systems view" be rigorously applied to evaluate the suitability of sites and repository designs and to minimize the pos-sibility of future human intrusion. The review group called for combining multiple natural and engineered barriers to the release of radionuclides, and said that some uncertainty and risk would always remain, with society having to judge whether the risk is acceptable. Although the report implied demanding exercises in site selection, systems design, and performance assessment, the IRG indicated that the first repository could be built and licensed by the early 1990s.[24] But beneath the

[22]Letter of March 22, 1978, from E-an Zen to members of the National Academy of Sciences' sections on geology and geophysics; subsequently endorsed by Francis R. Boyd, Jr., Hatten S. Yoder, Jr., Robert M. Garrels, Edward Anders, Arthur H. Lachenbruch, Wallace S. Broecker, W. Gary Ernst, Preston Cloud, Donald E. White, George R. Tilton, Gerald J. Wasserburg, Leon T. Silver, Hans P. Engster, Paul B. Barton, Jr., and Luna B. Leopold.

[23]Interview by the author with E-an Zen, March 30, 1983.

[24]Interagency Review Group on Nuclear Waste Management, *Report to the President*, TID-2944Z (Washington, D.C., Department of Energy, March 1979).

compromise language of the report, which reflected tensions between the Department of Energy and other agencies, lay some ill-concealed conflicts. For instance, the Geological Survey believed it would be years before the technical questions pertinent to salt as a host medium could be adequately resolved.[25] David Stewart, then the survey's geologic division coordinator for radioactive waste management, later told Congress that a decade of laboratory evaluations and tests inside a repository might be necessary before a decision to dispose of either spent fuel or reprocessing wastes in salt could be supported scientifically.[26]

The strategy for repository siting recommended by the Interagency Review Group called for a departure from the past policy of using the first of the repository sites investigated and found to be acceptable. Instead, the site to be used would now be chosen from among several acceptable sites.[27] The majority of the review group's participants held that technical prudence demanded that the first site be chosen from among four or five acceptable sites in a variety of rock types. For its part, the Department of Energy wanted the selection made from two or three suitable sites in different kinds of rocks. The aim of the IRG majority was to widen the choice beyond the Hanford basalt and the bedded and domed salt sites,[28] even though this would delay establishing the first repository by at least a few years. In another hedge against uncertainty, the IRG looked to the Nuclear Regulatory Commission licensing process to close significant information gaps. Information available for the initial stage of proceedings might justify issuing a repository construction permit, but licensing to operate would come only after more information was available from extensive tunneling and testing at the repository horizon. The emplaced wastes would be closely monitored and subject to retrieval, though for how long was not specified.

The President's Policy Statement

Nearly a year passed before the Interagency Review Group's final report and later memoranda to President Carter finally led to the president's

[25]James F. McIntyre, Jr., director of the Office of Management and Budget, and Stuart E. Eizenstat, director of the White House domestic policy staff, confidential memorandum for the president on nuclear waste management, August 29, 1979, p. 17.

[26]Testimony of David B. Stewart, prepared for the Senate Subcommittee on Energy, Nuclear Proliferation, and Federal Services of the Committee on Governmental Affairs, March 13, 1979.

[27]Interagency Review Group, *Report to the President*, chap. 2, "Technical Strategies for High-level and Transuranic Wastes," pp. 35–76.

[28]Interview by the author with Gerald Brubaker, staff scientist at the Council on Environmental Quality and participant in the Interagency Review Group.

nuclear waste policy statement of February 12, 1980.[29] Considerable time had been taken up by struggles among the principal IRG participants to have the White House take their side on issues never resolved within the review group itself. Although the Department of Energy was generally the loser, the issues were for the most part not very large, one exception being the question of whether to abandon the Waste Isolation Pilot Plant project in New Mexico. The president's statement fully embraced the IRG philosophy of "technical conservatism" in repository siting, even though that philosophy went beyond what DOE and the nuclear industry would have preferred. The president, like the review group, supported a program of away-from-reactor surface storage of spent fuel, but not the relatively large away-from-reactor program announced by the Department of Energy in the fall of 1977.[30] The Interagency Review Group had stated explicitly that "the responsibility for establishing a waste management program shall not be deferred to future generations."[31] In the group's view, and in the president's, storage at the reactor sites should generally be the rule until repositories became available.

The president's statement gave much attention to the Department of Energy's dealings with potential repository host states. The governing principle that he embraced was described as one of "consultation and concurrence." This loosely defined principle or concept was developed by the National Governors Association (NGA) and incorporated into a very early piece of nuclear waste legislation, the Glenn-Percy bill of 1977–1978 (for Senators John Glenn of Ohio and Charles Percy of Illinois). This bill, which did not become law but which influenced later legislation, was supported by environmental groups as well as by the NGA. But consultation and concurrence could be variously interpreted, and according to Edward Heiminski, who was on the governors' association staff when this principle was conceived, the governors never saw a right of concurrence as a right of veto by the states.[32] Their view of

[29]President Carter's message to the Congress on radioactive waste management, February 12, 1980, released by the Office of the White House Press Secretary.

[30]At the White House press conference announcing President Carter's nuclear waste policy, Gus Speth, chair of the Council on Environmental Quality and an IRG participant, said: "The president's limited away-from-reactor storage policy is very different, both in its intent and its effect, from the proposal for long-term surface storage[,] which . . . basically . . . say[s] we are going to put the waste somewhere in a mausoleum-type arrangement and leave it there for future generations to decide what to do with, fifty to a hundred years from now. That is not the approach we are taking." Transcript of the press conference, Office of the White House Press Secretary.

[31]Interagency Review Group, *Report to the President*, p. 16.

[32]Helminski was interviewed by the author on May 13, 1986.

it was that a potential host state would be continuously consulted and that siting activities would not be carried from one stage to the next until the state was satisfied and had concurred in what had gone before. To the environmentalists, on the other hand, if the state could concur, it could also withhold its concurrence. This amounted to saying no, or exercising a veto, although not as part of a formal process coming at some predetermined stage. The Interagency Review Group interpreted consultation and concurrence in the same way as the governors, and so did President Carter. The president, according to one of his top aides, even believed in the right of federal preemption if relations between the government and a host state reached an impasse.[33] But the term "consultation and concurrence" nonetheless represented a fundamental ambiguity which sooner or later would have to be more explicitly resolved.

President Carter also announced that he was creating a State Planning Council on radioactive waste management, an advisory group to be made up largely of governors and state legislators. Governor Richard Riley of South Carolina was to chair this new body. Before disbanding the next year, the council was to have substantial influence in shaping national policy only with respect to disposal of low-level waste. It effectively promoted the concept adopted by Congress in late 1980 of dealing with the disposal of such waste on a regional basis, through programs established by interstate compact—a concept whose political feasibility, however, remains unproved to this day.

In the selection of high-level waste repository sites, President Carter promised full compliance with the National Environmental Policy Act, which could help the states and the public to hold the Department of Energy accountable for the quality of its siting studies and environmental impact analyses. Carter insisted too, in keeping with IRG recommendations, that repositories for all categories of long-lived wastes—military as well as commercial, transuranic as well as high-level—be subject to Nuclear Regulatory Commission licensing. The president wanted the Waste Isolation Pilot Plant project in New Mexico cancelled, as the review group majority had recommended (with DOE dissenting). The Carlsbad site for WIPP would be included in the larger Department of Energy waste program and investigated along with all the other sites as a potential candidate for a repository for spent fuel and military high-level and transuranic waste. But only a few months before, in December

---

[33]Stuart Eizenstat gave the author this interpretation of the president's views following the press conference on February 12, 1980, at which the nuclear waste policy was announced.

1979, Congress had passed an act which authorized WIPP to continue as an unlicensed project for military transuranic waste alone.

The Interagency Review Group had taken a position of neutrality on the issue of the future growth of nuclear power, but there were significant differences within the group as to what neutrality should mean. One view was that of Gus Speth, former environmental lawyer for the Natural Resources Defense Council, who was now chair of the Council on Environmental Quality and was its representative on the IRG. Before the review group was created, Speth, in a speech upsetting to officials at the Department of Energy, called for a moratorium on further reactor construction unless a widely accepted solution to the waste problem were agreed upon and implemented by certain "nuclear deadlines."[34] That the review group should be neutral on the question of nuclear expansion was proposed by John Deutch. In a memorandum to Deutch, Speth accepted neutrality as an "excellent way to proceed" for the time being, provided that the IRG make a comprehensive and detailed analysis of the cumulative impact on the waste problem of several alternative nuclear futures.[35] These ranged from a 150-reactor future for the year 2000, signifying no growth beyond construction permits already issued, to the 380-reactor future contemplated in the administration's National Energy Plan. Speth said it appeared that, with a nuclear commitment for the year 2000 of 300 gigawatts (equivalent to 300 large reactors), six repositories would be needed, with a new one coming on line every six years after the commissioning of the first at the turn of the century. But the Speth proposal for a comprehensive impact analysis, which would have reached into every aspect of the waste problem (including transport logistics), had not been followed. At Speth's urging, however, the White House did include in the president's statement a favorable mention of the then ongoing proceeding by the Nuclear Regulatory Commission to determine whether the waste from nuclear reactors could be managed and disposed of safely.

The general policy direction for nuclear waste management indicated by President Carter's statement would in the end continue to be followed by the Reagan administration. But in the first years of the new administration the Reagan appointees at the Department of Energy had by no means embraced the IRG view that the site for the first repository should be selected from several acceptable sites in diverse geologic media. To the contrary, the prevailing DOE philosophy at that time was

[34]Speech of Gus Speth before an American Bar Association group on September 29, 1977.

[35]Memorandum from Gus Speth to John Deutch on "IRG and Neutrality Regarding Nuclear Growth," August 29, 1978.

for the field managers or contractors in charge of the siting efforts in basalt, tuff, and salt to compete, like horses on a racetrack, with the prize going to the fastest horse.[36] This horse-race approach was part and parcel with the semiautonomous nature of the various field programs, and lent credence to criticism that DOE headquarters was exercising only loose authority and quality control over its contractors, such as Rockwell Hanford, a subsidiary of Rockwell International.

Tighter control and a conscious abandonment of the horse-race approach came when the department brought in Robert L. Morgan, its top official at the Savannah River plant, to serve temporarily as waste program manager after the Nuclear Waste Policy Act took effect in early 1983. As Morgan knew, the horse-race approach was inconsistent with the terms of the new act, which required that the site submitted for licensing be chosen from among three sites that had undergone thorough investigation or characterization. This congressional requirement, moreover, followed a similar requirement that had been promulgated in early 1981 by the Nuclear Regulatory Commission: the NRC had called for site characterizations to include exploration and testing of the rock formations from deep shafts and tunnels at repository depths.[37] Morgan moved quickly to put the DOE siting effort in conformity with these congressional and NRC requirements. In particular, he reined in the horse at Hanford and called off plans to begin immediate construction of an exploratory shaft—an undertaking which, had it been allowed to proceed, would have put the basalt project far out in front of work elsewhere, and even far ahead of the congressionally prescribed process by which the top three candidate sites were to be chosen. Later, Rockwell Hanford's technical director for the basalt project was replaced following severely critical reviews by the U.S. Geological Survey and other agencies of the latest project report.[38]

The next chapter will examine the troubled real world of nuclear waste politics in the context of repository siting efforts and field investigations in potential host states.

[36]Interview conducted by John Graham of the American Nuclear Society with Ben C. Rusche, the waste program's present director, in *Nuclear News*, November 1984, p. 179.

[37]Nuclear Regulatory Commission, "Disposal of High-Level Radioactive Wastes in Geologic Repositories—Licensing Procedures," 10 CFR, pts. 2, 19, 20, 21, 30, 40, 51, 60, and 70, in *Federal Register*, February 25, 1981, p. FR13971.

[38]Interview by the author with Morgan, September 13, 1986.

5
Conflict in the Host States

As the search for a nuclear waste policy and strategy went on from the mid 1970s to the early 1980s in Washington, first the Energy Research and Development Administration and then the Department of Energy was engaged in a largely frustrating effort to identify and investigate potential repository sites in the field. This effort was taking place in ten states in several widely scattered regions, from the Great Lakes to the Gulf Coast and from the Southwest to the Northwest. The rock formations targeted in these investigations have been principally salt beds, salt domes, and two kinds of volcanic rock, basalt and tuff. Major problems have been encountered, problems arising from actual and perceived resource and land-use conflicts and frequently exacerbated by the presence or perception of technical difficulties and uncertainties.

The politics of repository siting have in some cases conformed poorly to stereotyped thinking and clichés about the waste problem. For instance, there is not invariably a not-in-my-backyard (NIMBY) response locally. On the contrary, in two of the salt states, Utah and New Mexico, the opposition has come not from the communities nearest the potential sites but from more distant parts of the state, and in Utah particularly from a governor who saw unacceptable resource conflicts. In these instances a repository is taken locally to mean additional jobs and federal dollars. For some of the same reasons the basalt program at Hanford has enjoyed a friendly political environment locally, but has met with strong criticism and opposition elsewhere in Washington state.

The experience in the generally troubled salt and basalt programs suggests that the Department of Energy's focus on finding sites in specific kinds of rock—and, at Hanford, in a specific place—has not represented the best way to go about repository siting. Later in this chapter I shall take up the experience of the department and its predecessor agencies with the Waste Isolation Pilot Plant project for military transuranic waste, now under construction in New Mexico. Taken together, the lessons from these experiences strongly suggest a need for new approaches to the problem of siting a repository for spent fuel and high-level waste. Such lessons are explored at the chapter's end (and in chapter 13) after an examination of the investigative effort and problems at each of the potential sites.

The Salt Beds

It was in the bedded salt of the huge Permian Basin, which reaches from southeast New Mexico across the Texas High Plains and Oklahoma panhandle into Kansas, where the geologic disposal program began with the Salt Vault research project at Lyons in the 1960s. Salt beds continued to have a central place in the program as it underwent a major expansion in the mid 1970s, and these formations remain important in the program today. They were formed when shallow inland seas or confined coastal embayments became increasingly briny, with the salt and other minerals present in solution reaching the point of oversaturation and precipitating out. This sedimentary or depositional process typically was accompanied by a slow subsidence of the shallow basin, thus allowing room for the salt, gypsum, and other minerals (such as dolomite and anhydrite) to accumulate in successive layers or beds. The total accumulation of these various evaporites ran to a thickness of thousands of feet and the beds in a particular basin might extend for many, many miles. Often during and following the deposition of evaporites from the briny seawater the rocks in the upland areas surrounding the basin were eroding. The gravel, sand, silt, and mud resulting from erosion were washed down into the basin, causing layers of this debris to be interspersed among the evaporite beds and ultimately to cover them over. Sometimes the entire evaporite sequence was covered to a great depth, only to be subsequently uplifted, permitting some of the overlaying deposits to erode away. In other instances the uppermost salt beds remained within a thousand feet or so of the earth's surface. In either case, to be considered a candidate repository site the salt must lie at a depth accessible by conventional mining techniques. The Permian Basin salt beds were formed from these depositional and sedimentary processes more than 200 million years ago.

Besides the Permian, the bedded salt basins of interest to the nuclear waste program have been the Paradox Basin in southeastern Utah and the Salina Basin in the lower Great Lakes region. The Salina Basin was where the Energy Research and Development Administration and the Department of Energy first got into trouble with salt-state governors.

The Salina Basin: Opposition in Michigan

The Salina Basin sweeps in a wide arc from the Finger Lakes region of western New York down across western Pennsylvania and northeastern Ohio, then up into Michigan's lower peninsula, ending where Lakes Michigan and Huron join at the Straits of Mackinac. The thickest and most extensive salt beds in the Salina Basin are said to be in Michigan.[1] Extensive but thinner deposits occur in Ohio, Pennsylvania, and New York. In late 1975 the Energy Research and Development Administration and the Oak Ridge National Laboratory undertook to start their Salina Basin field study in Michigan. The place first considered for borehole drilling was the largely rural northeast corner of the lower peninsula that fronts on Lake Huron and consists politically of the counties of Alpena, Montmorency, and Presque Isle. The largest town is Alpena, on Lake Huron's Thunder Bay.

The Energy Research and Development Administration needed the state's cooperation for this investigation. The Michigan Geologic Survey could provide helpful information and advice, and no borehole drilling could begin without a permit from the state Department of Natural Resources. ERDA and Oak Ridge representatives had already met several times with officials of these two state agencies when, in late May 1976, Governor William G. Milliken and Congressman Philip R. Ruppe, representative for the Alpena area, intervened. Ruppe was afraid that ERDA was much farther along in choosing a site than the agency would acknowledge and that a "bureaucratic sneak play" was afoot.[2] Ruppe persuaded the House Interior Committee, of which he and Michigan Democrat Robert Carr of Lansing were both members, to conduct public hearings in Michigan. Governor Milliken, for his part, put out word that there were to be no negotiations, discussions, or state commitments related to the proposed geologic investigation until he, as governor, was assured in writing that no disposal site would be selected in Michigan without the state's approval.[3]

[1]Office of Nuclear Waste Isolation, *Regional Summary and Recommended Study Areas for Ohio and New York Portions of the Salina Basin*, ONWI-29 (Columbus, Ohio, Battelle Memorial Institute, September 1979) p. 8 (hereafter cited as ONWI Salina Basin Study).

[2]*Proposed Nuclear Waste Storage in Michigan*, Committee Print 3, Subcommittee on Energy and the Environment, House Committee on Interior and Insular Affairs, 95 Cong. 1 sess. (March 1977) p. 4.

[3]Letter of May 26, 1976, from Governor William G. Milliken to Howard A. Tanner, director of the Michigan Department of Natural Resources, reprinted in *Nuclear Waste Disposal in Michigan*, Hearings before the Subcommittee on Energy and the Environment, House Committee on Interior and Insular Affairs, 94 Cong. 2 sess. (July 6, 1976) p. 276.

Congressman Ruppe and others in his district suspected strongly that ERDA and Oak Ridge were predisposed to select a site in the Alpena area. The *Alpena News* had quoted an Oak Ridge geologist as saying that this region "appeared to have more promise than other areas in the Michigan salt basin."[4] Furthermore, *Nucleonics Week*, a leading trade publication, had reported on May 20, 1976, that "the Salina Basin in Michigan and the inland Gulf coast salt domes [were] leading the list" of candidate areas. Confronting a waste program leader at an Interior subcommittee hearing, Ruppe told him, "You took the furthest, most remote—if that is the right word—point [within the Salina Basin] to begin the test. It just seems to me that this indicates a desire to move into an area like this and leave untouched the more populous, more urbanized, more politically powerful, perhaps, areas of the country."[5]

For three months, the Energy Research and Development Administration waffled on the states' rights issue that Governor Milliken had raised. Initially, the agency took a hard line, with a spokesperson informing the *Detroit Free Press* on June 4, 1976, that disposal of high-level waste was a "federal function . . . not subject to state control or veto." But later—on June 22—Robert Seamans, administrator of ERDA, was quoted in the *Free Press* as saying that, no, nuclear wastes would not be disposed of in Michigan "if the people don't want them there." Only two weeks after that, at the Interior subcommittee hearing, ERDA's general counsel, William L. Brown, said that the agency was "reserving the right" to press on with a repository project despite state objections if such action seemed justified.[6] Not until Seamans wrote Governor Milliken on September 17, 1976, was the veto question resolved to the governor's satisfaction. Seamans said "the project will be terminated in Michigan if the state raises issues on the project connected with [site selection] criteria, and their application, that are not resolved through a mutually acceptable procedure."

The Interior subcommittee later dismissed this pledge as "misleading doubletalk,"[7] while Congressman Carr, worried lest the governor surrender his political judgment to the opinions of federal and state technocrats, commented at the subcommittee hearings, "Congressman Ruppe and I have gone to bat . . . to protect the governor's veto. I am wondering, though, if we are not painting ourselves into a box by saying we will

[4]"Pinpoint Proposed Nuclear Dump Site West of Alpena," *Alpena News*, May 28, 1976.

[5]Testimony of Congressman Ruppe, Hearings on *Nuclear Waste Disposal in Michigan*, p. 266.

[6]*Proposed Nuclear Waste Storage in Michigan*, p. 5; Hearings on *Nuclear Waste Disposal in Michigan*, p. 264.

[7]*Proposed Nuclear Waste Storage in Michigan*, p. 8.

veto [a repository] unless ... it is proven to us that these things are safe. The people who have the power and money to make this demonstration of safety oftentimes can overwhelm us with information and overwhelm those who would offer some contrary advice."[8]

Governor Milliken appeared to be thinking much the same way. He chose not to follow the advice of a radioactive waste task force he had established, which had concluded that siting studies, and even construction of a repository, should be allowed to proceed provided ERDA agreed to a number of special conditions—one being that a repository also should be under development in another state to make abandonment of the Michigan site easier if its suitability were found questionable.[9] After reviewing the record of the public hearings held across the state by the task force, Milliken wrote ERDA asking that all its "activities in Michigan in connection with this project cease." He said that Michigan might suffer "negative economic impacts" if the project went ahead, and noted the "proximity of the Great Lakes" to any lower peninsula repository site.[10] The governor also told citizens at Alpena that the legislature would be asked to forbid the disposal of nuclear waste in Michigan, a promise fulfilled the next year when such a measure was enacted.

Governor Milliken's assessment of the risks squared fully with the intuitive judgment of most citizens and local officials in northeastern Michigan, where opposition to a repository had been strong from the beginning. Keith Titus, an Alpena County commissioner, had told the Interior subcommittee, "When the beast in us is aroused—and I assure [you] that it is—the jackpine savage is a ferocious animal. Despite hearings, conferences, and volumes of bureaucratic impact statements, we will not allow placement of this waste disposal facility here."[11]

The Salina Basin: Rebuff by Ohio and New York

No test boreholes were ever drilled anywhere in the Salina Basin formation. The Department of Energy, as ERDA's successor, kept its foot in the door for a while in Ohio and New York, but the governors of those states eventually did as Governor Milliken had done and asked that the department give up plans for exploratory drilling. Nothing beyond the preparation of literature surveys was ever accomplished.

---

[8]Ibid.; Hearings on *Nuclear Waste Disposal in Michigan*, p. 25.

[9]"Report of the Governor's Nuclear Waste Disposal Task Force," December 1976.

[10]Letter of May 11, 1977, from Governor William G. Milliken to Robert W. Fri, ERDA acting administrator.

[11]*Proposed Nuclear Waste Storage in Michigan*, p. 285.

   The initial Ohio survey itself turned out to be a factor in bringing on
intense criticism of the waste program in that state. Taking no account
of political considerations, this survey indicated that the most promising
geology for a site was in densely populated northeastern Ohio. An area
of particular interest was just east of Cleveland and Akron in Cuyahoga,
Summit, Lake, Geauga, and Portage counties.[12] Even to bring up the
possibility of radioactive waste disposal in this part of Ohio was guaranteed to provoke strong resistance. But if, politically, the place was
poorly chosen, so was the time. The question of test drilling came up
in 1978 when Republican governor James A. Rhodes was running for
another four-year term, along with George Voinovich, a Cuyahoga County
(Cleveland) commissioner and candidate for lieutenant governor (later
to be Cleveland's mayor). Rhodes and Voinovich were not going to be
left politically exposed on an issue so volatile as the disposal of radioactive waste, which already had led to the formation of several citizens'
coalitions in opposition.[13] Rhodes, like Governor Milliken in Michigan,
had at first sought expert advice on the problem. Several members of
his cabinet, including Robert S. Ryan, the state director of energy, urged
that the siting investigation be allowed to proceed.[14] But in the end
Rhodes, too, told the Department of Energy that he wanted no further
geologic studies made in Ohio.
   New York was no more receptive than Michigan or Ohio, and the
existing waste problem at West Valley made Governor Hugh Carey and
other New York officials particularly cool toward DOE. In fact, any
suggestion that New York accept more radioactive waste while the West
Valley problem remained unresolved was deemed obnoxious. "They [the
New York officials] didn't want to discuss anything related to radioactive
waste until they got an agreement on the cleanup at West Valley," says
Colin Heath, who at the time was directing the DOE waste isolation
program.[15] When the Deutch task force suggested that negotiations with
the state about the West Valley cleanup also cover the possibility of a
repository in the Salina Basin salt,[16] an outraged protest ensued. James

   [12]ONWI Salina Basin Study, figure 1. This document took note of the urban development
and discouraged further interest in the area, but an earlier survey by a subcontractor had
not.
   [13]Bob von Sternberg, "U.S. Drops Ohio Nuclear Waste Site," *Akron Beacon Journal*,
May 5, 1979.
   [14]Interview by the author with Ryan, April 30, 1983.
   [15]Interview by the author with Heath, April 18, 1983.
   [16]Department of Energy, *Report on the Task Force for Review of Nuclear Waste Management* (Washington, D.C., February 1978) p. 22 (hereafter cited as the Deutch report).

Larocca, Governor Carey's energy commissioner, went immediately to the press, crying "nuclear blackmail."[17] When Governor Carey later wrote Secretary Schlesinger to say that no geologic studies should be conducted in New York,[18] DOE complied with the governor's wishes.

The federal plans to investigate the Salina Basin salt in New York almost certainly would have met with strong opposition even had there been no West Valley. The places that DOE contractors had identified as having special interest were in the heart of New York's wine-growing Finger Lakes country. One of them, for example, was the Beaver Dam area only five miles southwest of Seneca Lake and the tourist mecca at Watkins Glen.

The Department of Energy was now stymied throughout the Salina Basin.

The Paradox Basin: Moratorium in Utah

When the Department of Energy first began its investigation of the Paradox Basin, Utahans gave it little notice and there was no trouble obtaining the state's cooperation. Colin Heath recalls that once when he went to Capitol Hill to brief the staffs of Utah senators and representatives there was so little interest that nobody came.[19] By the 1980s this had changed. The department's interest in the basin had become intensely controversial, but the controversy was in part between Utahans as well as between Utahans and DOE.

In discussions of the nuclear waste problem there is no phrase more often heard than "not in my backyard." In Utah the backyard in the most literal sense was in San Juan County, which takes in the state's entire southeast corner and much of its spectacular red-rock canyonlands. Yet the Department of Energy found this county eagerly receptive to the possibility of a repository. The decline in uranium mining and milling was putting many people there out of work, and the prospect of construction jobs was enticing. The growth-oriented San Juan County Commission, led by its chairman Calvin Black, a former uranium mine operator and a major political force in the county, not only supported the siting

---

[17]Reported to me by Colin Heath, interview of April 18, 1983, and confirmed by Frank Murray, New York State's Washington liaison on energy and environmental issues, in an interview on April 28, 1983.

[18]Interview with Frank Murray.

[19]Interview with Heath, April 18, 1983.

investigation but was (and is) highly impatient with Utahans who complained that a repository in the canyonlands country would be an unacceptable intrusion.[20] The state government maintained a kind of neutrality during the first years of the investigation, seeking to keep intimately informed but not prejudging the question of whether a repository should or should not be built. But Governor Scott M. Matheson became increasingly doubtful about the project after it became clear, by the summer of 1982, that the two sites identified as most promising were immediately adjacent to Canyonlands National Park and only about ten miles from the Colorado River.

The Department of Energy had not set out to find sites next to the park and near the river, but this was the luck of the draw. Three other areas within the Paradox Basin had been investigated. Two of them were clearly unacceptable, while the third was deemed to be only marginally acceptable at best.[21] The geology of a bedded salt formation in its idealized, or highly generalized, form resembles that of a layer cake. But the reality is never so simple. For example, in the Paradox Basin's Salt Valley, where the salt takes the form of a diapiric (or rising) structural fold that seeks to push up through the overlying rock, DOE found the geology to be quite complex. Moreover, the potential for oil, gas, and potash development appeared high. In the Lisbon Valley, displacements in the basement rock and compression of the regional crust had produced severe deformation of the salt beds; oil and gas, potash, copper, uranium, and vanadium were all known to be present. The Elk Ridge salt formation at the Paradox Basin's southwest edge, where the salt beds feather out, was also investigated. Test drilling there showed the beds to be of a thickness only marginally adequate for a repository.

Only at the Gibson Dome did DOE find a formation at minable depth and believed to have the desirable structural regularity, thickness of salt, and absence of recoverable mineral resources. To call this structure a dome is something of a misnomer because it is actually a very gentle structural fold or arch that extends laterally for several hundred square miles. Its center is a few miles east of Canyonlands National Park. There the salt is within two to three thousand feet of the surface. The Davis Canyon and the Lavender Canyon sites which the department identified on the flank of the dome are just beyond the park boundary in a valley that is part of a major scenic approach to the park—and that was within

---

[20]Black was interviewed by the author on November 1, 1982.

[21]Office of Nuclear Waste Isolation, *Paradox Area Characterization Summary and Location Recommendation Report*, ONWI-291 (Columbus, Ohio, Battelle Memorial Institute, August 1982); Utah Nuclear Waste Repository Task Force, "Summary of Public Meetings on Studies in Utah" (August 1982) p. 2.

the boundaries first proposed when the park was created in the early 1960s. A railroad might have to be built the entire length of this valley to haul spent fuel or waste to the repository.

None of this was objectionable to Calvin Black and his colleagues on the San Juan County Commission. Black had opposed the creation of the park and was one of those who had fought successfully to reduce its boundaries. But Black could not speak for Utah, or even for the Paradox Basin as a whole; at Moab, the largest town in the basin and county seat of Grand County, immediately to the north of San Juan County, opinion about the DOE project was sharply divided. Moreover, Utah's political center of gravity is along the Wasatch Front in the Salt Lake City–Ogden–Provo metropolitan region, some 200 miles northwest of thinly populated San Juan County. For many Utahans the redrock canyonlands are a distant but treasured recreational backyard.

From the time the Gibson Dome sites were first identified the Department of Energy could expect its Paradox Basin project to be opposed by a great many Utahans and by national environmental groups, led in particular by the National Parks and Conservation Association. The land-use and resource conflict was indisputably real, and added to it was an unfounded but perceived potential for a major radiological hazard arising from the proximity of the Colorado River. This perception had the potential of spreading from Utah to other states within or near the Colorado basin. A headline in the *Los Angeles Times* on September 19, 1983, proclaimed: "A Nuclear Row in Utah May Bear on the Water we Drink."

In mid 1982, Governor Matheson of Utah imposed a moratorium on further site investigations by refusing DOE the cooperation of state agencies, which meant among other things that requests for rights-of-way across state land would be denied. This moratorium grew out of a dispute between the state and the department over the timing and scope of environmental impact studies. In the end Matheson was to come out in flat opposition to a repository and to any further geologic investigation, declaring in a letter to the secretary of energy that the "canyonlands site is inherently unsuitable."[22]

The Texas High Plains and the Ogallala

The Department of Energy's siting investigation on the Texas High Plains has been taking place in Swisher and Deaf Smith counties in the Permian

[22]Letter of May 4, 1984, from Governor Scott M. Matheson to Secretary of Energy Donald P. Hodel.

Basin's Palo Duro subbasin. The land here is nearly flat except where a sharp escarpment divides the High Plains from the Rolling Plains to the east, and where the Canadian River valley separates the Palo Duro from the Dalhart Basin to the northwest. Strikingly characteristic features of this semi-arid region, especially when seen from the air, are the numerous playas—shallow depressions which fill with water after heavy rains but then usually soon dry out. The lifeblood of the region, as the people there know it, is the Ogallala aquifer, the prolific aquifer encountered 100 feet or so beneath the land surface. This aquifer, which is 300 feet thick or greater, is an immense resource, lying beneath not only the Texas panhandle but beneath the larger High Plains region, from Texas and New Mexico across Colorado, Oklahoma, and Kansas up into Nebraska. Without irrigation the Texas High Plains would be too dry to support today's intensive farming. By pumping from the Ogallala to irrigate the region's rich soils, an exceptionally productive farm economy has been created, the panhandle's annual output of wheat, corn, potatoes, soybeans, and other products being valued at more than $2 billion.[23] To support irrigated agriculture the aquifer has had to be "mined," or drawn down faster than the very slow rate of recharge, which is less than one inch per year.[24] A shift to a more sustainable but less productive regimen of dryland farming seems inevitable eventually; but while the water lasts the present agricultural bounty continues.

To build and operate a repository in the Palo Duro subbasin salt would mean sinking shafts through the Ogallala aquifer for emplacement of the radioactive waste. It was therefore predictable that sooner or later the Department of Energy would find itself accused of putting the people and farm economy of the Texas High Plains at risk. The investigation of the Palo Duro salt got under way in early 1977. The department's contractor was the Bureau of Economic Geology at the University of Texas, which serves as the Texas geological survey and was at that time accountable to a cabinet-level Texas Energy and Natural Resources Advisory Council, called TENRAC. TENRAC looked benignly on the salt investigation, especially in the early years. Furthermore, DOE officials such as Critz George, then director of geologic siting, had some success in reassuring people in Deaf Smith and Swisher counties that there was no reason for alarm. "In my opinion, they've done a good job of trying to inform people," County Judge Glenn Nelson, head of the Deaf Smith

---

[23]"Proposed Nomination of Repository Sites," Hearings before a U.S. Department of Energy panel, transcript, vol. 1, Hereford and Tulia, Tex., May 16–17, 1983, pp. 34–36, testimony of Congressman Kent Hance.

[24]Allen Kneese, Resources for the Future, personal communication to the author, 1985.

County board of commissioners told me in the fall of 1982.[25] "Nobody really wants it [nuclear waste], but many people feel a patriotic obligation," he added, saying too that "if you can't depend on your government to do a good job, we don't have anything to look forward to anyway." But there was more than faith in DOE and patriotism behind the fact that in the county seat towns of Hereford and Tulia the siting investigation was getting a relatively sympathetic reception. In early 1982 Wendell Tooley, editor and publisher of the *Tulia Herald*, told a Dallas reporter how the attitude in his town was shifting: "Now that we're more educated, we know about the safety aspects. I think the Chamber of Commerce and the industrial board want it [the repository] for the billions of dollars it would pump into here. Farming's been kinda slow, and there's not that much money circulating. Of course a lot of people are still afraid. Most are taking a wait-and-see attitude."[26]

Yet County Judge Nelson and publisher Tooley would later be on record as opposing the siting project. Expressed sentiment in the two counties and in the larger Texas panhandle region was to switch to strong opposition to any possibility of a repository being built in the Palo Duro salt. An opposition group was organized in each county. In Deaf Smith, Tim Revel, a local physician, and his wife were leaders of POWER, or People Opposed to Wasted Energy Repositories. In Swisher, Delbert Devin, a farmer and rancher, organized STAND, or Serious Texans Against Nuclear Dumping. Petitions were circulated among the general public and affidavits collected from bankers, grain elevator operators, health food processors, and others willing to declare their opposition to nuclear waste disposal in the region. The transcripts of Department of Energy public hearings show that the opposition included such politically or economically important entities and individuals as the Deaf Smith and Swisher county commissions, the Texas Farm Bureau, several county committees of the High Plains Underground Water Conservation District, the Texas Corn Growers Association, the Texas-New Mexico Sugar Beet Growers Association, and the Bishop of the Catholic Diocese of Amarillo (who linked the nuclear waste problem to the nuclear arms race).[27]

[25]Interview by the author with Nelson, November 16, 1982.

[26]Jere Longman, "Tulia Losing Fear of Nuclear Waste," *Dallas Times Herald*, January 21, 1982.

[27]"Proposed Nomination of Repository Sites," vol. 1; vol. 2, Austin, Tex., May 18, 1983. POWER and STAND have since received a $25,000 grant from the American Catholic Bishops' Campaign for Human Development, for use among the region's large low-income, Spanish-speaking population. "Dump Opposition Gets Grant from Catholic Bishops," *Hereford (Tex.) Brand*, August 6, 1986.

One Department of Energy official familiar with the Texas scene has confided to me his view that a critical factor in turning public attitudes against the siting effort has been the "unrelieved negativism" of state officials. "They have reinforced and confirmed the worst dreads expressed by any of the local opposition," he said. "With no holding back, very quickly a frenzy of demagoguery develops, much like the theatrics of domestic war propaganda except in this case DOE and nuclear waste are the evil. Fear and negativism are easily reinforced without a continuous voice of moderation." But in truth, this does the state officials less than justice. According to his aides, Governor Mark White, knowing that Texas may need the help of Congress to keep a repository out of the Palo Duro subbasin, was somewhat restrained for tactical reasons and chose to find fault with the siting process rather than to declare DOE's interest in the Palo Duro flatly unacceptable. The Texas commissioner of agriculture, Jim Hightower, has not felt so constrained. But even his comments have centered on what he not unreasonably perceives as a major land-use conflict. For instance, Hightower told a DOE hearing panel that, "for me, and for the agricultural community of this state, location of the nuclear dump site in the Panhandle really is not a technical issue at all. It's a human issue, an economic, cultural, and even a moral issue ... you could save us a lot of money, trouble, and time if we can just agree right now that prime agricultural land and a major fresh water aquifer is not a suitable site for a nuclear waste repository."[28]

The repository, if built, actually would be more than 2,000 feet beneath the Ogallala aquifer and separated from it by more than 1,000 feet of salt, not to mention numerous layers of other evaporites and interbedded clay and other material. According to DOE consultants, if someone drilled through the repository to the briny and nonpotable Wolfcamp aquifer deep below it, the water in that aquifer would not be under enough hydraulic head pressure even to approach the repository horizon, much less to rise above it and transport radionuclides up to the Ogallala.[29] Such technical facts can be marshaled for an argument in support of a Palo Duro site. Yet the Department of Energy must not only persuade the public in the region directly affected that the repository will be safe, but also convince that public that it has persuaded everybody else as well.

In 1983 Kent Hance, then the congressman representing Deaf Smith County, reported receiving a number of letters from High Plains food processors warning that they could be badly hurt if the public that buys their products perceives a contamination problem. One letter was from

---

[28]"Proposed Nomination of Repository Sites," vol. 2, pp. 428–433.
[29]Critz George, personal communication to the author, 1986.

Frito-Lay, a company that purchases up to $23 million worth of corn and potatoes a year from High Plains growers. The letter said, "We all know that nuclear waste is an emotional issue, and it doesn't take much imagination to conjure up the eventual hue and cry of the public about food crops being irrigated by water that flows over nuclear waste. Under such a situation our alternative would be to move into other corn and potato producing areas for raw materials, seriously disrupting the economy of your district. It seems that a far better alternative for every party concerned would be to locate nuclear waste disposals in areas of federally owned wastelands where even sage brush has a hard time growing."[30] More recently, the Holly Sugar Corporation, a major producer of beet sugar on the Texas High Plains, has given a $10,000 donation to the Nuclear Waste Task Force, a regional umbrella group for local groups such as POWER and STAND.[31]

The Texas legislature in 1983 enacted a bill, sponsored by a state senator from the panhandle and favored by TENRAC, to require anyone proposing to sink a shaft through the Ogallala or any other aquifer first to obtain a permit from the Texas Water Commission. The permit could only be issued upon a determination by the commission, following a formal hearing, that the shaft is needed in the public interest and will cause no pollution and impair no one's water rights.[32] If the Department of Energy were to attempt to sink an exploratory shaft in the Palo Duro subbasin, this new state permitting procedure would present a possibly insurmountable political and legal obstacle. In sum, the presence of the Ogallala aquifer above the Palo Duro salt in politically powerful Texas could put the potential repository sites there beyond practical reach, whatever their technical merits.

The Salt Domes

The Department of Energy and its predecessor, the Energy Research and Development Administration, began field investigations in salt domes in the three Gulf Coast states of Texas, Louisiana, and Mississippi in 1977 and 1978. Structurally, salt domes are quite different from salt beds, yet it is from such beds that the domes arise. As described earlier, the beds

[30]Letter to Congressman Kent Hance from Charles H. Murphy, vice-president of government and consumer affairs for Frito-Lay, in "Proposed Nomination of Repository Sites," vol. 1, pp. 36–38.

[31]"Holly Issues Statement Against Dump," and "Sweetens Opposition Coffer with $10,000 Lump," *Hereford Brand*, August 7, 1986.

[32]Texas Water Code, chap. 28, subchap. C (Drilled or Mined Shafts).

of salt and other evaporite minerals (such as gypsum and anhydrite) lie more or less flat, and in idealized form resemble a layer cake. Salt domes have formed in geologic basins where sediments have accumulated over deep layers of bedded salt until the overburden pressure has forced the plastic and relatively bouyant salt up through weak places in the overlying rock. The domes have thrust upward in enormous columns, some of them up to several miles across, and some vaguely mushroom-shaped. Although typically there is no surface expression of their presence, occasionally the upthrusting dome has caused the overlying surface material to bulge upward, as at Avery Island in Louisiana, where the visitor traveling across an otherwise flat coastal plain is surprised to see a small mountain rising in isolation from the surrounding marshes.

The process by which salt domes are formed creates an extraordinarily complex geology. Harry Smedes, a respected field geologist formerly with the U.S. Geological Survey and now a Department of Energy consultant, vividly described it to me:

A good way to visualize it is to take a bunch of colored cloths or old rags and let each color represent a different rock type and have them stacked up like a layer cake, and then draw this entire mass up through a hole you've made by thumb and forefinger. You'll notice there are very complex vertically oriented folds, highly convoluted. Then if you snip part of that off and make a horizontal section, you'll see how complicated these crenulations are, largely vertically oriented but really like a marble cake. To the extent that these different rock types have desirable and undesirable hydrologic or mechanical properties, you can see the difficulty, the near impossibility, of trying to predict or determine ahead of time what the configuration is.[33]

The Geological Survey in 1981 warned DOE that "the chances for predicting subsurface geologic and hydrologic conditions are significantly better in bedded salt than in domed salt."[34]

Besides their complexity, salt domes present certain other drawbacks as potential repository sites. For one thing, they are a resource in themselves for the salt they contain and for their potential for the storage of petroleum, chemicals, and the like in solution-mined cavities. For another, they are exploration targets for wildcatters drilling for oil because of the structural traps created along their flanks when the domes were formed. The attractiveness of the domes for resource use, extraction,

[33]Interview by the author with Smedes, November 8, 1982, quoted in Luther J. Carter, "The Radwaste Paradox," *Science*, January 7, 1983, pp. 35–36.

[34]Letter of March 17, 1981, from George D. DeBuchananne, chief of the USGS Office of Radiohydrology, to Colin A. Heath, director of the Division of Waste Isolation in DOE's Office of Nuclear Waste Management.

and exploration poses two problems for repository siting. First, exploration boreholes drilled into a dome in the past may have compromised its integrity, especially if the holes were never properly recorded and sealed. Second, in the distant future, when administrative control over the site may well be lost, explorers for resources might unwittingly invade and breach a repository, at great danger to themselves and perhaps considerable risk to the surrounding populations. For example, if solution mining accidentally occurred, contaminated brine might even be pumped to the surface. With respect to these potential problems salt domes differ from salt beds in that, unlike the beds (which extend over hundreds of square miles), the domes each represent specific, isolated targets.

The Department of Energy found itself constrained early on in its search for suitable domes. In the Gulf Coast basin there are about 500 salt domes, about half of them on shore, the rest off shore. Of the 115 onshore domes that are at practical depths, 79 were eliminated at the start because of competing uses. The remaining 36 were winnowed down for various reasons to only 7, but even these few showed substantial evidence of past exploratory activities. Indeed, a total of more than 180 known boreholes had been drilled into these domes, better than a third of them penetrating beyond the caprock into the salt itself. The penetrations into one of the east Texas domes were so numerous and poorly recorded as to cause the Office of Nuclear Waste Isolation, DOE's contractor at the Battelle Memorial Institute, to indicate that the site might be unlicensable for these reasons alone.[35] But in truth, under the site selection criteria eventually adopted by the Nuclear Regulatory Commission, the Environmental Protection Agency, and the Department of Energy itself, almost any salt dome would appear to be a candidate with a strike against it.[36]

Mississippi: The Dome in Richton's Backyard

In addition to the problems posed by salt domes as a rock type, there are the problems peculiar to particular domes. As luck would have it, of the seven salt domes which the Department of Energy investigated intensively, the one that appeared to be the best, all things considered,

[35]Office of Nuclear Waste Isolation, "Evaluation of Area Studies of the U.S. Gulf Coast Salt Dome Basins," ONWI-109 draft (Columbus, Ohio, Battelle Memorial Institute, March 31, 1981) pp. 45, 137–138 (hereafter cited as ONWI Salt Domes Evaluation Study).

[36]See, for example, NRC's final regulations, "Disposal of High level Radioactive Waste in Geologic Repositories, Technical Criteria," 10 CFR, in *Federal Register* vol. 48, no. 120 (June 21, 1983) pp. 28224–225.

was right next to a town. This was the Richton dome at Richton, Mississippi, a community of about 1,100 people some 20 miles east of Hattiesburg. It is an unusually large dome, roughly 4 miles by 2 miles in its horizontal dimensions at the proposed repository depth of 1,900 feet. There appeared to be room for a repository of almost 3,800 acres, with space left over for an 800-foot buffer extending in every direction.[37] With 2,000 acres considered adequate for a repository, this dome would allow generous latitude for selection of the areas best suited for the emplacement of waste. The geohydrologic characteristics of the dome also were believed to be superior. Several of the other domes had shown evidence of salt dissolution. This was true, for instance, of the two domes investigated in Louisiana and of a second one in Mississippi, the Cypress Creek dome, some 15 miles south of Richton, which was overlain by a depression and a swamp. There was a possible sign of dissolution at the Richton dome, too, but the evidence for this was relatively weak.[38]

While superior to the other domes in most respects, the Richton dome promised to be clearly the most unsettling in its impact on people. Moreover, press reports about the project's possible impact made the prospects appear worse than they actually were. "N-Waste Site Could Force Entire Town to Relocate," declared the *Jackson Clarion-Ledger*, Mississippi's largest newspaper, on May 24, 1981. Richton town attorney Tom Sims, at a DOE hearing, would ask, "Will there still be a Richton?" The rumor that there would not, he said, "persists and causes great anxiety among many of our citizens, particularly the elderly."[39] A DOE spokesman dismissed the rumor as "absolutely fallacious," but he acknowledged that as many as forty families might be displaced.[40] Mississippi's senior senator, John C. Stennis, in response to many complaints from constituents in Richton and elsewhere in the state, was assuring them that he was doing his best to persuade the department to site the first repository in a western desert, far away from towns and where people are few.[41]

There were certain other peculiarities about the situation confronting the department in Mississippi. The department had enjoyed easy access there to drill boreholes in the early years. Colin Heath thinks, as he told me, that this may have been an indirect result of Mississippi's traumatic

[37]ONWI Salt Domes Evaluation Study, pp. 119–168.

[38]Ibid., pp. 126–128.

[39]Department of Energy hearings on proposed site nominations, Richton, Miss., April 28, 1983, transcript, vol. 1.

[40]Ibid., response of J. O. Neff, manager of the DOE National Waste Terminal Storage Program office, Columbus, Ohio.

[41]Senator John C. Stennis, "Report to Mississippi," column in the *Richton Dispatch*, September 10, 1981.

experience with federal preemptive power during the showdown over court-ordered school integration in the 1960s. State officials were now assuming, Heath believed, that they were powerless to prevent nuclear waste disposal in Mississippi and that unless they cooperated with DOE they might lose all control over the situation.[42] The reality was, and is, that federal preemption successfully exercised in support of a great moral and legal principle such as racial equality before the law tells nothing about the political feasibility of making a state accept a repository for something as frightening to people as nuclear waste. As time went on, Mississippi politicians could more easily appreciate this distinction, particularly inasmuch as the governors of the Salina Basin states had told the Department of Energy to leave and had gotten away with it.

In 1981 Governor William Winter demanded and got a moratorium on further DOE field studies to allow the state to complete a review of data from earlier investigations. The state attorney general, Bill Allain, filed two Freedom of Information Act requests with the department, through which Mississippi received many thousands of documents—so many that most were stored away, unread.[43] The Mississippi legislature, for its part, established by statute a permitting procedure for fieldwork related to repository siting that was every bit as labyrinthine and baroque as any of the laws enacted during the heyday of massive resistance to school integration.

Potential repository host states typically suspect that other, more politically powerful states want to dump their nuclear waste on them. Department of Energy officials found this attitude to be especially pronounced in Mississippi. Allegations that Mississippi is despised and persecuted have long been a staple of Mississippi politics. Attorney General Allain, speaking before some 5,000 people at an anti-nuclear forum at Gulfport, Mississippi, in November 1981, plucked on that string to loud applause by observing, "They have been dumping on Mississippi so long, I guess now they feel they can dump in Mississippi."[44]

---

[42]Interview with Heath, April 18, 1983.

[43]Robert Ourlian, "Next Year May Be Too Late to Stop N-Dump—State," and Ourlian, "Allain Urges Organization of Campaign to Oppose Nuclear Waste Disposal Site," in the *Jackson (Miss.) Clarion-Ledger*, May 22, 1984, and June 5, 1984.

[44]Laurence Hillard, "Nuke Dump Site 'A Political Question,' " *The Hattiesburg American*, November 9, 1981. A further peculiarity of the Mississippi situation has been the lingering memory many citizens have of an underground nuclear test that was conducted at the Tatum salt dome in southern Mississippi in 1964. The AEC and its scientific advisers were astonished when this explosion produced an earth shock strong enough to break windows and crack masonry in Hattiesburg more than twenty miles away; one result was more than a thousand successful damage claims that totaled over $650,000.

Louisiana: Distrust and Alarm

The Vacherie dome in northwest Louisiana, about thirty miles to the east of Shreveport, was the second choice among the Gulf Coast salt domes. In that state the situation faced by the Department of Energy was, if anything, even less favorable than in Mississippi. As in Mississippi, the department and its predecessors enjoyed easy access at the start of the geologic investigation. Here the explanation seems to have been that Louisiana State University's Institute for Environmental Studies was the DOE contractor; the institute's director, Joseph D. Martinez, was a respected geologist and a native son who shrewdly argued that it was in the state's interest to have the geologic studies done by its own university.

The first significant political opposition was encountered when drilling of the first deep test boreholes was about to begin in 1977. Early that year the *Bienville Democrat* and the *Minden Press-Herald*, two widely read local newspapers, prominently displayed photographs of a locked gate and Keep Out signs on a road leading to the Vacherie dome site. These generated strong feelings of distrust and alarm.[45] Clayton Zerby, director of what was then the Office of Waste Isolation (OWI) run by Union Carbide at Oak Ridge, came down to try to explain the siting program at public meetings. But his explanation involved an obvious contradiction. Whereas Zerby emphasized that the project was in "the earliest stages" and that "no candidate sites [had] been selected,"[46] under the waste program's schedule selection of the first salt sites was supposed to occur in the next budget year, which was near at hand.

J. Robert Kemmerly, a local physician, and R. H. Manning, a retired Webster Parish schools superintendent, were among the prime movers behind Citizens Against Radioactive Storage, a recently formed group that had succeeded in arousing northern Louisiana and turning the state's political establishment against the salt dome project. Kemmerly and Manning were very different personalities, yet each had proved effective in raising the alarm.

Manning, now deceased, was a tall, imposing man and his opinions were no doubt widely respected. But when I called on Manning one afternoon in March 1981 at his home on a lonely back road of Webster Parish, I found him driven by deep and pervasive fears of nuclear waste,

[45]Interviews by the author in March 1981 and April 1983 with Charles Kilgore, a retired dean and professor of chemical engineering at Louisiana Tech University and a consultant to DOE.

[46]"Same Song: Nuclear Experts Go Home," *Minden Press Herald*, April 28, 1977.

of poor blacks, of the press, of communist governments, of his own government, and, in general, of the entire world of conspiracies known to the extreme political right wing. His house was surrounded by a high, man-proof fence which was electrified and topped with barbed wire, and had mounted along it more than a score of floodlights. Every afternoon at 4 o'clock the front gate was shut and a Doberman pinscher was turned loose in the yard. The "big ear" listening device mounted on the roof of his house was sensitive enough to pick up voices two miles away. "I think once they cut off welfare payments and food stamps there will be a crime wave around here," Manning told me. Manning had built two bomb shelters, one at his home and another just down the road at his mother-in-law's place. He also had a powerful radio antenna that allowed him to pick up broadcasts anywhere in the world. "It's amazing how much of our own news is censored," he remarked. Manning's interest in things nuclear went back to 1945 when, as a Marine Corps officer, he visited Hiroshima a few weeks after the bombing. "About thirty years [later] to the day, I got cancer," he said. "I don't know whether that had anything to do with it or not."[47]

Manning had been speaking before civic groups at Shreveport and at towns in Webster and Bienville parishes, attracting quite a bit of attention from the Shreveport television stations. He was warning everyone of the hazards of radiation. "I stressed that you can't see it, hear it, taste it, or feel it," he told me. "It's a silent killer. You don't know it's there until you're gone. I think my most effective argument is that [radioactive waste has] leaked every place they've put it and that [the repository] would be experimental."

Kemmerly was probably even more effective than Manning in arousing opposition to the salt dome investigation. With a flourishing practice in obstetrics and gynecology, he was, and is, one of the more prominent, affluent, and respected individuals in Minden, a small city about twenty-five miles east of Shreveport. Kemmerly, as I found in a long interview with him one evening, is deeply religious and is active in church mission work, having served two summers as a physician among the poor of Central America.[48] It is clear that his considerable effort to warn the public about radioactive waste was for him another mission. His argument, which he hammered away at constantly at public meetings, was that the concept of committing radioactive waste to salt dome repositories was unproved and experimental. His view of the worst-case repository failure was apocalyptic and not well supported scientifically,

[47]Interview by the author with Manning, March 1981.
[48]Interview by the author with Kemmerly, March 1981.

but his insistence on the experimental nature of what was afoot was not inconsistent with the then ongoing scientific reappraisal of geologic disposal in general and of disposal in salt formations in particular.

Kemmerly contended that both morality and common sense precluded disposal of nuclear waste in northern Louisiana, with its abundance of oil and gas, water, arable land, and many cities and towns. His refrain was much the same as that voiced by Senator Stennis in Mississippi. "The risks involved dictate that this experiment be carried out in a desolate area, just as were the initial nuclear explosions," Kemmerly said.[49] His warnings on this point would be echoed in the Louisiana press. The *Shreveport Times* later was to point out, for example, that some spectacular accidents had taken place in the mining of Louisiana salt domes, as in 1965 when the Carey Salt Company mine at Winnfield was lost when a powerful jet of water broke through a side wall and flooded the some fifteen miles of tunnels and galleries within a day's time.[50]

Test borehole drilling was to proceed in Louisiana, but very likely only because in early 1978 the Department of Energy promised Governor Edwin Edwards, in a signed agreement, that no waste would be committed to a Louisiana dome without the state's consent.[51] This concession was made by the department in its negotiations with the state to gain access to the salt domes needed for the Strategic Petroleum Reserve. But many Louisianans found the situation still worrisome. They reasoned this way: "DOE knows we won't accept a repository, yet the geologic testing still goes on. If they mean to allow us a veto, why are they drilling?" Their apprehensions were heightened when the U.S. Comptroller General, in an opinion requested by a congressional committee, said Louisiana's purported right to veto had no legal standing.[52] The Louisianans' fears and suspicions persisted even after a new governor, Republican David Treen, extracted from Ronald Reagan during the 1980 presidential race a promise that Louisiana's right to say no would be honored.[53] In 1981 the legislature passed a bill to ban geologic

[49]J. Robert Kemmerly, "The Search for Safe Disposal of Radioactive Nuclear Waste—A Series of Trial and Error Experiments, May 1978 Update," memorandum.

[50]Keenan Gingle, "Winnfield: North Louisiana's Mine Disaster," *Shreveport Times Enterprise,* January 4, 1981.

[51]"Principles of Understanding," signed on February 27, 1978, by Governor Edwards and Deputy Secretary John O'Leary of the Department of Energy.

[52]Letter of June 19, 1978, from R. F. Keller, acting comptroller general of the United States, to Congressman John D. Dingell, chair of the House Interstate and Foreign Commerce Committee.

[53]Letter of November 8, 1983, from Robert L. Morgan, acting director of DOE's Office of Civilian Radioactive Waste Management, to Governor Edwards.

testing, but Governor Treen vetoed it, believing that his state's best chance of keeping nuclear waste out of its salt domes still lay in the agreement his predecessor had negotiated.[54]

The Hanford Basalt

The Hanford reservation where the Department of Energy was looking for a repository site in basalt rock lies within the elongated Pasco Basin on the Columbia Plateau. It begins just southeast of Washington state's tri-cities of Richland, Kennewick, and Pasco, and extends to the northwest on either side of the Columbia River up to the foothills of the Cascade Range. This basin is said to include some of the thickest, flattest basalt flows of the Columbia Plateau Basalt Group, which is how geologists refer to the lava fields that cover more than 78,000 square miles of Washington, Oregon, and northern Idaho. The lava, or basalt, typically came forth not as a fiery torrent rushing down the slopes of mountainous volcanic cones, but rather as a flood extruded in enormous volume from long, deep fissures in the earth. This occurred during a period of intermittently active vulcanism that lasted for 11 million years and finally ended about 6 million years ago. The successive flows, typically from 90 to 120 feet thick but sometimes thicker, spread far from their source, with the depth of the entire sequence in some basins going to as much as 16,000 feet, or greater than three miles.[55]

The difficulties and uncertainties confronting any effort to explore and develop the Hanford basalt for a repository are extraordinary. But what the Hanford site has had going for it is strong local support for hosting a repository. For the leaders of the tri-cities, of which Richland is the largest and most politically important, Hanford is both their backyard and their bread-and-butter. They have been all for any new nuclear development on the reservation, including the development of a waste repository. Glenn C. Lee, formerly editor and publisher of the *Tri-City Herald* and for years the unofficial voice of the local business and political leadership, has been accurately described as having the appearance and manner of a Marine Corps general. "Nuclear waste is a peanut problem, a piss-ant problem," Lee said when I saw him one morning at breakfast several years ago. "It's a problem that's blown all out of proportion, just like Three Mile Island." With us was Sam Volpentest, executive

---

[54]Robert Moore, "Treen Vetoes Dome Bill," *Shreveport-Bossier City (La.) Times*, August 5, 1981.

[55]Department of Energy, "Draft Environmental Assessment: Reference Repository Location, Hanford Site, Washington," DOE/RW-0017 (Washington, D.C., December 1984) p. 2-1.

vice-president of the Tri-City Nuclear Industrial Council. Since its establishment in 1963, the council has been a force in keeping Richland, Pasco, and Kennewick economically alive despite the phase-out of plutonium production that began at Hanford in the mid 1960s. "We would like to have a national repository," Volpentest told me flatly.[56]

But Washington state's population and electorate are mostly to the west of the Cascades, in Seattle and the rest of the area around Puget Sound. There, as in most of the state, public attitudes about Washington taking in nuclear waste from other parts of the country have been no more positive than those noted in other potential host states.[57] Yet the thrust of state policy on the Department of Energy's basalt investigation has never been to oppose the project out of hand, but to make sure that the judgments made on the suitability of the basalt site by DOE and its Hanford contractor, Rockwell Hanford, are technically sound and supported by scientists independent of the department, such as those of the U.S. Geological Survey and the Nuclear Regulatory Commission.[58]

As time has passed the problems associated with the Hanford site have become increasingly evident to independent reviewers. In 1981 Rockwell's Hydrology and Geology Overview Committee, a group made up largely of academics, cautioned Rockwell about representing the Hanford site as geologically "favorable." The committee observed that the contrary was true, although adding that an acceptable site perhaps could be identified through careful data gathering and engineering design.[59]

As a geologic medium, basalt flows present inherent difficulties for an exploratory and development effort carried on at repository depths. The flows usually lie one on top of another, but not always. When the vulcanism of the Columbia Plateau was interrupted for a considerable period, sand, gravel, and other erosional debris washed down from the higher terrain surrounding the basin. This material would form a highly permeable "interbed" and aquifer when later covered by the succeeding basalt flow. The Pasco Basin's Cohassett flow, which the Department of

[56]Interviews by the author with Lee and Volpentest, February 1981; interview with Volpentest, July 1, 1983.

[57]In 1980 the electorate overwhelmingly passed Initiative 383, prohibiting the disposal in Washington of most nuclear waste from out of state, the principal exception being low-level waste from states that join Washington in a Northwest regional compact. The initiative law was later struck down as unconstitutional.

[58]Interview by the author on September 23, 1984, with David. W. Stevens, program director, Office of High-Level Nuclear Waste Management, Department of Ecology, State of Washington.

[59]Rockwell International, Hydrology and Geology Overview Committee, Reports and Responses from the Basalt Waste Isolation Project, RHO-BWI-LD-50, Informal Report, September 1981.

Energy has selected as a repository candidate,[60] lies at a depth of more than 3,000 feet and is beneath a score of flows and almost half that many interbeds, several of them prolific aquifers. Indeed, it has been estimated that if a 15-foot-diameter repository access shaft were sunk and left uncased, water from these interbed aquifers would pour into it at the rate of 140,000 gallons a minute.[61]

The hydrologic situation of the Hanford basalt dictates an unusual shaft construction method. The large, 15-foot-diameter access and ventilation shafts required for a repository, however, are beyond the established state of the art. Normally when large shafts are sunk for any purpose they are not drilled, but are dug out by the drill-and-blast method. Where prolific aquifers must be penetrated, the first step is often to freeze these water-bearing zones by drilling numerous closely spaced holes around the circumference of the proposed shaft and injecting chilled brine. This is in fact what would be done first at the Texas salt bed site. But at the basalt site the flow of water is so great that the freezing method is unworkable. The only choice there is to attempt to drill the shafts by using huge bits. To drill the 10-foot-diameter shafts needed for the deep exploratory investigation is clearly possible. But to drill the 15-foot-diameter shafts needed for the repository in rock of the hardness of basalt would be unprecedented. In 1983, at the Agnew nickel mine in the Australian outback, a 14-foot-diameter shaft was sunk in hard rock to 2,460 feet, when drilling had to be stopped some 900 feet short of the intended depth. The huge bit had begun running roughly in brittle fractured rock, and the drill rig mast was shaking from side to side.[62] According to current plans for Hanford, nine shafts would have to be drilled for the repository, seven of them to the 15-foot-diameter size. These specifications are dictated in part by the need for a powerful ventilation system to cope with the high humidity and 125° F ambient temperature that would otherwise prevail in the repository's 2,000-acre maze of tunnels—and to keep the methane gas present from reaching potentially explosive concentrations.[63]

Rockwell has deemed it likely that the Cohassett flow would offer an acceptably slow radionuclide release rate and a long groundwater travel

[60]Department of Energy, "Draft Environmental Assessment, Hanford Site," pp. 3–20.

[61]Interview by the author with DOE geological consultant Harry Smedes, November 8, 1982.

[62]Paul Richardson, "Australia's Largest Blind-drilled Shaft," paper presented before the American Society of Civil Engineers, Atlanta, May 16, 1984. See also Department of Energy, "Draft Environmental Assessment, Hanford Site," pp. 4-10 to 4-12 and 6-172 to 6-179; and Nuclear Regulatory Commission, "Comments on Draft DOE Environmental Assessment," pp. 9–10 and 96.

[63]Department of Energy, "Draft Environmental Assessment, Hanford Site," pp. 5-7 through 5-33.

time. But there are in fact many uncertainties. Once any repository built in the Cohassett flow is filled with waste and decommissioned, the uncontrolled seepage of water into the underground workings would lead to saturation. Eventually, perhaps within a thousand years or less, leaching of the waste packages would result in the contamination of groundwater passing through the repository. One crucial question at Hanford, then, has been how long the groundwater would take to travel from the repository in the Cohassett flow upward 3,000 feet through the geologic profile, and laterally no more than about five miles, to a discharge point along the Columbia River.

But the predictability of the groundwater regime at the basalt site has been in doubt from the start. Horizontal pathways can follow the zone of brecciated basalt at the top of the flows as well as along the interbeds of sediments between flows. (The breccia, or sharply fragmented basalt, was formed when the surface of the molten lava cooled and hardened, only to be broken up by the hot lava still flowing beneath it.) Vertical pathways can follow faults, shear zones, and columnar joints. The multisided columnar joints were formed as the lava cooled and contracted, the total effect being that of a densely arrayed bundle of vertical columns, rather resembling a huge pipe organ.

Together, the columnar joints and brecciated zones should permit easy movement of groundwater, except that overburden pressure and minerals deposited in the joints and the breccia zones have made most of them tight and relatively impermeable. But not all have been tightened, so pathways for relatively rapid groundwater movement are a possibility. Moreover, wherever such pathways exist, the water will move upward under differential pressure unless less resistance is encountered laterally in the interbeds or along the basalt flows. Groundwater recharge for the Pasco Basin takes place in the Cascades to the west and in the hills to the east, at elevations hundreds to thousands of feet above the basalt flows and interbedded aquifers. Thus, for the groundwater regime relevant to a 2,000–acre repository to be accurately characterized from the information gained from a necessarily limited number of test wells is extraordinarily difficult. Yet this has been the challenge confronting the Department of Energy and Rockwell Hanford.

On the fundamental issue of groundwater travel time, Rockwell reported in 1982 that it would take no fewer than 10,000 years for water from the repository to reach the "accessible environment," which is to say the springs and usable aquifers beyond the site boundaries, and no fewer than 20,000 years to reach the Columbia River.[64] But reviewers

---

[64]Rockwell International, Hanford Operations, "Site Characterization Report for the Basalt Waste Isolation Project," DOE/RL 82-, vol. 1 (November 1982) p. 7.

at the U.S. Geological Survey and elsewhere rejected these calculations, questioning the data and assumptions that went into them. The Geological Survey concluded that travel time to the accessible environment, especially in light of vertical pathways that might be present, was possibly less than 1,000 years, which if true would not meet licensing criteria. Later the reviewers observed that to judge from past Geological Survey studies in the Pasco Basin and in the larger plateau region, "the Columbia River Basalt ... should be generally thought of as a series of semi-confined aquifers that are hydraulically connected to each other by semi-permeable confining beds. Overall, the system appears to be very leaky."[65]

The Geological Survey reviewers warned that construction workers and repository operating personnel would face the potential hazard of catastrophic flooding. Water pressure at the repository horizon would be as much as 1,500 pounds per square inch—or about fifteen times the pressure of water from a firehose which takes three men to hold. At such pressures, any "fractured, brecciated and pillowed zones" encountered during construction "could wash out, resulting in disastrous (perhaps even explosive) flooding in the mined passages which could be impossible to control initially by pumping."[66] The possibility of mining accidents and floods could actually be all the greater because the Pasco Basin is within a region known to be under high compressive stress and to be seismically active. The highly critical comments by reviewers led to a management shake-up at the Rockwell Hanford operations office, and to greater caution and tentativeness in Rockwell's conclusions. But as I shall point out in chapter 13, the Department of Energy was to continue looking to the Hanford basalt as one of the preferred sites.

Of all these issues, the risk of contaminating the Columbia River most bothers citizens and elected officials in Washington state (and, increasingly, in Oregon). In this connection, a problem unique to the Hanford situation is also of concern: Hanford's 149 old and corroding military high-level waste tanks are less than two miles from the basalt site. Washington state officials worry that, some generations hence, it may not be possible to distinguish radionuclides from the repository from radionuclides from the defense waste burial and storage sites. The state Nuclear Waste Board has put it this way: "If succeeding generations are unable to discriminate between sources of measured radionuclides in the land and water environments, and if original records are lost, then the burden on future generations becomes immense: they will have to

[65]U.S. Geological Survey, "Review Comments on the Site Characterization Report for the Basalt Waste Isolation Project," May 6, 1983, pp. 5, 13.

[66]Ibid., pp. 2, 3.

assume a worst case and react accordingly. This would mean almost incalculable costs and losses of productive time to ensure a habitable and useful environment."[67]

But of all the Hanford site's problems, probably the worst is that the repository would have to be built deep below several highly prolific aquifers in a medium that the Geological Survey characterizes as "leaky." Ernest L. Corp, supervisory mining engineer at the U.S. Bureau of Mines Research Center at Spokane, has devoted about 15 percent of his time to advising the Nuclear Regulatory Commission on the Hanford site. He says, "It's always amazed me how they could have selected a site like this one, with all its water problems."[68] He acknowledges that mining companies in South Africa and elsewhere are accustomed to coping with such difficult conditions. But to face the risk of catastrophic flooding for the sake of recovering gold or diamonds is entirely different from assuming such risks in order to bury nuclear waste. There is, Corp says, "no sound technical or economic basis to even consider locating a repository in an area where ground water is known to be a problem."[69]

Yucca Mountain and the Unsaturated Zone

The search for suitable rock for a repository at the Nevada Test Site began focusing within a few years on the southwest corner of that site, some thirty miles away from the nuclear weapons testing, the activity in which this unusual facility has been engaged since its creation in 1950. This focus led the investigation to Yucca Mountain, an unimposing flat-topped ridge that runs six miles from north to south and rises some 1,000 to 1,500 feet above the surrounding terrain. On its west side the mountain looms as an irregular escarpment that overlooks Crater Flat and faces southwest toward the Amargosa Desert, the Funeral Mountains along the Nevada–California border, and Death Valley beyond. To the east the mountain drops off less precipitously and is dissected by side canyons and deep washes. On this side Forty Mile Wash parallels the mountain, and to the east lie the sandy wastes and sagebrush of Jackass Flats. It is a dry, eroded, hard-bitten country that has seen little settlement or human activity of any kind.

Yucca Mountain is tuffaceous, and the rock of particular interest is known as welded tuff. Tuff is volcanic in origin but is quite unlike lava rock or basalt. Rather, tuff is formed when the magma inside a volcano

---

[67]Washington State Nuclear Waste Board, "Final Comments on the [DOE] Draft Environmental Assessment," May 20, 1985, p. 8.

[68]Corp was interviewed by the author on April 4, 1984.

[69]Letter of April 5, 1984, from Corp to the author.

erupts explosively, with ash, pumice, and other material blown out of the crater, or caldera. The eruption at Mount Saint Helens in Washington in 1980 was such an event, but it was a minor one compared to the enormous eruptions that took place in Nevada 10 to 14 million years ago. The ejecta from those eruptions could be measured in cubic miles. They were charged with hot gases and flowed until finally coming to rest at the bottom of a canyon or other low place as a thick, hot mass of volcanic ash. There, as the mass became increasingly compressed or "welded" by the weight and heat in its deeper parts, it formed welded tuff.

The singular thing about Yucca Mountain, and indeed about much of the Nevada Test Site and the surrounding region, is the great depth from the ground surface to the water table. From the ridge top down to the water table the distance is nearly 1,800 feet, and at some places in the region the water table lies even a few hundred feet deeper than that. By contrast, in the humid regions of the eastern United States the water table typically lies within a few tens of feet of the surface, if not just below it; even in many arid or semi-arid regions the water table is down not more than several hundred feet.

As technically defined, the water table marks the point in the geologic profile below which all the pore spaces in the rock are filled with water—in the jargon of the hydrologist, the "saturated zone." Above the water table, both water and gases are present in the pore spaces in varying proportions, but always at less than saturation; this is the "unsaturated zone." One essential difference between the two zones is that in rocks in the saturated zone the groundwater moves continuously through the pore spaces; in the unsaturated zone, on the other hand, the water is held in place by capillary tension and either does not move at all through the pore spaces or moves extremely slowly. Except for those few radionuclides such as iodine-129 which assume a gaseous form, radionuclide transport within a geologic formation usually is only by groundwater movement; if the water does not move, neither do the radionuclides. In the unsaturated zone rainwater from the surface does percolate downward by gravity, especially beneath major arroyos. But in the view of proponents of waste disposal in the unsaturated zone there is so little water available for downward percolation in an arid region such as Nevada that waste canisters should remain intact over many millennia. The onset of a rainy climate in the future would cause the water table to rise; but if the repository is situated in a place of high topographic relief, as at Yucca Mountain, the waste canisters should still be hundreds of feet above it.[70]

[70]Eugene H. Roseboom, Jr., "Disposal of High-Level Nuclear Waste Above the Water Table in Arid Regions," U.S. Geological Survey Circular 903 (1983); interviews by the author with, and personal communications from, Roseboom, Newell J. Trask, and Isaac J.

When the investigation at Yucca Mountain began in 1979 the rock layers of principal interest were not in the unsaturated zone. All previous investigations in the geologic disposal program had been below the water table, and while the idea of going above it had received some mention in the technical literature, it was still one in which the Department of Energy was showing little interest. The department had not, for example, followed up with field investigations a National Academy of Sciences recommendation that priority be given to the possibility of siting a repository in the unsaturated basalt of the Rattlesnake Hills at Hanford.[71] Disposal in the unsaturated zone was not yet well developed as a concept, and it departed from convention in that disposal would be at depths considerably less than the 2,000- to 3,000-foot depths usually envisioned for deep geologic repositories.

The Department of Energy's candidate site at Yucca Mountain might well have been below the water table except for the fact that both of the horizons investigated there, in the saturated zone, presented major problems. The tuff of the Bull Frog horizon at about 2,000 feet beneath the surface was found to have poor mechanical strength. The tuff of the Tram horizon at 3,500 feet was competent rock, but to install and operate a ventilation system large enough to cope with the high temperature (up to 130° F) and high humidity expected at that depth would pose problems similar to some of those at the Hanford basalt site.[72]

As these problems were becoming increasingly apparent, the Geological Survey was beginning to play a key role at Yucca Mountain. Founded in 1879, the USGS is one of the government's oldest and strongest scientific agencies, with a reputation for high professional competence and integrity. At Hanford and in the Gulf Coast salt dome region the

---

Winograd of the Geological Survey over the period 1982–1985. For early development of the concept of disposal in the unsaturated zone, see I[saac] J. Winograd, "Radioactive Waste Storage in the Arid Zone," *Eos Transactions* vol. 55, no. 10 (American Geophysical Union, 1974) pp. 84–94; Interagency Review Group on Nuclear Waste Management, "Subgroup Report on Alternative Technology Strategies for the Isolation of Nuclear Waste," TID-28818, draft (October 1978) appendix A, pp. 80–81.

[71]National Academy of Sciences, *Radioactive Wastes at the Hanford Reservation* (Washington, D.C., 1978) pp. 3–4 and 108–110. In 1983 Larry R. Fitch, manager of research and licensing for Rockwell Hanford, told a reporter, "Rattlesnake Mountain has never been nor is it currently being seriously considered" for a repository. He explained that this possibility was excluded under site-selection criteria which Rockwell had adopted with respect to earthquake hazards. Hill Williams, "Tunnels Endorsed as Best N-waste Repositories," *Seattle Post-Intelligence*, October 16, 1983. But both the academy panel and U.S. Geological Survey scientists had taken the earthquake hazard into account—the Rattlesnake Hills lie along a major structural lineament or fault—and regarded it as acceptable.

[72]Interview by the author with Don Vieth, DOE manager for the Yucca Mountain project, February 21, 1984.

agency either already was or soon would be looking critically over DOE's shoulder and warning of possible trouble ahead. But at Yucca Mountain the Geological Survey was calling attention to a new approach to repository siting that could ease the vexing problems of uncertainty that were contributing to technical and political stalemates. Twice in early 1981 the survey's Isaac J. Winograd appeared before Department of Energy officials to describe how the concept of disposal in the unsaturated zone might be applied at Yucca Mountain.[73]

Meanwhile, the DOE waste program managers had redirected the investigation at Yucca Mountain to horizons above as well as below the water table. A horizon of particular promise was found at 1,200 feet in the Topopah Spring member; this exceeded the minimum depth of about 1,000 feet that the Nuclear Regulatory Commission would require for a repository, and was more than 600 feet above the water table.[74]

In February 1982 the chief of the Geological Survey's Office of Hazardous Waste Hydrology and two other USGS officials dispatched a letter to DOE's Nevada operations office, saying they felt that the concept and advantages of disposal in the unsaturated zone were still not well understood within the Department of Energy. Their purpose was to review the hydrogeologic characteristics of such a disposal system and to point out its merits. They observed in particular that some things that make for a problem in the saturated zone are no problem at all in the unsaturated zone, and in some instances are a major advantage. If the highly fractured tuff of the Topopah Spring member were beneath the water table, the fractures would represent potential pathways for radionuclide migration. But above the water table the fractures were expected to function usefully by allowing such rainwater as did percolate down from the surface to pass quickly through the repository to the water table below, with little or no contact with the waste canisters. Similarly, the letter observed that whereas in the saturated zone the sealing of access shafts and boreholes represented a "major unknown" in terms of repository performance over millennia, in the fractured rock of the unsaturated zone this was much less important.[75]

The amount of water expected to reach and pass through a repository in the unsaturated Yucca Mountain tuff would be small, the letter pointed out. Annual precipitation there is about five inches, of which all but a

[73]Winograd's role is noted in letter of February 5, 1982, from John B. Robertson, Gary L. Dixon, and William E. Wilson of the USGS to Mitchell Kunich of DOE's Nevada operations office, Las Vegas.

[74]Department of Energy, "Draft Environmental Statement, Yucca Mountain Site," DOE/RW-0012 (December 1984) fig. 2-15, p. 2-44, and pp. 3-1 through 3-106.

[75]Robertson, Dixon, and Wilson to Kunich, February 5, 1982.

few tenths of an inch is either lost to runoff or evaporation, or is taken up by the desert vegetation. The repository would be designed to divert this little bit of water away from the canisters and hasten its descent to the horizons below. "Even if we make the outrageous assumption that an entire year's average precipitation is instantaneously transported to the repository horizon as a layer of water ankle deep, it would drain in hours to days," the USGS officials wrote.

They did not maintain that a site in the unsaturated tuff would eliminate uncertainty. They argued, rather, that by making the first and most effective natural barrier to radionuclide migration the minimal contact of waste by groundwater, the uncertainties present would be much less significant. The present inability of mathematical models to describe the flow of dissolved substances through unsaturated fractured rock was said to be "perhaps the main source of confusion" about the unsaturated zone. They added that there was little real need for a detailed characterization of this system if it could be shown that little water would pass through the repository and that the waste canisters would be affected very little, if at all, by that water.

The letter noted two drawbacks to disposal in the unsaturated zone at Yucca Mountain: the possibility of difficulties in mining in the fractured tuff, and the possibility that some gaseous radionuclides would escape from spent fuel and migrate upward through the rock fractures. (Later it would appear that neither of these was to be a major problem. Tests made in fractured welded tuff at Rainier Mesa have indicated that such usual mining safety precautions as roof-bolting and the use of steel mesh will suffice.[76]) Iodine-129 is the gaseous radionuclide of principal concern. Virginia Oversby, a materials scientist at the Lawrence Livermore Laboratory, has told me that the release of iodine from spent fuel and its subsequent upward migration through the fractured rock are never likely to result in unacceptable concentrations of iodine at the surface. Most of the iodine, she says, will remain as iodide inside the ceramic uranium oxide pellets.[77]

A potential problem not mentioned in the Geological Survey letter, but which has since been much investigated and discussed, is the earthquake hazard. The general region is seismically active and Yucca Mountain itself is bounded by faults. But inasmuch as ground motion is attenuated at depth, the principal concern would not be the stability of the deep-mined maze of tunnels that make up a repository; rather, it would be to

[76]Interview with Don Vieth, February 21, 1984. Roof bolts are inserted into the ceilings of tunnels to help hold the rock in place; steel mesh is fastened to the ceiling to hold any rock that might be dislodged.

[77]Oversby was interviewed by the author on January 19, 1984.

build a repository's surface facilities to adequate earthquake-proof standards.[78] (See the discussion of the earthquake hazard in chapter 13.)

In light of the Geological Survey's recommendations and the difficulties being encountered in the saturated zone, Don Vieth, the Department of Energy manager for the Yucca Mountain project, decided in July 1982 to change the major focus of the investigation from the saturated to the unsaturated zone, with the Topopah Spring horizon at a depth of 1,200 feet being of primary interest. Vieth was influenced in this decision by his own experience as a veteran of the salt program.[79] Although candidate horizons in salt are invariably below the water table, the salt itself, though it dissolves readily, is impermeable and dry at its interior except for tiny inclusions of brine. Vieth had found this a major advantage and was therefore well attuned to the arguments made for a dry site in tuff. But only by further detailed investigation at Yucca Mountain could it be determined that the advantages claimed for waste disposal in the unsaturated zone would hold for this specific site.

Land-use and resource conflicts of the kind present at other sites are conspicuously absent at Yucca Mountain. Yucca Mountain is 90 miles from Las Vegas and some 15 miles from Beatty, the nearest town or hamlet. It is on public land, and is on or adjacent to a large (1,350 square miles) federal reservation which for nearly thirty-five years has been dedicated to the testing of nuclear weapons.[80] Further, the Yucca Mountain site is part of a large desert region that is for the most part economically unproductive. There are no known mineral resources on the site. There is no important regional aquifer (like the Ogallala) present, although beneath the site there is an aquifer that flows southward and supplies irrigation water to a small farming community in the Amargosa Desert about 30 miles away. The Colorado River is more than 100 miles distant, and there are no other water courses near Yucca Mountain other than the ephemeral streams which flood the arroyos and washes after heavy rains. Yucca Mountain is in fact in a hydrologically closed basin, out of which no surface water flows. Moreover, the proposed site is not on the edge of a national park or other important natural or scenic area.

In spite of these advantages, state and local acceptance of a repository project at Yucca Mountain was not assured. Sparsely settled Nye County, where the Nevada Test Site and Yucca Mountain are located, favored the project, as did some major business and political leaders in southern

[78]Interview by the author with Don Vieth, June 12, 1986.

[79]Interview with Vieth, February 21, 1984.

[80]Yucca Mountain straddles the boundaries between the Nevada Test Site, Nellis Air Force Base, and federal public domain land under the jurisdiction of the U.S. Bureau of Land Management.

Nevada who wanted to be sure that the Nevada Test Site would have a future, whether weapons testing was continued or not. But a bad political legacy had been left from the unfortunate early history of atmospheric weapons testing, as chapter 2 has shown. Moreover, the notoriously poor management, during the 1970s, of the commercial low-level waste site at Beatty, Nevada, had made nuclear waste disposal a much-publicized issue.

Nevada's governor, Richard H. Bryan, who as attorney general from 1978 through 1982 tried unsuccessfully to shut down the Beatty site, was promising to oppose the siting of a repository at Yucca Mountain. He insisted that the transport of nuclear waste into Nevada might discourage further growth of Las Vegas as a tourist center. Accordingly, if Yucca Mountain were ultimately to be chosen as the preferred site, efforts would have to be made to ease such fears and assure Nevadans that a repository project would bring them not trouble, but benefits.

The WIPP Experience in New Mexico: An Important First

When the Waste Isolation Pilot Plant, or WIPP facility, now being built in Permian Basin salt near Carlsbad, New Mexico, begins receiving waste packages in late 1988 or early 1989, an important first will have been scored for the nuclear enterprise. WIPP will be the first geologic repository constructed by the United States or any other country expressly for the isolation of nuclear waste. The Germans' Asse repository is a salt mine that was adapted for the disposal of low- and intermediate-level waste, and as this facility has been shut down since the late 1970s to await a possible relicensing, WIPP may also be the world's only repository for a considerable time. Accordingly, the WIPP experience is an important one to consider even though this repository has been authorized by Congress as an unlicensed and hence limited-purpose facility. It will receive only military transuranic waste for disposal; modest amounts of military high-level waste may be received also, but only for research purposes and ultimate retrieval.

The WIPP experience has been in some ways so remarkable with respect to the resource conflicts, the hydrogeologic complexities, and the state and local politics involved that it offers lessons relevant to repository siting not only in New Mexico but in other potential host states as well.

The Siting Problem: From Lyons to Carlsbad

The WIPP project, which goes back to the early 1970s, grew out of the failure of the Atomic Energy Commission's attempts to establish a repository in the salt mine at Lyons, Kansas. After the commission's belated discovery that past oil and gas exploration and solution mining called into question the Lyons site's integrity, the agency had to find some other site for the repository. The prospects were found to be best in New Mexico.

If the search was to remain in the Permian Basin—and this was the AEC preference—the choices, as seen by the agency's technical advisers at the Oak Ridge National Laboratory, came down to either another site in Kansas or a site in or near southeastern New Mexico's Delaware Basin. According to Thomas Lomenick, one of the Oak Ridge people principally involved, not much was known at that time about the thickness and evenness of the salt beds in the Texas and Oklahoma panhandles.[81] On the other hand, a lot was known about the beds in central Kansas from the extensive salt mining there, and about the beds in New Mexico, from extensive potash mining. That oil and gas development had come relatively late to the Delaware Basin was seen as a particular advantage, because there would be fewer exploratory boreholes to contend with.

By the early 1970s, the responsibility for nuclear waste management had been split off from the AEC's reactor development division and placed under a new division of waste management and transportation. Owen Gormley, a top engineer in the new office, recalls that the initiative that led the Atomic Energy Commission to the Carlsbad site in New Mexico came from the New Mexicans themselves.[82] A potash company wrote twice to the commission suggesting that its old mine galleries in the Carlsbad area might be used for a repository, and that if put to such use would not only help solve the waste problem, but would also provide jobs for unemployed potash miners. The agency was interested in New Mexico, but having been stung politically in Kansas, "the state people were told we were not coming to New Mexico until invited by the governor," Gormley recalls. "And we said that, if invited, we would be coming with geologists, not bulldozers." The upshot, Gormley says, was that Governor Bruce King (a Democrat who later would be dealing with WIPP issues during a second term) invited the AEC to come to New

---

[81]Lomenick was interviewed by the author on October 1, 1984.
[82]Gormley was interviewed by the author on October 4, 1984.

Mexico to look for a repository site. As it turned out, the potash mines were not considered suitable for anything except geologic testing, but the commission's advisers at Oak Ridge and the U.S. Geological Survey were confident that an acceptable repository site could be found.

From the outset the WIPP project enjoyed strong support from political and business leaders in Carlsbad and Eddy County. For these leaders the backyard in question was a windblown, semi-arid plain of scrub brush and red sands twenty-five miles to the east of Carlsbad. This was no agricultural cornucopia of the kind found on the Texas High Plains. The soils were poor and there was no prolific aquifer to irrigate them. About the only thing out there that Carlsbad's mayor Walter T. Gerrells and the other local leaders felt was worth much were the potash and oil and gas reserves. These leaders had, moreover, only a few years before received a brutal reminder that the potash represented a dwindling resource: in early 1968 the U.S. Potash Company, a big employer, had shut down its Carlsbad operations.

"We were shocked," Mayor Gerrells told me. "We had had a monopoly of potash in the Western Hemisphere."[83] A year later, Carlsbad had 1,100 empty homes and commercial buildings, an enormous setback for a community of 25,000. Furthermore, local leaders were afraid that other potash companies eventually would be shutting down their operations in response to changing market conditions and depleted and less easily recoverable reserves.

The local base of support for WIPP was to weather all of the project's vicissitudes, and it remains strong today. Some opposition later developed in Carlsbad, but neither Mayor Gerrells nor the two state legislators from Eddy County who were prominently identified with WIPP were ever to be challenged on the WIPP issue at election time. Why was there acceptance of WIPP in Carlsbad, when in such places as southern Mississippi and northern Louisiana repository siting investigations were giving rise to strong opposition? The answer is not entirely clear. But for the most part the strong support for WIPP appeared to be a function of circumstances: hard times and joblessness, the absence of real or perceived land-use or resource conflicts, and positive local leadership (including consistent editorial support by the *Carlsbad Current-Argus*). Local support for WIPP was to be of critical importance during the late 1970s and early 1980s when the fate of the project was being decided.

"The power structure has stuck together like glue on this issue," Roxanne Karchner, wife of a potash company worker and leader of a largely inactive anti-WIPP group, told me when I called on her several

[83]Gerrells was interviewed by the author in March 1981.

years ago.[84] According to Karchner, her group collected 2,000 signatures against WIPP but had found people generally apathetic and believing that to buck the establishment would be futile. Some local physicians were known privately to be against WIPP, she said, but no one like Robert Kemmerly in northern Louisiana had emerged.

According to Gormley, at the start WIPP was to have been a pilot facility for commercial high-level waste, with the disposal there of military transuranic waste from Idaho and elsewhere to come only later. Experimentation and research were to receive substantial emphasis, and the pilot phase was to continue indefinitely, no particular time having been set for conversion of the facility to a permanent repository. It was not seen, Gormley says, as a complement to, or substitute for, the then proposed Retrievable Surface Storage Facility that was expected to provide interim storage for the bulk of the high-level waste generated by commercial reprocessing plants. But as a result of controversy over the RSSF program (among other reasons), the Energy Research and Development Administration decided in 1975 to withdraw the environmental impact statement for the commercial waste program which the AEC, its predecessor, had issued. As Gormley saw it, this meant at least a two-year delay in a commitment to build WIPP. On Gormley's recommendation, then, ERDA removed WIPP from the commercial program, redefining its mission as that of an unlicensed facility for military transuranic waste. This policy was to remain undisturbed until the Department of Energy succeeded ERDA in 1977.

Three places in southeastern New Mexico were at first investigated for a site, but two were rejected, one because of extensive oil and gas exploration and development, the other because salt of the desired purity was too deep. The third area was the Delaware Basin, to the east of Carlsbad. This basin, extending about 150 miles from north to south and about 100 miles from east to west at its widest point, is encircled by the Capitan Reef, a large limestone formation shaped like a giant horseshoe, its open end lying to the south in Texas. The reef, most of which is buried, outcrops to the west in the Guadalupe Mountains, where its famous Carlsbad Caverns bear spectacular witness to the power of underground water flows to carve out huge passages and rooms.

In Permian time, more than 200 million years ago, the salt and other evaporite beds reached across the entire Delaware Basin, but in later geologic ages a massive and continuing process of salt dissolution began,

---

[84]Karchner was interviewed by the author in March 1981.

probably triggered by a tilting of the basin to the east.[85] The salt beds thus exposed to the surface in the western part of the basin were dissolved away by the flow of surface water and shallow groundwater. There is also geologic evidence that deeper-lying layers of salt were partly dissolved away by water from the Capitan Reef aquifer.

As a result, the salt over the entire western half of the Delaware Basin was lost to dissolution. In the middle of the basin, over an area extending ten miles or so to either side of the Pecos River, the salt is not yet gone but has long been in an active stage of dissolution, as is the salt along part of the reef on the basin's east side. The largest part of the Delaware Basin believed by WIPP geologists not to have been extensively and massively affected by shallow or deep dissolution, or by both, is at the north end, which includes the Carlsbad potash district. So it was here, in the northern part of the basin, that the search for the WIPP site was to take place.

The first serious difficulties to confront the WIPP project in the Delaware Basin had to do with three things: boreholes from oil and gas exploration; brine pockets or reservoirs; and the presence of significant potash reserves where they were not expected to be. Although oil and gas exploration had indeed been less extensive here than elsewhere in the larger Permian Basin, WIPP investigators found it exasperatingly difficult to avoid exploration boreholes by at least two miles, as required by the AEC in light of the Lyons experience. Twice the site had to be moved as new oil and gas wells were drilled nearby. Part of the problem was that the chair of the commission, Dixy Lee Ray, refused to seek the federal land withdrawals necessary to maintain adequate buffer zones. According to Gormley, this was shortly after the first OPEC oil embargo, and Ray did not want to appear to be favoring nuclear energy over oil and gas development.

The WIPP investigation was still at an early stage in 1975 when ERDA succeeded the AEC and the Sandia corporation took over the technical direction of the project. The chosen location for WIPP at that time was at ERDA-6, a drilling site only about three miles to the south of the Capitan Reef. This was not a good place to be, as Sandia was shortly to discover. The salt body of interest was the Salado formation, which is encountered at about 1,000 feet below ground surface and goes down to 2,800 feet or more. The Salado salt overtops the Capitan, but where the evaporite beds below meet the reef they typically have undergone

---

[85]The description of dissolution phenomena in the Delaware Basin is derived from Lokesh Chaturvedi and Kenneth Rehfeldt, "Groundwater Occurrence and the Dissolution of Salt at the WIPP Radioactive Waste Repository Site," *Eos Transactions* vol. 65, no. 31 (July 31, 1984) pp. 457–459.

severe deformation. This zone of deformation is now believed to extend at least several miles from the reef, taking in the area where Sandia was about to drill.[86]

The core samples from ERDA-6 brought bad news. They showed the evaporite beds dipping down at sharp angles, and part of the Castile anhydrite formation directly below the Salado was missing. Still worse, a reservoir of pressurized brine was encountered in the Castile at a depth of 2,700 feet, only 300 feet below the Salado. Unlike the heavy rigs routinely used in oil and gas exploration, the light drilling rig used by ERDA at the time was not equipped with a blowout preventer to shut in the well. Brine, together with methane gas and deadly poisonous hydrogen sulfide, rushed up the borehole and gushed out, endangering the drilling crew (which, however, escaped harm).[87] The origin of such brine reservoirs remains a mystery; deep dissolution of salt in the geologic past is only one possible explanation. But the brine encounter at the Salado could clearly be taken as warning of a possible hazard in repository development, even though all such encounters in the basin had been below the Salado salt. To punch into a large pressurized brine reservoir accidentally during the mining of a repository could be catastrophic, with brine gushing into the underground workings in torrents. Once when an oil company hit a brine pocket in the Delaware Basin, brine flowed at the rate of 36,000 barrels a day.

The result was that Sandia, in a decision concurred in by the U.S. Geological Survey, moved the WIPP site about six miles to the southwest, well away from the Capitan Reef. In the process the criterion on proximity to oil and gas exploration boreholes was relaxed from two miles to one mile in order to widen the selection of sites.[88] A recent analysis had indicated that for water flowing through an inadequately plugged borehole to dissolve the salt over a distance of a mile would take a quarter of a million years. Yet while this relaxation may have been justified, the circumstances invited suspicion that criteria were being picked to accommodate the available sites rather than vice versa.

Two other criteria for selection of the new site called for staying outside the Carlsbad potash district, and away from places attractive for

---

[86]Interview by the author with Lokesh Chaturvedi, geologist with New Mexico's Environmental Evaluation Group; Chaturvedi and Rehfeldt, "Groundwater Occurrence and the Dissolution of Salt."

[87]Interview by the author on October 4, 1984, with George Griswold of the New Mexico Institute of Mining and Technology, who at the time brine was encountered was a Sandia geologist-engineer on the WIPP project.

[88]Personal communication to the author, May 6, 1985, from Wendell D. Weart of Sandia National Laboratories, technical director of the WIPP project for DOE.

oil and gas exploration. But 160 million tons of potash turned out to be present above the Salado within the boundaries of the site and the two-mile control zone, most of it langbeinite potash, which occurs nowhere in the United States except in the Carlsbad district.[89] New Mexico's WIPP review group would conclude that the mining of these significant reserves—the langbeinite present represented 7 percent of the known U.S. supply of this mineral—should be banned indefinitely to reduce risk to the repository.[90] Some 24 billion cubic feet of natural gas was also presumed (from statistical analysis) to be present at a depth of 10,000 feet or more beneath the Salado. If actually present, this resource, representing a field of small to moderate size, would be exploitable only at the additional cost of directional or slant drilling from a well started outside the WIPP project boundaries.

From 1978 on, the siting investigation was to be carefully reviewed by the State of New Mexico's WIPP Environmental Evaluation Group. The EEG was, and is, funded by the Department of Energy (at this writing at a level exceeding $600,000 a year), and from the outset the understanding was that neither the department nor the state would attempt to bias or interfere in the group's technical conclusions.[91] The group was answerable to a special state cabinet task force under the governor, through the head of the New Mexico Environmental Improvement Division. EEG's director, Robert H. Neill, was to be called on frequently to appear before a special radioactive waste committee established by the state legislature. By virtue of its professional staff of seven, together with its use of outside consultants, the EEG proved competent to press both Sandia and the scientific critics of WIPP to defend and substantiate their positions.

The principal uncertainties that the Environmental Evaluation Group had to deal with pertained to salt dissolution, the regional hydrology, and the presence of pressurized brine in the Castile anhydrite layer beneath the Salado salt.

Salt dissolution.   On this issue, the Environmental Evaluation Group was in both agreement and disagreement with WIPP's principal scientific critic, Roger Y. Anderson, professor of geology at the University of New Mexico. Anderson's general hypothesis for the presence of an active

---

[89]Later alterations in the location of the site, together with a reduction in the control zone from two miles to one mile, reduced the potash present to 14.5 million tons of langbeinite.

[90]Robert H. Neill, James K. Channell, Lokesh Chaturvedi, Marshall S. Little, Kenneth Rehfeldt, and Peter Spiegler, *Evaluation of the Suitability of the WIPP Site*, EEG-23 (Santa Fe, Environmental Evaluation Group, May 1983) p. 142.

[91]Letter of December 24, 1984, from Colin Heath to the author.

deep dissolution phenomenon in the northern Delaware Basin was found to be a strong one, but from the borehole testing done by Sandia the EEG concluded that the WIPP site itself would not be affected over the period of hazard. On Anderson's advice the EEG recommended, and the Department of Energy agreed to carry out, deep drilling of a possible point-source dissolution feature two miles north of the site. The evaluation group has since concluded that no dissolution has taken place there.

Regional hydrology.   Here the questions were principally two.[92] First, were the deep-lying sandstone aquifers 2,000 feet below the WIPP repository site under sufficient hydraulic head pressure, from recharge occurring at higher elevations outside the basin, to cause the water in any well drilled into those aquifers to rise up through and above the Salado and the repository? If so, water contaminated by repository wastes might enter the relatively shallow aquifers in the Rustler Formation above.

Second, were solution cavities present in the Rustler aquifers? If so, radionuclides might move much more rapidly than had been previously assumed to springs at Malaga Bend on the Pecos River, some seventeen miles from the WIPP site. If this question were not laid to rest by further studies, the EEG would recommend more robust engineered barriers in the repository system, such as mixing radionuclide-retarding bentonite clay in with the rock salt used in backfilling tunnels.[93]

Brine reservoirs.   The evaluation group regarded the presence of any pressurized brine reservoirs beneath the WIPP site as a hazard for the very long term, looking to the time when administrative control over the site might be lost. The drilling of oil or natural gas exploration holes into such a reservoir could eventually cause brine to rise up through the Salado, enter the repository, then continue on up to the Rustler aquifers and to the surface.[94]

Of the nine deep boreholes drilled by the Department of Energy and its predecessors within about an eight-mile radius of the WIPP project's present central shaft, two had hit brine reservoirs below the Salado salt. There was the first encounter in 1975 at ERDA-6, as described; the second came in late 1981 when a much larger reservoir was encountered at WIPP-12, a mile to the north of the WIPP central shaft. After this second encounter, DOE, acting on the evaluation group's recommendation and the state's request, agreed to build the repository to the south

[92]Interview by the author with Lokesh Chaturvedi, September 14, 1984.

[93]Interview by the author with Lokesh Chaturvedi, January 2, 1986.

[94]Neill and coauthors, *Evaluation of the Suitability of the WIPP Site*, pp. 25–39.

of the central shaft rather than to the north. Even so, there was the very real possibility that the reservoir encountered by WIPP-12 extended beneath the site as relocated. The evaluation group had estimated, though none too confidently, that the WIPP-12 reservoir contained from 5 to 10 million barrels of brine and might be up to four miles across. Whether brine was present beneath the site could not be confirmed short of drilling boreholes through the Salado and the repository block.

"Consequence analysis" scenarios done by the evaluation group as well as by a WIPP contractor indicated that if brine were to flow up through WIPP to the surface, the consequences for public health would be within acceptable limits.[95] The concentration of radionuclides in a repository containing only transuranic waste would be low to start with: less than 6 million curies would be present, compared to the several *billion* curies in a repository for high-level waste or spent fuel. The concentration reaching the human environment would be much lower still. The EEG and Sandia scientists agreed that for a repository containing high-level waste, the health consequences would be different, and possibly much worse.

The Politics of WIPP

The strong local support for WIPP served not only as the political bedrock on which the project rested, but also contributed to the relatively independent and objectively critical role that the evaluation group was able to play—a role particularly important for WIPP in that the project was not subject to Nuclear Regulatory Commission licensing and had no other independent overseer. Carlsbad has been potently represented in the state legislature in Santa Fe, and in 1979, when the Radioactive Waste Consultation Committee was established, two Carlsbad legislators were named as chair and vice-chair. Between this pro-WIPP legislative committee and the not anti-WIPP but at least skeptical WIPP task force under the governor, there was a degree of political tension that worked in the EEG's favor. Robert Neill, the health physicist who has directed the evaluation group from its inception, told me, "We haven't been pressured by anyone, not by the governor or his task force and not by the legislature."[96]

Nonetheless, the politics of WIPP in New Mexico reflected a situation in delicate balance. People living in the populous Albuquerque and Santa Fe areas had nothing to gain from WIPP, and were being regularly exposed to news accounts of the problems and uncertainties associated

[95]Ibid., p. 38.
[96]Interviews by the author with Neill, March 1981 through October 1984.

with the project. One of the New Mexico environmental groups, the Southwest Research and Information Center in Albuquerque, was devoting much of its attention to WIPP, even publishing a WIPP newsletter. Thus the Department of Energy had reason to proceed cautiously. Yet on two occasions the department made moves that caused alarm in Santa Fe and among members of the New Mexico congressional delegation in Washington.

The first such occasion was in the fall of 1977, when reports circulated that DOE officials wanted to change the project mission and have WIPP accommodate military high-level waste. Indeed, the department's plans for this appeared to be farther along than its officials were acknowledging.[97] Signs that DOE was planning to expand the project's scope without taking the state into its confidence contributed to increasing opposition to WIPP in the legislature. In January 1978 the legislature seriously entertained a ballot proposal for a constitutional amendment to keep nuclear waste from being brought into New Mexico for disposal; the proposal failed in the House by only three votes.[98]

Shortly thereafter, when Senator Pete V. Domenici and other members of the New Mexico congressional delegation called on Secretary of Energy Schlesinger, they were promised that WIPP would not be built over the state's objection. Schlesinger's deputy, John O'Leary, has since told me that making such concessions to New Mexico, Louisiana, and other states amounted simply to a recognition that a repository cannot be built over determined host-state opposition. "When you think of all the things a determined state can do, it's no contest," O'Leary told me, citing by way of example the regulatory authority a state has with respect to its lands, highways, employment codes, and the like.[99] The federal courts, he added, would strike down each of the state's blocking actions, but meanwhile years would roll by and in a practical sense DOE's cause would be lost.

The second Department of Energy provocation came when the Deutch report, made public in March 1978, recommended two changes in the WIPP project that turned out to be highly controversial. One was for a major spent fuel disposal demonstration, strongly favored by Deputy Secretary O'Leary. The other was for committing the military transuranic waste to WIPP without provision for retrieval; the reasoning here was that inasmuch as such waste generates little heat, there would be no interactions between the waste and the host rock to observe even if a period of retrievability were provided. New Mexicans objected strongly

[97]Luther J. Carter, "Trouble Even in New Mexico for Nuclear Waste Disposal," *Science*, March 10, 1978, pp. 1049–1050.
[98]Ibid.
[99]Interview by the author with O'Leary, May 9, 1983.

to these recommendations at public hearings. What previously had been described as a repository for relatively low-level military waste was now seen as one that would receive high-level commercial waste. Moreover, the 1,000 fuel assemblies that would be used in the demonstration were perceived as possibly the vanguard for spent fuel from throughout the country. The retrievability issue was important to the New Mexicans because, according to one observer, the WIPP facility had originally been "portrayed as a 'pilot plant' from which wastes would be removed if something went wrong."[100] The Department of Energy beat a quick retreat on the retrievability issue, but sought to press on with the spent fuel demonstration. According to Deputy Secretary O'Leary, the department saw this demonstration as particularly important because it was expected to satisfy the California nuclear moratorium law by showing that the means for spent fuel disposal were clearly in hand.[101]

Although DOE continued to promise that WIPP would not be built if the state objected, the U.S. Comptroller General and the department's general counsel were taking the position that these assurances were not legally binding. New Mexico's attorney general, Toney Anaya, a Democrat running for the Senate, was insisting that Congress should enact a law giving New Mexico an explicit legal right of veto. Republican Senator Domenici, a member of the Subcommittee on Nuclear Regulation, was himself seeking such a right for repository host states, but was not finding much support among his congressional colleagues. In demanding a veto right the state was saying, in effect, that it wanted at least enough control over WIPP to prevent unexpected and undesirable changes in the project mission. It was saying, too, that it wanted to make sure the repository would be safe and that the state would be compensated for any new burdens associated with the repository. The surprise Deutch proposal had pointed up the state's lack of control, putting WIPP at risk by challenging state officials to take action.

WIPP may well have been saved by the developments that took place in 1978 and 1979. The establishment of the state Environmental Evaluation Group, while perhaps not critical, was important, for it contributed to defining WIPP as an issue that could be dealt with on its technical as well as its political merits. A second and clearly critical development came in late 1979 when Congress passed the WIPP authorization act, which blocked both DOE's initiative to include a spent fuel disposal demonstration in WIPP and President Carter's initiative to cancel the project.[102]

---

[100]Randall Francis Smith, "Strategies for Siting Nuclear Waste Repositories" (Ph.D. dissertation, Kennedy School of Government, Harvard University, 1981) pp. 114–121.

[101]Interview with O'Leary, May 9, 1983.

[102]Public Law 96-164, sec. 213, 93 Stat. 1265, December 29, 1979.

The House and Senate armed services committees felt that WIPP could be fully justified as a repository to receive the large and still growing inventory of military transuranic waste stored in Idaho and elsewhere. And Senator Henry M. Jackson of Washington, a powerful figure on the Senate committee, believed that the best chance for a nuclear waste repository to be built in the near future lay with WIPP as an unlicensed military waste facility. According to James C. Smith II, a committee staff member, Jackson believed that WIPP would indirectly help the commercial nuclear power program by providing some early answers to the geologic disposal problem.[103]

The WIPP legislation gave the State of New Mexico a considerable voice in the project, although not a right of veto. Its provision for the Department of Energy and the state to negotiate a "consultation and cooperation" agreement put New Mexico in a position where DOE had to give great weight to the state's demands. But WIPP went through another period of political danger before an agreement acceptable to the state was achieved. In negotiations with the state the department took a hard line, rejecting the state's position that the agreement should be a legally enforceable contract, with issues in dispute to be resolved by the secretary of energy in written decisions. The department's attitude was encouraged by Chairman Melvin Price of the House committee (a former chair of the Joint Committee on Atomic Energy), who in reply to an inquiry from DOE wrote that the act conferred "no additional rights" on the state and that lack of an agreement should not delay the WIPP project.[104]

In January 1981 state officials were outraged when the Department of Energy, behaving like the AEC at Lyons a decade earlier, formally announced without prior consultation with the state that WIPP would be built unless the site was discovered to be unsuitable.[105] Not long thereafter New Mexico's attorney general at the time, Jeff Bingaman, brought suit against DOE. Some key people in DOE's Albuquerque operations office are said to have been ready to battle it out in court. But Secretary of Energy Edwards, who as a former governor of South Carolina had the credentials of a practical politician, met with Governor King to smooth things over, and the department subsequently acceded to the state's demands. As part of the court-approved settlement,[106] the de-

[103]Smith was interviewed by the author on October 10, 1984.

[104]Letter of October 6, 1980, from Congressman Melvin Price to Duane C. Sewell, assistant secretary for defense programs, Department of Energy.

[105]Department of Energy, "Waste Isolation Pilot Plant (WIPP) Record of Decision," January 23, 1981.

[106]Approved by the U.S. District Court, the agreement was signed on December 27, 1982, by Governor King of New Mexico and Raymond G. Romatowski, manager of the Department of Energy's Albuquerque operations office.

partment agreed to a series of state-designed experiments to test the suitability of the WIPP site and repository design; to state and public review of any proposals for changes in the scope of the project; and to the creation of a state-federal task force to address state concerns about waste transportation, emergency preparedness, accident liability, federal compensation for financial burdens the project would impose, and other issues.

Had Secretary Edwards not yielded to New Mexico's demands, the entire state political establishment might have turned against WIPP. Moreover, Senator Domenici had now become a powerful senior member of Congress—he was chair of the Senate Budget Committee, among other things—and alone might have been able to kill the project. Admittedly, considerable uneasiness about WIPP would persist among New Mexico environmentalists and on the part of Governor Anoya, whom they helped elect in November 1982. Governor Anoya seemed to worry in particular lest WIPP prove to be a foot in the door for a licensed repository intended for spent fuel and high-level waste.[107]

But the WIPP project would soon be yielding data about salt creep and other phenomena which indicated that the Carlsbad site could never be a plausible or defensible choice for a commercial repository. In fact it appears today that WIPP has been saved—technically—by its limited mission as an unlicensed repository for military transuranic waste that generates little heat and that is not subject to retrieval except during an initial five-year trial period. WIPP almost certainly could not be licensed as a repository for heat-generating high-level waste or spent fuel in light of the waste retrieval requirements of both the Nuclear Regulatory Commission and the Nuclear Waste Policy Act of 1982.[108] NRC regulations specify that it must be possible to begin retrieval of emplaced spent fuel or high-level waste at any time during the first fifty years in the life of a repository, and that this capability must extend for thirty years, or as long as it takes to retrieve all of the emplaced canisters. For reasons explained below, it is becoming increasingly clear that any attempt to exercise the retrieval option at the WIPP site—or possibly at any of the salt sites—would confront overwhelming difficulties.

Salt Creep, Cracking, and Other Disturbing Phenomena

It was not until late 1985 that findings pertaining to salt creep, salt cracking, and brine occurrences and migration began coming fully to

[107]Luther J. Carter, "WIPP Goes Ahead Amid Controversy," *Science*, December 9, 1983, pp. 1104–1106.

[108]Nuclear Regulatory Commission, "Disposal of High-Level Radioactive Waste in Geologic Repositories Technical Criteria," 10 CFR, pt. 60, in *Federal Register* vol. 48, no. 120 (June 21, 1983) pp. 28197–198. For the congressional retrieval option, see Public Law 97-425, sec. 122, 96 Stat. 2228, January 7, 1983.

light in the WIPP investigation. Although incomplete and in some cases not well understood, the new data seem compelling. Even before the new data were in hand, it was evident from calculations done with theoretical models that to maintain a capability for retrieving high-level waste over several decades could present a major problem. According to Lynn Tyler, the Sandia scientist in charge of WIPP experiments that would involve temporary emplacement of up to forty canisters of such waste, the model indicated that by the end of the first ten years total salt creep in the 18-foot-high emplacement chamber would be about 8 feet.[109] That is, in no more than a decade this chamber would lose almost half its original height: the floor, into which the canisters would be inserted in vertical holes, would heave up and the roof would creep down. The heat from radioactive decay hastens creep, but the "heat loading" used in the model was lower than the near-field temperatures that would be present in a repository for commercial spent fuel.

The actual measured rate of creep at the WIPP site has confirmed Tyler's belief that the assumptions used in the theoretical calculations understated the reality. Indeed, in a 13-foot-high, 33-foot-wide room used for design verification, with no waste and no heat present, the creep has been three times what was expected. While the very high early creep rate was not maintained beyond the first several months, the creep three years later—in 1986—was still progressing vertically at about 3 inches a year and was decreasing ever more slowly. Closure of the room could be far advanced, even without the presence of heat-producing waste, in less than fifty years. In late 1984 Lokesh Chaturvedi of the New Mexico Environmental Evaluation Group cautioned me not to conclude on the basis of the early WIPP creep data that fifty-year retrievability was an impossibility.[110] But after another year of data collection Chaturvedi felt that to entertain the possibility of waste retrieval for up to fifty years from a repository in salt was "unrealistic."[111] He had thought that the creep might stabilize at a much lower rate, but it did not.

There is now also reason for concern at WIPP over the large (up to 4 inches wide), discontinuous but interconnected horizontal cracking that has occurred several feet beneath the floor of design verification room number 3, one of four rooms where creep measurements have been made. Discovered by chance in September 1985 in the course of drilling boreholes for construction, the cracking was subsequently found to extend the entire 300-foot length of the room and all the way across its 33-foot width before pinching out at or near the walls. To say that

[109]Tyler was interviewed by the author on October 10 and 23, 1984, and on January 3, 1986.

[110]Letter of December 4, 1984, from Chaturvedi to the author.

[111]Interviews with Chaturvedi, January 2 and 16, 1986.

the first 4 feet or so of salt beneath the room's floor is now a huge slab held up chiefly by its tie into the walls might not be an oversimplification.[112] Sooner or later extensive vertical cracking will occur and the slab will begin to buckle unless the room is backfilled and the rate of creep is slowed.

Most of the cracking has taken place in the middle of a 3- to 5-foot-thick layer of anhydrite, an impurity within the salt. As the salt has heaved upward, hairline cracks previously existing within the anhydrite have widened. But whether the cracking, which has assumed a saucer shape, has propagated upward from the anhydrite into the salt or downward from the salt into the anhydrite is not known. Cracking associated with anhydrite or other layers of impurities could be avoided by building the repository at a horizon within the bedded salt where no such layers are present. But this can present its own dilemmas. The WIPP contractors have recommended to the Department of Energy that the site not be moved up or down. To find much purer salt the repository would have to be moved at least another 400 feet down, which would be closer to the Castile formation, where the deep boreholes drilled outside the repository block have hit pressurized brine.[113]

The creep and cracking phenomena would complicate waste canister retrieval in several ways. If canisters are emplaced vertically into the chamber floor, the rooms could be kept open only by periodic remining to raise the ceiling. The emplacement holes could be sleeved with cast iron to facilitate retrieval, or an attempt could be made to retrieve the canisters by "overcoring"—that is, by boring around each canister and removing it along with the salt in which it is encased. But in neither case would the canisters be found where first emplaced. The canisters or emplacement sleeves would move upward with the salt, though probably not at the same rate (the salt having less density). Furthermore, their orientation could be a confused jumble because of tiltings and deflections resulting from lateral creep, the presence of impurities, and the buckling of the repository floor.[114]

Adding to the complications for waste retrieval is the fact that any voids around the canisters or emplacement sleeves would be occupied

[112]Wendell D. Weart of Sandia National Laboratories disputes this characterization (personal communication to the author, April 2, 1986), but Lokesh Chaturvedi of the WIPP Environmental Evaluation Group believes it to be borne out by the fracture mapping done from six 36-inch-diameter boreholes across the width of the room. See "17th Quarterly Presentation to EEG/WIPP," WIPP-DOE-229 (March 27, 1986), figure labeled "Fracture mapping in 36-inch holes."

[113]Lokesh Chaturvedi, personal communication to the author, April 22, 1986.

[114]Interviews by the author with Lynn Tyler of Sandia on October 10, 1984, and Lokesh Chaturvedi in March and April 1986.

by a hot and briny slurry, possibly under high pressure.[115] In a heater experiment at WIPP that began in April 1985, the brine which had accumulated around the simulated waste package by the end of the year measured 15 liters, whereas only a liter or so had been expected.[116] The Sandia investigators are at this point uncertain whether their experimental method has distorted, and exaggerated, the results; but as Chaturvedi points out, free-flowing brine has been found to occur even without the influence of heat in the salt of the WIPP repository. Twenty-two boreholes drilled in the floor of the WIPP excavations filled up with brine shortly after drilling. These holes were drilled for various purposes, from instrumentation to stratigraphic control, and are scattered over the entire excavation. The brine encountered apparently originates in the same anhydrite layer in which the 4-inch-wide crack appeared in response to the upward creep of the salt.[117]

All of this is bad news for the Department of Energy siting effort in salt. A decision by the Nuclear Regulatory Commission and Congress to withdraw their present waste retrievability requirements would appear necessary to give this effort a reprieve, but such a decision would be unwise, not least because the political problems of a waste program already suffering from public distrust could be further aggravated. The creep, cracking, and brine migration phenomena observed at WIPP obviously are of great importance inasmuch as seven of the nine existing sites in the commercial repository program are in salt. The extensive cracking is perhaps significant chiefly as a revelation that the rock mechanics associated with establishing a deep-mined repository in salt are still poorly understood.[118] Much bigger surprises may lie ahead if a commercial repository is actually built in salt and heat-generating waste is introduced.

---

[115]Chaturvedi has pointed out that in one borehole drilled into the layer of impurities beneath the tunnels and rooms at WIPP an outflow of gas initially measuring 1,000 cubic centimeters (or 1 liter) a minute was encountered; the gas was 90 percent nitrogen and 8 percent methane. Personal communication to the author, March 11, 1986. According to Wendell Weart, the outflow of gas from this borehole declined over a five-hour period to a baseline condition at which, when the borehole is capped, the pressure builds up again and stabilizes at 90 pounds per square inch, or at about six times atmospheric pressure; the pressure encountered when the borehole was first drilled may have been about 90 psi. Interview of February 12, 1986.

[116]Interview with Lynn Tyler, January 3, 1986.

[117]Interview with Chaturvedi, January 16, 1986.

[118]Chaturvedi points out that the presence of the kind of anhydrite layer associated with the cracking at WIPP is by no means unique to the Carlsbad site and the Delaware Basin salt. Layers of anhydrite and other impurities are common in the bedded salt of Texas's Palo Duro subbasin and Utah's Paradox Basin. According to Chaturvedi, the same cracking phenomenon is a distinct possibility in the event a repository is built in either

Conclusions from Field and Host-State Experience

Most of the Department of Energy siting experience in New Mexico and other states has been so recent, so concentrated, so controversial, and so given to self-serving interpretations by the various players involved that to draw conclusions from it is to invite an argument. Nonetheless, a few conclusions do emerge strongly.

One is that real or strongly perceived environmental and land-use conflicts in the choice of candidate sites should be avoided. The not-in-my-backyard stereotype obscures this lesson by suggesting that state and local opposition is inevitable and predictable. The truth is much more complex, as the experience in New Mexico reveals.

A second conclusion is that by getting down to the task of thoroughly investigating one or more candidate sites much indispensable information can be gained. However one feels about the WIPP project, it would be hard to deny the importance of the information coming from the measurements now being made in exploratory tunnels.

Still another conclusion is that there is an enormous advantage, both technically and in terms of establishing public trust, in having a competent, objective, vigorously independent, and adequately financed group to follow and critique every detail of the siting investigation. The New Mexico Environmental Evaluation Group, which has been filling this role with respect to WIPP, has benefitted from an unusual constellation of political forces and circumstances which probably would not be present in another state. Ingenuity and inventiveness might be necessary to establish effective independent project oversight in any other state.

A final conclusion is that local and possibly state stakeholders with an interest in having a siting effort succeed may be established by the promise of substantial economic rewards. At Carlsbad, a town in the economic doldrums when the Atomic Energy Commission first arrived on the scene, the prospect of the new jobs and spending associated with WIPP brought a sympathetic local response—and the local legislators had enough clout in Santa Fe to give the project some standing there. But in another state something more might be necessary.

In their work shaping nuclear waste policy, President Carter's Inter-agency Review Group, the Department of Energy, and the Congress seem to have been influenced remarkably little by these lessons from experience in the field and the host states. In part this was because the

---

of those basins. Unlike the salt beds, salt domes were not formed in a way that left regularly configured layers of impurities; but, Chaturvedi observes, whether the diapiric stresses present in these domes would make the rate of creep faster or slower than at WIPP is not yet known.

lessons were lost in the confusion of events. But another explanation has been that the DOE people in the field typically lack the rank necessary to advance their views effectively within the Washington headquarters bureaucracy. From his struggles with the repository siting problem, Colin Heath concluded early that to offer economic incentives to the potential host states and localities was essential. "I tried to get a study of the legal precedents [for such incentives] ordered by Secretary Schlesinger," Heath says, "but I could never get this proposal through the system to Schlesinger's desk. Others who had to approve it felt that to offer such incentives would be tantamount to a bribe."[119]

Another reason why policy deliberations have been poorly informed by experience is that to find out what's going on and to search for its meaning can take a lot of legwork. It is easier to approach policy analysis as an exercise in abstract formulations that may have little relevance to the real world. An instance of this can be seen in the 1984 report by the National Academy of Sciences' Panel on Social and Economic Aspects of Radioactive Waste Management. The panel held all of its nine meetings in Washington and ignored the wealth of experience in the host states.[120]

The next chapter examines how Congress, whose members cannot escape the real world for very long, addressed the repository siting problem—dealing with it in a manner which can be seen from hindsight to have been surprisingly lacking in political realism.

[119]Interview with Heath, February 1981.

[120]Luther J. Carter, review of National Research Council, *Social and Economic Aspects of Radioactive Waste Management: Considerations for Institutional Management*, in *Environment* vol. 27, no. 2 (March 1975) pp. 25–29.

6

The Nuclear Waste Policy Act

The Nuclear Waste Policy Act was largely a product of the Ninety-sixth and the Ninety-seventh Congresses, gradually taking shape over the period from early 1979 to late 1982. The recommendations by President Carter, coming in his last year in office when Carter was in a losing struggle for political survival, were given no priority and were not pushed by the administration on Capitol Hill. But in their emphasis on deep geologic isolation as the first order of business and on investigating several potential repository sites thoroughly and then picking the best of them, Carter's recommendations did influence the legislation that eventually emerged. However, the legislative effort was an untidy process and much time would elapse before even a fragile consensus was arrived at. It should be pointed out that for most members of Congress the waste issue was strictly a domestic concern. An early promise of success in demonstrating safe geologic disposal of spent fuel could be important to U.S. nuclear nonproliferation policy, but this consideration was not to figure as an imperative in the congressional deliberations over nuclear waste policy.

The Interests at Play

The four major interests at play in the shaping of the waste legislation were the nuclear industry, the potential host states for geologic repositories and surface storage facilities, the environmental and anti-nuclear groups, and the Department of Energy.

The nuclear industry.   All parts of the nuclear industry saw a need for Congress to act on the waste issue, and there were no major disagreements as to the form the legislation should take. The utilities felt an immediate need for the government to provide interim storage for their growing accumulations of spent fuel, but shared in the general view that deep geologic disposal was the long-term solution most likely to satisfy the public. The feeling was that the sooner this solution could be effectively demonstrated, the better for the industry politically.

Spearheading the industry lobbying effort was the American Nuclear Energy Council (ANEC), an umbrella group made up of such disparate elements as uranium mining companies, reactor vendors, architecture-engineering firms, and utilities, both public and investor-owned. Closely allied with the council were other industry groups such as the Edison Electric Institute and the Utility Nuclear Waste Management Group, as well as sympathetic organizations like the U.S. Chamber of Commerce, the National Manufacturers Association, and the building trades unions. Industry lobbyists were typically out in force for important sessions of the key legislative committees and for major floor votes, but usually the industry's lobbying effort was carried on by just a few people, particularly John Conway, George Gleason, Thomas Kuhn, and Edward Davis, all with ANEC.[1]

The industry lobbyists enjoyed two major advantages. First, the utilities counted as major industrial enterprises in every congressional district and no congressman or senator could ignore them. Second, the industry, as represented by the American Nuclear Energy Council, was united in wanting a bill very, very badly. In fact, to get a bill the ANEC lobbyists eventually were to agree to many of the concessions demanded by the host states with respect to procedural safeguards and states' rights.

The host states. A dozen states that were potential hosts for a geologic repository or for interim spent fuel surface storage facilities took an active part in the congressional lobbying. These included Washington, Nevada, and the four salt states—Louisiana, Mississippi, Texas, and Utah—where the Department of Energy had sites under investigation. In addition, the upper Midwest states of Wisconsin, Michigan, and Minnesota, among the potential hosts for the second repository because of their Precambrian granite formations, became involved; one of their senators was to play a critical role in shaping the legislation's provisions on states' rights. Finally, there were three states that perceived themselves as the prime (if unwilling) candidates for an away-from-reactor (AFR) surface storage facility—South Carolina, Illinois, and New York, homes of the Barnwell, Morris, and West Valley reprocessing ventures. Although the reprocessing plants were not operating, their spent fuel receiving and interim storage facilities might be expanded and adapted for the Department of Energy's AFR needs.

[1] Interview by the author with Edward M. Davis of the American Nuclear Energy Council, June 10, 1983.

The host states all wanted the waste legislation to give them a strong voice, but in a number of instances their interests could be divergent or conflicting. For instance, the Gulf Coast salt dome states could see an interest in widening the search for repository sites to the granite states of the upper Midwest and the Appalachians. The granite states, on the other hand, would not want that at all. Also, potential away-from-reactor storage states could see an advantage in expeditious development of a repository for permanent disposal of spent fuel; the sooner this happened, the less chance there would be for an AFR to come their way. Most of the potential repository states, on the other hand, would want slower, more deliberate siting schedules. For these states the early availability of surface facilities for interim or indefinite spent fuel storage was to have its advantages, too. The concept of indefinite commitment of spent fuel to surface storage facilities was to be skillfully promoted by a senator from a Gulf Coast salt dome state.

The environmental and anti-nuclear groups.   The Washington environmental lobby, loosely defined, included membership groups such as the Sierra Club and Friends of the Earth, but also the Environmental Policy Center,[2] which has no members and was created for lobbying, and Public Citizen, Inc., which represents the ensemble of public interest activities initiated by Ralph Nader. Public Citizen includes both Critical Mass, an umbrella group for anti-nuclear activists, and Congress Watch, the Public Citizen lobbying arm on Capitol Hill.

In all of these groups, the "public" means people who are interested or alarmed. It can refer to any individual or group that takes the initiative to appear at a public hearing, prepare comments on an environmental impact statement, testify before a congressional committee, or file a court suit alleging failures to comply with the law. As a practical matter, only that small part of the interested public found in the larger or more resourceful environmental groups made a sustained commitment to lobbying on the nuclear waste legislation. The standing and influence of these groups were greatest with those members of Congress who, to one degree or another, shared their view of the world. One who did was Congressman Morris Udall of Arizona, chair of the House Interior Committee. Although not in agreement with environmental lobbyists on all issues, Udall was, and is, a dedicated environmentalist. In the special case of the nuclear waste legislation, the influence of the environmental lobbyists was the greater because they identified themselves with the potential repository host states by seeking a larger voice in siting decisions for both those states and the public at large.

---

[2]Renamed the Environmental Policy Institute in 1982. EPI both lobbies and does environmental research.

But the attitude of the environmental lobbyists toward the nuclear waste legislation was fundamentally ambivalent. They wanted a nuclear waste act that would help solve the waste problem, yet were afraid of being party to legislation which, while helping the nuclear industry politically, might serve only to create an illusion of progress and lead to the selection of unsatisfactory repository sites.

Brooks Yeager, lobbyist for the Sierra Club, perceived a major difference in point of view between his group and Congress Watch. "They [the Nader lobbyists] tended to view legislation as a congressional endorsement of a technical solution that actually does not exist. Our aim was to have Congress mandate a waste program with a real test in it to see whether there *is* a technical solution," Yeager told me.[3] This difference in outlook led to important differences in lobbying approach. Yeager and David Berick, lobbyist for the Environmental Policy Center, worked closely with the leaders and staff of the key committees, such as Morris Udall's Interior Committee and Congressman John Dingell's Commerce Committee, where the legislation was actually being shaped. The lobbyists for Congress Watch, on the other hand, found little rapport with the staff and leadership of those committees, and chose to work with members such as Congressman Edward J. Markey of Massachusetts, whose views on the nuclear waste legislation were similar to their own.[4] This put Congress Watch effectively outside the bargaining process. When the proposals that Congress Watch favored were rejected, its lobbyists simply withdrew instead of sticking with the legislation longer and trying to improve it.

The Department of Energy.   The Department of Energy was desperately eager for Congress to pass a nuclear waste policy act. Experience in the salt states had convinced department officials that unless the repository siting effort was backed by a strong and explicit congressional mandate, top executive branch officials were not going to assert federal rights of access over state opposition. As much as anything, the department did not want to venture boldly out on limbs that might break under the weight of political controversy. Passage of a waste act would shift at least part of the political burden from DOE to Congress.

Although department officials were amenable to state participation in repository siting, they did not want the host states accorded a right of nonconcurrence or veto. Those DOE waste program leaders who dealt with the states wanted above all to have the rules of the game clearly

---

[3]Interview by the author with Yeager on July 8, 1983.
[4]Interview by the author on July 11, 1983, with Caroline Legette, a Congress Watch lobbyist during the Ninety-seventh Congress, 1981–1982.

defined. Critz George, who continued to direct the site selection program during the first years of the Reagan administration, had broken his lance more than once because of the lack of mutual trust between the department and state and local officials.[5]

The Ninety-sixth Congress: Defining the Nuclear Waste Issues

In 1980, the last year of the Ninety-sixth Congress, the Senate and the House each passed nuclear waste bills, only to have conference negotiations between the two bodies end in failure just before adjournment. But the work done that year further defined several major issues, including the approach to be taken to repository siting, how big a voice potential host states should have in such siting, whether interim away-from-reactor spent fuel storage should be provided, and whether there should be long-term monitored retrievable surface storage of spent fuel as a complement to geologic disposal, or even as a substitute for it.

Congress addressed one fundamental question pertaining to repository siting which had also confronted President Carter's Interagency Review Group: Was deep geologic disposal a straightforward engineering task, or was it an unproved proposition to be approached with great caution? Congressman Mike McCormack of the tri-cities area near Hanford, a

[5]George was now picking himself up from what he regarded as an unpleasant and gratuitous encounter over a DOE proposal to conduct research at an old International Salt Company mine just outside Cleveland, and beneath Lake Erie. The research was to have involved no use of radioactive materials, but was deemed important for the study of salt creep, among other purposes. The Cleveland mine, 1,700 feet beneath the surface, would have afforded DOE its first chance to measure the response of salt to simulated spent fuel canisters (containing electric heaters) at a depth approaching that of an actual repository. Salt creep is in part a function of pressure from the weight of the overlying rock mass, which increases with depth. While the Cleveland mine was well suited for the intended research, to use it as a waste repository was out of the question: not only was it located in a major metropolitan area and beneath one of the Great Lakes, it had an entirely unsuitable geometry and was crossed by a major fault. George and others had emphasized these points in discussions with local officials and Ohioans in Congress, but the suspicions of most of them could not be dispelled. DOE's worst moment had come at a meeting at which its representatives had tried to explain the project to local officials and the press. The meeting had gone well until Congresswoman Mary Rose Oakar arrived to give the reporters present the angry confrontation they had come expecting. "They're absolutely lying," Oakar told reporters. "They know and we know that once they test in an area, that area comes under consideration for storage." At that point, several Cuyahoga County commissioners and state legislators present also told reporters that DOE was engaged in a sneak play. "They're after the real thing," said State Senator Timothy McCormack. "This is not a flirtation." Jerry Masek, "Salt Mine Tests for Atomic Dump Stir Furor Here," *Cleveland Press*, November 26, 1981.

subcommittee chairman on the House Science and Technology Committee, believed the former, and viewed the Carter policy of investigating four or five sites before selecting one for a repository as time-wasting and likely to result in endless debate over what goes on thousands of feet underground. McCormack favored early construction of four small, unlicensed demonstration repositories, and a system of waste isolation that would place heavy emphasis on engineered barriers. His bill to that effect was favorably reported by the Science and Technology Committee in July 1980,[6] but it got no further. Congressmen Udall of the Interior Committee and Dingell of the Commerce Committee won out over McCormack and the Science and Technology Committee by gaining House passage in late 1980 of legislation that reflected essentially the Carter philosophy. Moreover, acceptance of the proposition that repository siting should include elaborate, multiphased screening of several sites and rock types was to carry over into the next Congress and not be seriously challenged.

Another issue disposed of in the Ninety-sixth Congress was whether to impose a moratorium on further licensing and operation of reactors until the waste problem had been solved. In September 1979, Senator Gary Hart of Colorado, then chair of the Environment and Public Works subcommittee on nuclear regulation, proposed that continued use of nuclear power be made subject to Nuclear Regulatory Commission findings in 1985 and succeeding years that permanent disposal of radioactive waste was achievable. By "achievable" he meant that feasible disposal plans and a workable technology had to be well in hand. But most senators were reluctant to put this pistol to the head of the nuclear industry, and, according to a Hart aide, some environmental lobbyists were concerned that if the Department of Energy were faced with a moratorium law, repository siting might be pushed ahead precipitously and ill-advisedly.[7] The principle of a moratorium was retained in the bill reported by the Environment and Public Works Committee, but the deadline was moved forward to the year 2000. The Senate Energy and Natural Resources Committee wanted no moratorium at all, however, and there was to be none in either the nuclear waste bill passed by the Senate in 1980 or in the legislation that actually became law two years later.

---

[6]The Nuclear Waste Research, Development, and Demonstration Act of 1980, H. Rept. 96-1156, pt. 1, 96 Cong. 2 sess. (1980). McCormack was not to figure much longer in the legislative debate over nuclear waste; he was among the Democrats who lost their congressional seats in the Reagan landslide of November 1980.

[7]Interview by the author on June 24, 1983, with Keith Glaser of the Nuclear Regulation Subcommittee staff.

Most members of Congress were willing to give the potential repository host states a limited veto over siting decisions, subject to a federal override, but they disagreed as to where the balance between state and federal prerogatives should be struck. Some legislators, especially those from states that were likely candidates for a repository, argued for a veto that would stand unless overridden by a joint resolution of both houses of Congress. But what prevailed in the bills passed by both the Senate and House in 1980 was a veto that would stand only if sustained by at least one house of Congress. This was not to be the last word on the states' rights question before a nuclear waste act became law.

Although controversial, the issue of away-from-reactor storage was rapidly diminishing in substance and significance. At the start of the nuclear waste debate, industry lobbyists were deeply concerned lest some reactors be shut down for lack of space in their spent fuel storage pools. The Senate was particularly responsive to this concern. But the House made no provision at all for away-from-reactor storage in the 1980 bill, and would insist in the next Congress that the AFR program be narrowly limited. The technical means to allow spent fuel to continue accumulating at reactor sites were now clearly foreseeable, for the entire operating life of the reactor if necessary.[8] Modular dry storage units could be added as needed, for instance, and at a cost to the utilities comparable to or less than that of storage at a federal AFR facility.

Monitored retrievable storage, or MRS, was not different in concept from the Retrievable Surface Storage Facility pushed by James Schlesinger as chair of the Atomic Energy Commission. But Congress did not take up this concept until Senator Bennett Johnston of Louisiana, a gifted politician of subtle and complex motives, brought it to the fore. At the start of the Ninety-sixth Congress Johnston was assuming an important leadership role on the Energy and Natural Resources Committee. He liked to be thought of as the Senate's "Mr. Energy," as a resourceful problem-solver concerned with rapid development of all major energy sources, including nuclear. But the nuclear waste issue was a tricky one for Johnston, involving potentially conflicting interests. His home political base was northern Louisiana, a region politically up in arms since 1978 over the Department of Energy's exploration of the Vacherie and Rayburn salt domes (see chapter 5). By adopting the concept of long-term retrievable surface storage, Johnston could oppose disposal in salt domes yet put forward a positive alternative.

The senator had embraced this concept after discussions with Daniel A. Dreyfus, then the energy committee's staff director, and after having

---

[8]Office of Technology Assessment, *Managing the Nation's Commercial High-Level Radioactive Waste*, OTA-0-171 (Washington, D.C., 1985) pp. 55–61.

been impressed by the French nuclear center at Marcoule. There John-ston had seen how canisters of glassified high-level waste were inserted in holes in a massive, air-cooled concrete block or vault, then capped with thick concrete plugs in such a way that people could stand above them without fear of radiation exposure.[9] For his part, Dreyfus, trained as an engineer and political scientist, had for years heard talk of geologic disposal of radioactive waste, only to see this solution recede further into the future. The problem, as he has caustically described it, was that the Department of Energy was searching vainly for a "technically appro-priate subsurface with a politically compliant governor on top."[10] Drey-fus thought that the waste problem could be solved by turning to retrievable storage in surface facilities in a friendly state (such as Wash-ington) already accustomed to having nuclear waste in storage, espe-cially if the state and affected localities could count on economic benefits. Dreyfus also believed that states with nuclear power plants would find retrievable surface storage in DOE facilities preferable to a long, contin-ued accumulation of spent fuel at reactor sites, and that the political problem would ease after the first MRS facility went into service. "It could become a way of life," he told me. "Workers in clean white uniforms driving around in trucks, taking radiation readings. You do it once, and people say, 'It's not so bad.' "

The proposition that the siting of retrievable surface storage facilities could be readily accomplished was unproved and not undisputed, how-ever. At the Department of Energy, Colin Heath doubted from the first that such facilities would be any less controversial than geologic re-positories. And the reaction of potential host states to away-from-reactor storage was anything but friendly.

Monitored retrievable storage held priority over geologic disposal in the nuclear waste bill that Johnston and Senator Henry M. Jackson, chair of the energy committee, introduced in March 1979. The bill called for the Department of Energy to present to Congress, on a crash basis and within one year, detailed plans for the construction of an MRS facility, with a specific site proposed. This measure was characterized by John-ston himself as "radical legislation."[11] It represented a direct challenge to a widely shared assumption that went back more than twenty years:

[9]Interview by the author in the spring of 1983 with Proctor Jones of the Senate Appropriations Subcommittee on Energy and Water Development. The subcommittee visited Marcoule as part of a tour of French nuclear facilities, November 7–11, 1978.

[10]Interviews by the author with Dreyfus on April 20 and 28, 1983.

[11]Introduced as S.685, the Johnston bill became a part of S.2189, the measure reported by the Senate Energy and Natural Resources Committee on January 3, 1980. S. Rept. 96-548, 96 Cong. 2 sess. Johnston's remark was quoted in the 1979 CQ Almanac (Washington, D.C., Congressional Quarterly, 1980) p. 699.

that all long-lived, high-level waste would eventually be committed to geologic repositories and thus isolated so securely from the biosphere that, apart from protecting the site from deep drilling or mining, no custodial care would be necessary.

Like Dreyfus, Johnston saw monitored retrievable storage as offering the great advantage of freeing the siting of waste management facilities from geologic constraints. An MRS facility could be built almost anywhere. Senator Johnston himself regarded the Nevada Test Site—which is 1,300 miles from Louisiana—as one of his "particular favorites" for an MRS facility. As he envisioned it, existing tunnels at the test site could be adapted for the storage of spent fuel or high-level waste, which could be kept there indefinitely—for centuries if necessary.[12] The canisters would be under constant surveillance, with retrieval for re-canning or other maintenance always possible. At the same time, if geologic disposal were the ultimate objective, retrievable surface storage would allow more time for site selection and development of long-lived waste packages and for longer cooling of the waste.

The reaction to the monitored retrievable storage proposal by the various interests with a direct stake in the waste legislation was quite mixed. Several host-state legislators besides Johnston argued for the concept, but it was still new and speculative enough not to trigger much of a reaction from the states. Industry lobbyists viewed retrievable surface storage as perhaps a necessary step in nuclear waste management, but not as a sufficient one. "It wouldn't answer the critics' complaint that we don't know how to dispose of radioactive wastes," Edward Davis of the American Nuclear Energy Council told me.[13] On the other hand, according to Davis, ANEC recognized that if the effort to site and develop a geologic repository took longer than hoped and expected, an MRS facility could provide a needed stopgap. In his view, without a vigorous repository program, MRS would not be politically viable. The Department of Energy, in keeping with Carter administration policy, at first opposed monitored retrievable storage, but never made an issue of it. Later, under Reagan, the MRS concept was accepted by DOE in much the same way as it was by the industry. For his part, Colin Heath worried that retrievable surface storage might be seized upon as a politically convenient alternative before a new push to site repositories could get under way. Some key members of Congress, particularly Congressman Udall, were to share this concern.

[12]Senator Johnston's reference to the Nevada Test Site came on July 28, 1980, during Senate debate on S.2189, the nuclear waste bill brought to the floor under his sponsorship. See *Congressional Record* daily ed., July 28, 1980, p. S9972.

[13]Interview by the author with Davis, June 27, 1983.

The environmental and anti-nuclear groups were totally opposed to long-term retrievable storage, believing that for this generation to burden future generations with its wastes would be unethical and unsafe. As later conceptually defined in a Department of Energy study, an MRS facility might have a capacity of 48,000 tons of spent fuel.[14] After a century of storage and radioactive decay, there would remain more than one and one-half billion curies of radioactivity, including more than a quarter of a billion curies in long-lived transuranic elements (mostly americium and various isotopes of plutonium).[15] Thus more radioactivity would be contained in the MRS facility than is present today in the entire inventory of military waste. But such ironic implications of the monitored retrievable storage approach to national nuclear waste policy were not to be explored in the congressional debate, in either the Ninety-sixth or the Ninety-seventh Congress.

Hope that the Ninety-sixth Congress would pass a nuclear waste act vanished in December 1980, just before adjournment. The states' rights provisions of the House and Senate were comparable, except that the Senate bill provided for a considerable state voice even in the siting of repositories for military waste. But included in the Senate bill were two items conspicuously absent from the House measure. One was monitored retrievable storage, the other a program for interim away-from-reactor storage. The House bill included a key feature that the Senate measure lacked: insistence on geologic disposal as the first priority. These differences were to prove unbridgeable, and the House–Senate conference on the waste legislation ended hopelessly deadlocked.

The Ninety-seventh Congress: The Senate

By 1982 the nuclear industry was playing for higher stakes than ever in its efforts to have Congress pass a nuclear waste bill. Without a waste act, the Nuclear Regulatory Commission could conceivably refuse to license more nuclear reactors. In an interview in July 1982, Nunzio J. Palladino, the professor of nuclear engineering whom President Reagan appointed NRC chairman, told me frankly, "If Congress passes the waste

[14] Department of Energy, *The Monitored Retrievable Storage Concept* (Washington, D.C., 1981).

[15] Computed from the Department of Energy's Final Environmental Impact Statement, *Management of Commercially Generated Radioactive Waste*, vol. 2 (Washington, D.C., 1980) appendixes, tables A.2.1a and A.2.1b, pp. A. 31–32.

bill, that very much increases the confidence on the part of a number of the [NRC] commissioners."[16]

Moreover, the nuclear industry badly needed some good news on the waste front because certainly there was none coming from elsewhere. Utilities had already cancelled seventy-seven reactor orders, and in 1982 they would cancel another eighteen.[17] The decline in the growth of electricity demand was perhaps a sufficient reason for the lack of orders and the rash of cancellations. But in the minds of utility executives and the managers of the utility bond markets, the whole tangle of public controversies, regulatory uncertainties, liability questions, and rising and uncertain construction costs associated with nuclear power was something to be avoided. Passage of a nuclear waste policy act alone could not save the industry from its manifold troubles, but the industry could not be revived if the waste problem were left unresolved.

The McClure Nuclear Waste Bill

The first action on a waste bill in the Ninety-seventh Congress came in the Senate, in late April 1982. But most of the major decisions made in the Senate had come earlier, beginning the previous fall, in negotiations among key senators and key committees. On September 24, 1981, Senator James A. McClure of Idaho, who had become chair of the energy committee following the 1980 elections, introduced a nuclear waste bill on behalf of himself and several other Republican senators, including Robert T. Stafford of Vermont and Alan Simpson of Wyoming, chairpersons of the Environment and Public Works Committee and its Nuclear Regulation Subcommittee, respectively.

The McClure bill (S.1662), the vehicle for the ensuing action in the Senate, was meant to modify and give statutory sanction to the emerging DOE–NRC program. It would establish a series of procedural steps for locating, exploring, and choosing waste disposal sites; strengthen the

[16]Palladino, who came to the NRC from Pennsylvania State University in 1981, was interviewed by the author on July 14, 1982. In announcing the confidence rulemaking in October 1979 the commission said that its purpose was "solely to assess generally the degree of assurance now available that radioactive waste can be safely disposed of, to determine when such disposal or offsite storage will be available, and to determine whether radioactive waste can be safely stored onsite past the expiration of existing facility licenses until offsite disposal or storage is available." *Federal Register* vol. 24, no. 108 (1979) p. 61373.

[17]Atomic Industrial Forum, "Historical Profile of U.S. Nuclear Power Development," AIF background paper (Bethesda, Md., 1986). Another fourteen reactor orders were cancelled in 1983 and 1984; there were no cancellations in 1985.

rights of the states with respect to siting decisions; speed up the siting and licensing schedule; and levy a user fee on the utilities to pay for the program. Almost from the beginning the utilities had been expected to pay the cost of waste disposal and to pass those costs on to consumers of electricity, but the McClure bill spelled out how this was to be accomplished: a 1 mill or one-tenth of a cent per-kilowatt-hour fee (subject to adjustment) would be imposed on all nuclear-generated electricity; the proceeds, amounting to about $200,000 for every ton of spent fuel discharged from a reactor,[18] or roughly $5 million per reactor per year, would go into a nuclear waste fund. The fund was expected to be an ample source of appropriations from year to year, a critical need for a repository siting and development effort which was certain to extend over a period of decades.

The imposition of a fee on nuclear electricity to fund the waste program was to be accompanied by an obligation on the part of the federal government to begin accepting the spent fuel or high-level waste from the utilities by a certain date, later fixed in the act as January 31, 1998. This commitment could compel the government to push ahead, whatever the difficulties, and establish facilities of one kind or another to receive the fuel.

The McClure bill, unlike the energy committee bill of the previous Congress, all but embraced the principle that away-from-reactor storage would be only a "last resort."[19] In the end this storage would be defined so narrowly that few utilities could qualify for it, and would promise to be so expensive that probably few would seek to qualify. Nevertheless, the bill's away-from-reactor provision was to be hotly contested. Senators and congressmen from states where an AFR facility seemed most likely to be established wanted no federal AFR program at all.

In introducing his bill, McClure spoke scornfully of the Carter administration's waste policy as a "blueprint for delay, study, litigation, eternal research, and more and more delay." Under that policy, he said, the earliest possible date for an operational repository would be the year 2007. His bill would provide for a "much accelerated" schedule, with issuance of a Nuclear Regulatory Commission construction permit by the end of 1987 and full operational status well before the turn of the century. A critical step would be the early identification and detailed

[18]Testimony of Thomas A. Cotton, director of the Office of Technology Assessment radioactive waste study, before the Senate Committee on Energy and Natural Resources and the Subcommittee on Nuclear Regulation, 97 Cong. 1 sess. (1981).

[19]As finally passed, the Nuclear Waste Policy Act of 1982 authorized DOE to establish AFR storage capacity not in excess of 1,900 tons. Stephen Kraft of the Edison Electric Institute told me, in an interview on July 7, 1983, that the demand for such storage was unlikely to approach this ceiling.

characterization of three sites from deep exploratory shafts. The president, on the recommendation of the Department of Energy, would eventually select one of the three for the first repository. McClure was going to allow about two years for this step, plus about three years for the NRC licensing proceeding. During this five-year period there was also to be a "test and evaluation facility" built for packaging, handling, and emplacing spent fuel. In McClure's view, a "successful assault" on the waste problem required only an aggressive "battle plan."[20]

The next step in the waste bill's evolution in the Senate came in the two days of hearings held jointly in October 1981 by McClure's energy committee and Simpson's Nuclear Regulation Subcommittee. Witnesses from all of the agencies most knowledgeable about geologic disposal, except those from the Department of Energy, said that the McClure bill's schedule for siting a repository and licensing its construction was far too compressed. The principal Department of Energy witness, Shelby T. Brewer, assistant secretary for nuclear power, applauded the schedule and told McClure that it could be met and that DOE would have only a few changes of a "minor technical nature" to suggest.[21] A much slower schedule, however, had been developed by the nuclear waste program's professional managers at the department, such as director Colin Heath and his deputy Critz George. Indeed, later in the month the department sent McClure detailed comments on S.1662 that were by no means of a "minor technical" nature. The Department of Energy was now saying that the accelerated schedule "could jeopardize" the chances of a repository being built by a fixed date by possibly reducing the amount and quality of geologic data available and by impairing DOE's working relationships with the host states and the Nuclear Regulatory Commission.[22] The disparity between Brewer's testimony and the subsequent department comments revealed a pronounced difference in attitude toward and perception of the realities of geologic disposal between the waste program managers, on the one hand, and top DOE officials on the other. At the joint hearings the witnesses from the U.S. Geological Survey, the Environmental Protection Agency, the Nuclear Regulatory Commission, and the congressional Office of Technology Assessment had also

---

[20]Press release from Senator McClure's office, excerpting the *Congressional Record* daily ed., September 24, 1981.

[21]*Joint Hearings on Nuclear Waste Disposal* before the Committee on Energy and Natural Resources and the Nuclear Regulation Subcommittee of the Committee on Environment and Public Works, October 5 and 6, 1981, pp. 210–236 (hereafter cited as *Joint Hearings on Nuclear Waste Disposal*).

[22]Letter, with attachment, from Robert G. Rabben, Assistant General Counsel for Legislation, to Senator McClure, October 16, 1981, reprinted in *Joint Hearings on Nuclear Waste Disposal*, pp. 678–683. See especially the comments on "Milestones," p. 680.

made it clear that the risks posed by S.1662's accelerated schedule were substantial, and that the adverse consequences could be grave.[23]

The report on S.1662 issued jointly in late November 1981 by the Energy and Natural Resources Committee and the Environment and Public Works Committee was no consensus document, for in the rush to get the bill through committee for Senate action before the end of the year some major questions had received scant consideration. Environment and Public Works, for instance, felt that the accelerated siting schedule had not been justified and was unrealistic. But the compromise bill the Senate was to eventually agree on would relax the schedule only slightly; any further easing of it would have to come in the House.

In its part of the report, Environment and Public Works called for assurances that the host state for the first repository would not receive all of the nation's spent fuel and high-level waste. A new Republican member of the committee, Slade Gorton of Washington, had recently conducted a field hearing at Richland, Washington, at which an important statement was presented on behalf of Governor John Spellman, who was insisting on an equal voice for his state in the repository siting process. He also was suggesting new policy directions: in his view, high-level waste repositories should be established on a regional basis, as was already the policy for low-level waste sites.[24] Senator Gorton, promptly acting on this advice, persuaded his committee to call for a second repository site to be chosen within three years of the selection of the first, and to compel completion of the second repository by limiting the tonnage of spent fuel that could be received at the first one before the second was in operation. This concept of enforced regionalization held obvious and important implications for the waste program, but occasioned little debate. The energy committee leadership found the Gorton provision "an easy amendment" to accept.[25]

The monitored retrievable storage concept was accepted by both the Energy and Natural Resources and the Environment and Public Works committees, but in very different ways. Senator Johnston had not been satisfied with the MRS provision that was in S.1662 as Senator McClure had first introduced it. Whereas the Johnston bill of the previous Congress had given more emphasis to monitored retrievable storage than to geologic disposal, the McClure bill had shifted the emphasis, stating that MRS "shall not constitute an alternative to the disposal of high level

[23]*Joint Hearings on Nuclear Waste Disposal,* pp. 236–248, 290, 320–324.

[24]The Spellman testimony was reprinted in *Congressional Record* daily ed., April 29, 1982, pp. 4255–4257. Governor Spellman was interviewed by the author in February 1981.

[25]Interview by the author with Senator Bennett Johnston on July 25, 1983.

waste or spent fuel in a repository."[26] Again, the Department of Energy was to submit plans for an MRS facility within a year, but with use of the facility—if Congress chose to build it—to be integrated with short-term away-from-reactor storage and geologic disposal.

Johnston persuaded the energy committee to rewrite the McClure bill to put monitored retrievable storage and geologic disposal on an equal footing. The committee referred to them as "parallel programs" and said that "any express or implied preference or priority" for deep geologic disposal over MRS was to be removed. But when the Environment and Public Works Committee rewrote the McClure bill, the MRS provision was made still less to Senator Johnston's liking: MRS was defined as interim storage to be co-located with either a reprocessing or geologic disposal facility, whichever came first, and the Department of Energy proposal for the first MRS facility would not be submitted to Congress until after construction of the reprocessing plant or repository had been approved by the Nuclear Regulatory Commission.[27] In the end, Simpson and his colleagues on Environment and Public Works gave in to Senator Johnston's demands to drop their bill's restrictions on monitored retrievable storage. There were other things, such as retaining the concept of an away-from-reactor facility as a last resort, to which they were giving a higher priority.

Negotiations for a compromise Senate bill were completed in late March of 1982 when the Armed Services Committee agreed to a Simpson amendment calling for military high-level waste to go in the same repositories as commercial waste unless the president could justify a repository for defense waste alone.[28] Initially, Armed Services had wanted military wastes excluded from the bill's coverage for fear of regulatory or other interference in the nuclear weapons program.

The McClure Bill on the Senate Floor

Senate bill 1662 reached the floor on April 28, 1982. Although the committee negotiations had smoothed the way for Senate passage, consensus on certain major issues was still lacking. A substantial but losing effort was mounted on the floor to give repository host states a stronger

[26]S.1662, sec. 501 of Title V, "Long-term Storage of High-level Radioactive Waste and Spent Fuel from Civilian Nuclear Activities," as originally introduced by Senator McClure on September 24, 1981.

[27]*Joint Report by the Committee on Energy and Natural Resources and the Committee on Environment and Public Works*, S. Rept. 97-282, 97 Cong. 1 sess. (1981) pp. 10, 24.

[28]See statements by Senator Simpson and Senator Henry M. Jackson, *Congressional Record* daily ed., April 29, 1982, pp. 4270–4273.

veto, or a veto that could be overridden only by a joint resolution of Congress. Senator Howard Cannon of Nevada, a state with one of the smallest congressional delegations and probably the most promising of the repository sites, led this effort, which in the end failed by a vote of 52 to 40.[29] Senators from three potential host states—Louisiana, Washington, and Texas—did not support Cannon. The Louisianans already had their promise of a veto from President Reagan. The Washington senators had won significant concessions on the regionalization issue in the compromise bill. In the case of the Texans, nuclear waste disposal was at that time a hot issue only in the remote west Texas panhandle. Several senators from states considered unwilling potential hosts for away-from-reactor facilities—namely, Illinois, New York, and South Carolina—also chose not to support Cannon. An aide to Senator Stennis of Mississippi had tried to promote a coalition among senators from the potential repository and AFR host states, but without much success. When an amendment to eliminate all away-from-reactor storage, proposed by senators from the potential AFR states, failed by a 3-vote margin,[30] there were eleven repository state senators voting against them.

Monitored retrievable storage was not challenged on the Senate floor, and obvious conflicts of interpretation over the ultimate purposes or implications of MRS were left unresolved. Senator Simpson spoke in favor of MRS, but stressed that such storage should serve only interim needs. The bill "makes it clear," he said, "that all high level waste and spent fuel must ultimately be disposed of in a repository."[31] Johnston held tenaciously to the interpretation that the bill put monitored retrievable storage and geologic disposal on parallel and, in his view, competing, tracks. "I am not sure that you could ever have the geologists agree as to the geologic characteristics of a salt dome or, indeed, of any other geologic formation," he told the Senate.[32] The "two tracks" approach taken by the pending bill represented, to him, a "great compromise."

Conflicting opinions also were expressed as to the best policy for siting geologic repositories. The two Washington senators, Slade Gorton and Henry Jackson, argued for regional repositories and against having the burden of disposal fall heavily on one state; but Senator Stennis of Mississippi argued for the "common sense" proposition of putting repositories "where the people are not," which is to say in the arid West.[33]

[29]*Congressional Record* daily ed., April 29, 1982, p. 4261.

[30]Ibid., p. 4288.

[31]*Congressional Record* daily ed., April 28, 1982, p. 4158.

[32] Ibid., pp. 4137–4138.

[33]*Congressional Record* daily ed., April 29, 1982, pp. 4255–4257, 4264; April 28, 1982, p. 4129.

An amendment relevant to the risks of nuclear proliferation and nuclear terrorism was offered by Senator John Glenn of Ohio, ranking Democratic minority member of the Governmental Affairs Subcommittee on Energy, Nuclear Proliferation, and Government Processes. Sponsored by Glenn and Senator Charles H. Percy of Illinois (the subcommittee chair), the amendment stipulated that non-nuclear weapons countries could receive U.S. technical assistance for spent fuel storage and disposal. Not controversial, the amendment was routinely agreed to in the Senate by voice vote. The nuclear industry could see a possible commercial advantage in the amendment; the prospect of reactor sales abroad could only be improved if potential foreign customers mastered their spent fuel management and disposal problems. From its own perspective, the Glenn-Percy subcommittee could see the technical assistance effort helping to steer some other countries away from the separation and recycling of plutonium. As the subcommittee sized up the situation, many leaders of foreign utilities and atomic energy agencies still believed reprocessing to be essential to nuclear waste management, when in fact waste management was simpler and easier without reprocessing. As one subcommittee insider expressed it to me privately, "We saw the technical assistance amendment as a Trojan horse. Our feeling was that once the time became riper, when the unfavorable economics of reprocessing became clearer, the direct geologic disposal option would be attractive."

The Glenn-Percy amendment had not, however, provided for the United States to accept even small or token amounts of foreign spent fuel for storage or disposal, as President Carter had called for in his nonproliferation policy. According to a subcommittee source, it had become evident during the previous Congress that such a provision would be strongly opposed. Some senators felt it would weigh down the already intensely controversial nuclear waste bill with more "political baggage." Others, strongly pro-nuclear and pro-reprocessing, would not want to lend credence to the view that reprocessing abroad was generally to be discouraged. Then, too, most environmental groups were expected to oppose any proposal that foreign spent fuel be shipped to the United States.

The Glenn-Percy technical assistance amendment was later to be routinely approved by the House and to survive as the waste legislation's only provision addressing nuclear waste problems abroad. As of the fall of 1986, however, the Department of Energy's offer of technical assistance had been limited essentially to countries still early in their development of nuclear power; no major programs had been undertaken with any country.[34]

[34]Interview by the author with Roger Gale of DOE's Office of Radioactive Waste Management, October 1, 1986.

Late the evening of April 29, 1982, the McClure nuclear waste bill was headed for final Senate passage, by a vote of 69 to 9.[35] The Department of Energy was satisfied with the measure, and industry lobbyists were delighted. But the environmental and anti-nuclear lobbyists were anything but pleased. They objected to the away-from-reactor and monitored retrievable storage provisions; to the burden put on host states to persuade at least one house to sustain a veto; to some significant exemptions and restrictions placed on National Environmental Policy Act procedures; and, not least, to the bill's accelerated siting and licensing schedule. Officials in the host states liked parts of the bill but wanted a stronger veto and a full NEPA review, particularly one that would offer an early comparative evaluation of alternative sites. Nonetheless, passage of the bill came routinely, without fuss or drama.

But had senators known the true nature of a "clarifying" amendment that Senator McClure slipped into the bill at the last minute, there would have been a donnybrook. According to this amendment, passage of the act was to be construed by the courts as satisfying any law requiring either the existence of a government-approved technology for disposal of spent fuel and high-level waste or the timely availability of storage and disposal facilities. It was meant to vacate the Nuclear Regulatory Commission waste confidence proceeding and the laws of California and a half-dozen other states that had imposed a moratorium on reactor licensing pending a solution to the nuclear waste problem. This clearly was no routine clarifying or technical amendment, but it was as such that McClure had asked the Senate to accept it, unread, on a voice vote.[36]

This episode ended badly for McClure. A *Washington Post* story on July 6, 1982 headlined "Idaho's McClure ... Is a Capitol Hot Potato," reported that McClure was losing the trust of his colleagues. Senator Patrick Leahy of Vermont, whose state was among those whose laws would be affected by the McClure amendment, had complained to the Democratic minority leader of a "serious violation of senatorial comity."[37] As one Senate aide told me in confidence, "The McClure amendment was dead as soon as it was passed," and it would indeed be among

[35] *Congressional Record* daily ed., April 29, 1982, p. 4325.

[36] The amendment is in Ibid., p. 4310. A waste confidence proceeding is carried out by the NRC "to assess generically the degree of assurance now available that radioactive waste can be safely disposed of, to determine when such disposal or off-site storage will be available, and to determine whether radioactive wastes can be safely stored on-site past the expiration of existing facility licenses until off-site disposal or storage is available." *Federal Register* vol. 44 (1979) p. 61372.

[37] Letter from Senator Leahy to Senator Robert C. Byrd, Democratic minority leader, June 8, 1982.

the first things dropped in negotiations with the House. But for McClure the worst was to come a year later. When a U.S. Supreme Court opinion upheld the California moratorium law, the House's refusal to accept the McClure amendment was cited as evidence that Congress did not favor federal preemption.[38]

The Ninety-seventh Congress: The House

Circumstances that gave the nuclear industry and the Department of Energy an advantage over the environmental lobby and the host states in the Senate were not present in the House. The Democratic-controlled House was generally less sympathetic than the Republican-controlled Senate to the positions taken by the Reagan White House and the department on nuclear waste issues. Moreover, whereas the Senate passed S.1662 in the early spring, the House committees were still struggling in late September to arrive at a compromise bill. The fact that the clock was running out on the Ninety-seventh Congress gave negotiating leverage to any committee or any member capable of causing delay. Four House committees—Interior, Commerce, Science and Technology, and Armed Services—were asserting legislative jurisdiction. Thus, four distinct bills would have to be reconciled and fitted together. Furthermore, the Rules Committee would have to be persuaded to grant a "rule" allowing the legislation to be brought up on the floor. The three Rules Committee members from potential away-from-reactor or repository host states—Butler Derrick of South Carolina, Gillis Long of Louisiana, and Trent Lott of Mississippi—would be in a position to get further states' rights concessions. Morris Udall and John Dingell, the chairs of the Interior and Commerce committees which had general jurisdiction over the waste issue, wanted the House to go further than the Senate in responding to host-state concerns. Dingell would side for the most part with Richard Ottinger, the environmentally oriented New Yorker chairing the Commerce subcommittee which was handling the bill. To one degree or another, Udall, Dingell, and Ottinger were all nuclear critics, with Ottinger the most critical of the three.

But set against the foregoing were circumstances working in favor of the Department of Energy and the industry. The House is not on the

[38]In his opinion for the Court in *Pacific Gas and Electric v. State Energy Resources Conservation and Development Commission,* Justice Byron White specifically noted that a manager of the nuclear waste legislation in the House had explained on the floor that the McClure amendment was deleted in committee "to insure that there be no [federal] preemption." White wrote that it would "appear improper for us to give a reading to the act that Congress considered and rejected." 465 U.S. 190 (1983).

whole a liberal body, and in the Commerce Committee lobbyists for the
department and the industry generally could count on at least a slender
bipartisan majority made up of Republicans and conservative (or "Boll
Weevil") Democrats. Heading this majority coalition were James T.
Broyhill of North Carolina, the committee's ranking Republican, and Phil
Gramm of Texas, leader of the Boll Weevils. Also in the industry's favor
was the fact that Udall, Dingell, and Broyhill all felt congressional action
on the nuclear waste problem to be long overdue. In the judgment of
one Department of Energy lobbyist, this was more important to the
ultimate passage of the waste act than any of the deals struck on particular features of the bill.[39]

Struggles in the Interior and Commerce Committees

The first legislative "markup" sessions by a House committee to review
and accept or reject the detailed provisions of the nuclear waste legislation came in the Interior Committee in the fall of 1981, and by late
April 1982 Interior had reported its bill. The Interior markup, though
by no means free of conflict, did not involve the intense infighting that
was to characterize the markup in Ottinger's Commerce subcommittee.
Interior had decided without much argument on a repository siting
schedule somewhat less accelerated than the one in the Senate bill, but
still very tight. The major milestones in the schedule were: detailed
characterization of at least three sites was to start in 1985; the president
was to submit to Congress in March 1987 the site chosen for the first
repository; and the Nuclear Regulatory Commission was to grant or
deny a construction permit for the first repository no later than 1992.
The projected operational date was "around 1995."[40] As in the Senate,
there was expert testimony before the House committees recommending that time be allowed for a more ambitious program of in situ testing,
for the identification of more sites before the first repository was actually
chosen, or simply for contingencies. But the schedule described here
was to become part of the bill finally enacted.

When policy disputes arose during Interior Committee discussions,
the lineup of members varied from issue to issue. Oddly, the important
proposition that large positive incentives or rewards should go to a
repository's host state and locality was almost a nonissue because only

---

[39]Interview by the author with Michael T. Kelley, formerly deputy assistant secretary
of DOE for congressional liaison, summer 1983.

[40]H. Rept. 97-491, pt. 1, House Committee on Interior and Insular Affairs, 97 Cong. 2
sess. (1982) pp. 30–31.

Morris Udall, the committee chair, seemed to advocate this. Specifically, Udall advocated a kind of bidding process. If three states all had qualified sites, each would be invited to draw up a package of conditions—including substantial economic benefits—under which the repository would be acceptable and on the basis of which DOE would negotiate.[41] But Udall's colleagues were not receptive to the incentives concept. "He never got any takers. The idea never went anywhere," says Andrea Dravo, the Udall aide who handled the waste bill.[42]

One reason for this was that the National Governors Association, after taking soundings among the potential host states, was advising Congress against offering economic incentives. Holmes Brown, the association's liaison with these states, says that while the question was not treated as a big issue in his meetings with state representatives, "their attitude was that this [an offer of incentives] would really hurt the credibility of public officials who would even tolerate the DOE siting program. It would look like they are after the money."[43]

In the Commerce subcommittee, where Ottinger was greatly outnumbered by the Republicans and Boll Weevils, Broyhill got his way without difficulty. But in full committee the votes were often exceedingly close, some ending in a tie. On behalf of away-from-reactor storage Broyhill gained the support of two Illinois Republicans, both AFR opponents, by agreeing that the country's three defunct commercial reprocessing plants at Morris, West Valley, and Barnwell should be excluded as potential AFR storage sites. Broyhill recruited an ally from another quarter by agreeing to support a monitored retrievable storage amendment by Louisianan W. J. "Billy" Tauzin if Tauzin would support away-from-reactor storage.[44] Among other things, Tauzin wanted co-location of MRS facilities and geologic repositories to be flatly prohibited. Another AFR vote was picked up when Broyhill struck a deal with Ohio Democrat Ronald Mottl, who would vote with Broyhill if Broyhill would intercede to have the Department of Energy give up the Cleveland salt mine research project.[45]

The old adage that there are two things best not directly observed, one the making of sausage, the other the making of laws, is particularly

---

[41]Udall described his concept of a bidding process in an interview with the author in January 1983 shortly after the Nuclear Waste Policy Act was passed. He believed that in due course such a process actually would be used. "It may take some fine tuning of the legislation," he said. "Right now it is not explicit in the act. But that's how it's going to work out."

[42]Interview by the author with Dravo, September 27, 1985.

[43]Interview by the author with Brown, October 3, 1985.

[44]Interview with Christopher Warner of the Broyhill staff, January 5, 1983.

[45]Interview by the author with Mottl, August 5, 1983.

apt with respect to the Commerce Committee deliberations over repository siting guidelines and population density criteria. In the Interior Committee, John F. Seiberling, Democrat of Ohio, knew that as recently as 1979 a Department of Energy contractor had identified part of the Cleveland–Akron metropolitan region as geologically promising, so Seiberling had the guidelines in Interior's bill written to keep repositories out of areas of even moderately high population density.[46] But in the Commerce Committee the population density issue presented itself differently. Trent Lott of Mississippi, the House Republican Whip and member of the Rules Committee, wanted a population criterion that would block construction of a repository in the salt dome just outside the town of Richton, population 1,100. The way Lott and other Mississippians such as Senator Stennis looked at the matter, for DOE even to consider siting a nuclear waste repository on the edge of a town, whatever its size, was an awful idea. But the "Richton amendment" which the committee first adopted was so poorly drafted that, had it become law, all of the sites DOE then had under investigation might be excluded. When this was discovered new language was offered to disqualify only the Richton site. The proposal failed on a tie vote, 21 to 21.[47] In the end, the population density criterion adopted as part of a compromise Interior–Commerce committee bill fell short of disqualifying the Mississippi site, and the Richton amendment became something of a laughingstock.

The very idea of a population test could be challenged. To disqualify sites near existing communities in favor of remote sites in deserted areas ignores geologic disposal's vast time dimension, for over the period of hazard, running to at least 10,000 years, areas now densely inhabited could be abandoned while some areas now remote could become thriving urban centers. But for all the clumsiness and apparent cynicism of his efforts to disqualify the Richton site, what Lott had been trying to do was to have the proposed waste act reflect the commonsense notion of his constituents that things which are thought to be dangerous should not be put next to or under a town, even a small one.

The Commerce Committee bill was reported in early August, and by mid September Commerce and Interior had produced a common discussion draft. But this draft did not command a solid consensus. The

---

[46]The Seiberling population density criterion would disqualify any site which, according to the latest census, was "located in any metropolitan statistical area, county, urbanized area, or place, having both (1) a population of not less than 2,500 individuals; and (2) a population density of not less than 1,000 individuals per square mile." H. Rept. 97-491, pt. 1.

[47]H. Rept. 97-785, pt. 1, House Committee on Energy and Commerce, 97 Cong. 2 sess. (1982) pp. 100–101. The tie vote resulted because Congressman Ottinger and others wanted the population test eliminated entirely, rather than rewritten to disqualify only the Richton site.

environmental lobbyists and certain host states were particularly unhappy with it. Whether the committees now had in hand a bill that could or should be enacted was in doubt.

Seeking Consensus: Concessions to Environmentalists

The day after the discussion draft was first circulated, six environmental groups (the Sierra Club and the Environmental Policy Center among them), plus a few allied organizations such as the Union of Concerned Scientists, wrote Congressman Richard Bolling, chair of the Rules Committee, to say that the draft failed to resolve or even address a number of major issues. Fault was found particularly with the exemptions from the National Environmental Policy Act and the failure to call for full environmental review of the screening of sites for characterization.[48]

Bolling took the environmentalists' objections seriously but with reservations. In his view the legislation was at an impasse because lobbyists for both the environmental groups and the nuclear industry were taking extreme positions. "The environmentalists want it perfect and would like to kill nuclear power," Bolling told me.[49] "The industry wants what will serve the interests of the stockholders, and to hell with everything else." This was, he conceded, an "oversimplification," but he stood by his basic point that the situation called for a moderating influence. In his judgment, Morris Udall was the one who could best bring about sensible compromise.

By now Udall, Dingell, and Ottinger wanted concessions made to the environmentalists and the host states. They got together with the ranking Republicans on Commerce and Interior, and an agreement was reached with surprisingly little difficulty.[50] One significant concession came in response to the demand by Governor Matheson of Utah and others for an early comparative evaluation of nominated sites. Another major concession was the agreement that the secretary of energy must make a "preliminary determination" that the three sites from which the final site is to be selected are all suitable for repository development. The environmental lobbyists saw this as a way of making sure that the sites investigated are all truly promising and likely to be licensable. These

---

[48]Letter to Bolling, September 15, 1982, signed by representatives of the Sierra Club, the Environmental Policy Center, Friends of the Earth, the Union of Concerned Scientists, Environmental Action, the Nuclear Information and Resource Service, the Natural Resources Defense Council, and the National Audubon Society.

[49]Interview by the author with Bolling, September 1982.

[50]Interviews by the author with Andrea Dravo, July and August 1983, and with Jeanine Hull, counsel of the House Subcommittee on Energy Conservation and Power, August 1983.

conceded provisions were to become part of the waste act. The wording of the latter provision was ambiguous, however, and DOE would invite litigation by choosing to interpret it as a mandate for preliminary determination before, rather than after, site characterization (see chapter 13).

Input by the Science and Technology and the Armed Services Committees

To the legislation worked out by the Interior and Commerce committees now had to be joined the provisions developed by the Science and Technology and the Armed Services committees. The bill reported by Science and Technology provided for a test and evaluation (T&E) facility, but the ever-suspicious host states viewed such a facility as a foot in the door for a repository. Ultimately, the principals from Science and Technology, Interior, and Commerce agreed that surface facilities for a T&E facility could not be built at a candidate repository site until the Nuclear Regulatory Commission has authorized construction of the repository itself.

The Armed Services Committee, to everyone's surprise, had decided that all state participation provisions should apply if the Department of Energy moved to develop a separate repository for military waste. Included would be the right to a strong veto, or one that only a joint resolution of Congress could override. Some senior members of Armed Services were aghast at this, still viewing the prospect of any state control over a defense activity as intolerable. But a conservative Mississippi Democrat, G. V. Montgomery, whose district lies near the Richton salt dome, had joined with a Philadelphia liberal, Thomas M. Foglietta, to have this states' right provision overwhelmingly approved.[51] The Armed Services amendment was readily accepted by the Interior and Commerce committees.

For the waste legislation to make it through the Rules Committee, Udall, Broyhill, and the other principals had to accommodate the demands made by Rules members Long and Derrick—and supported by chairman Bolling—who wanted the states' rights provisions pertaining to the siting of repositories and AFR facilities to be strengthened. The compromise Interior–Commerce bill would have given the host states not a strong veto but a weak one, requiring sustaining action by either the House or the Senate. But now Long and Derrick were able to have the bill include a strong veto for any state asked to host either a geologic repository or a major away-from-reactor facility.[52]

[51]Interview by the author with Celane McWherter, aide to Congressman Montgomery, January 17, 1983.
[52]Interview with Andrea Dravo, August 4, 1983.

On September 23, Bolling, realizing that time was running out for the Ninety-seventh Congress, asked Udall and the other principals to prepare a finished bill without further delay. This was done by the next morning, and on September 28 the bill was reported from Rules to the floor.[53] It was not to come up for major House action until Congress reconvened after the November elections.

The Environmentalists Abandon the Bill

The House bill—H.R.7187—had now been considerably improved from the standpoint of states' rights and environmental protection, as some of the leading environmental lobbyists have acknowledged.[54] But by mid November all of the environmental groups lobbying on the waste issue had decided that the effort to pass a bill in the Ninety-seventh Congress should be abandoned and taken up again the next year. The Sierra Club and ten other groups, mostly national environmental organizations, sent a letter to Udall, Dingell, and House Speaker Thomas P. O'Neill, Jr., on November 19, 1982. They said that the weaknesses of H.R.7187 would be compounded by the necessary reconciliation with the Senate bill, which, they added, "falls short of even minimal acceptability in many critical areas." There was concern that Udall and the other House leaders wanted a waste act so much that they might accept one at almost any price.

The environmental lobbyists objected principally to the House bill's provision for a hurried and constrained monitored retrievable storage study, its failure to ensure a common program and a single set of rules covering both defense and commercial wastes, and its provisions for an unlicensed test and evaluation facility and AFR storage program. As events were to show, the time pressures of the short lame-duck session could be made to work more for the environmentalists than against them.

---

[53]"Comprehensive Nuclear Waste Plan Enacted," *1982 CQ Almanac* (Washington, D.C., Congressional Quarterly, 1983) p. 309.

[54]In a letter published in the May 1983 issue of Friends of the Earth's *Not Man Apart*, David Berick of the Environmental Policy Center and Brooks Yeager of the Sierra Club said that negotiations conducted with Udall, Dingell, and Ottinger had led not to weakening changes in the bill—as an article in the January 1983 issue had stated—but to "major improvements," which they proceeded to cite. The letter also said that the earlier article had misinterpreted the bill in various ways, by suggesting, for instance, that environmental assessment would not be reviewable in court. But *Not Man Apart*, in an editor's note in the May 1983 issue, put its finger on the ambiguity in Berick's and Yeager's position: "We ... find it curious that the two writers so vigorously defend a bill that their organizations eventually opposed."

Congress Watch had been the first to express its disillusionment with the legislation. It had broken ranks with the other groups back in mid September, before the final House negotiations, by dismissing all pending legislation as defective beyond remedy.[55] Later, Friends of the Earth, which had been occupying an uneasy middle ground in the testy relationship between Congress Watch and the other groups, also abandoned the effort to pass a bill. It did so along with a half-dozen local or regional groups such as New York City Audubon, New York SANE, and an Illinois group called Pollution and Environmental Problems, Inc.[56]

Udall unwittingly increased the pressures on the environmental lobby to take a harder line by claiming, in a floor statement on September 30, that the new compromise bill had the support of "major environmental groups," together with that of the Reagan administration, the nuclear industry, and the host states. Udall genuinely believed that he had now met the environmentalists' concerns, but his statement caused no little discomfort for some of the lobbyists. David Berick of the Environmental Policy Center, Brooks Yeager of the Sierra Club, and some of the other lobbyists were beginning to think that the environmental lobby should strongly express its dissatisfaction with the bill, and Berick also felt that the environmental lobby's position was fuzzy.[57]

This attitude was reinforced in October when some grass-roots environmental and anti-nuclear activists in the Northeast wrote a letter to the Washington lobbyists, telling them, in effect, that they were out of step and out of touch. The letter was prepared at a meeting of people from a variety of public interest research groups as diverse as the Council on Economic Priorities, the Vermont Yankee Decommissioning Alliance, and the Sierra Club's New York-based Radioactive Waste Campaign. Mina Hamilton, a Sierra Club activist, had called the meeting at Blue Mountain Lake in the Adirondacks in response to news that spent research reactor fuel from Chalk River, Ontario, would soon be shipped to South Carolina for reprocessing at the Savannah River plant. The State of Michigan had blocked such shipments, so they would now be coming through the Northeast. Hamilton says the "Dear Washington Lobbyists" letter "sprang out" of the meeting spontaneously.[58] The activists were disturbed to see environmental lobbyists in Washington cooperating in the development

[55]Congress Watch, "Nuclear Legislation: Now or Next Year," paper dated September 14, 1982.

[56]Friends of the Earth and the other groups sent House Speaker Thomas P. O'Neill, Jr., a mailgram on November 8, 1982, urging him to withdraw and rewrite the pending bill. A Friends of the Earth press release of November 9 noted that the groups were now making common cause with Congress Watch to oppose the existing legislation.

[57]Interview by the author with Berick, September 2, 1983.

[58]Interview by the author with Hamilton, August 26, 1983.

of legislation that would provide for at least some central AFR storage even though this meant fuel elements would have to be shipped twice, first to the AFR facility, then later to the place of final disposal. The letter said, "We are concerned that there is a dire need for a greater communication between Washington lobbyists and safe energy groups in the country." Its signers were opposed not only to away-from-reactor storage, but to all forms of at-reactor storage as well. "We support the rapid phase out of all nuclear reactors as the solution of the problem of irradiated fuel build up at reactors," they said.

Here was a classic conflict. If the Washington lobbyists had adopted the activists' absolutist positions they would have lost their place at the negotiating table. But for the activists, conflict and confrontation with the Department of Energy was the weather of their lives. They distrusted the department and doubted that the legislation would offer the benefits claimed. Mina Hamilton, for instance, felt the bill's siting schedule was "preposterous." As she saw it, the states, although offered an oversight role, would never catch up with DOE decision making.

Debate on the House Floor

The waste bill was taken up again on November 29, and passed on December 2, 1982, after three days of debate.[59] The compromises and understandings reached in September had proved durable. Only two major substantive amendments were adopted, Udall's on the monitored retrievable storage program and Broyhill's on the state veto provision. Both were controversial, but in neither case did the outcome occasion much surprise.

The first important business was the monitored retrievable storage amendment, allowing the Department of Energy up to five years instead of only two to submit a plan for such storage, and eliminating the restrictions on National Environmental Policy Act review and on the scope of Nuclear Regulatory Commission licensing proceedings. Udall stated that MRS could serve as a back-up if technical or institutional failure to develop permanent deep geologic facilities left the country without a better answer to the waste problem. But he argued that to have the DOE study and plan was not urgent, "since an MRS decision will not have to be made until well into the next decade." He indicated that inasmuch as MRS facilities could become "indefinite repositories," unrestricted NEPA and licensing reviews were needed. To Ottinger,

[59]*Congressional Record* daily ed., November 29, 1982, pp. 8523–8552; November 30, 1982, pp. 8580–8598; December 2, 1982, pp. 8777–8800.

monitored retrievable storage was "the way to go," but it should not be
hurried and its procedural safeguards should not be truncated. Markey
predicted that unless geologic disposal remained the centerpiece of the
nuclear waste program, without competition from MRS, it would be
displaced by the latter. "The political will to accomplish the goal of
pressuring a state to accept the nation's nuclear waste will end,"
he said.[60]

Most of the members speaking against the Udall amendment were
from potential repository states. The nub of their argument was stated
by Tauzin: the amendment would stretch out the Department of Energy
study too long, denying the country the timely availability of the monitored retrievable storage option, and forcing a choice between proceeding with deep geologic disposal despite questions that remain
unanswered or shutting down the nuclear reactors generating the waste.[61]
But in the end Udall prevailed, his amendment winning approval almost
routinely, on a voice vote without a roll call.

The key vote on the Broyhill amendment to require the support of at
least one house for a state's veto of a repository site was a close one,
coming out 190 to 184 in Broyhill's favor.[62] Although Udall spoke against
the amendment in keeping with the position of his committee's majority,
he saw no "great cosmic differences" between a state veto that stands
absent a joint House–Senate resolution overriding it and a veto that
must be sustained by one house. But Montgomery of Mississippi said
that a repository host state faced a "tough time" at best, with "forty-nine
other states ... ganged up against it."[63] There was also concern among
environmental lobbyists and the host states that a veto which only one
house could sustain might be found unconstitutional, in which case host
states would have no veto at all.

Trent Lott's Richton amendment got little support. Udall argued that
adoption of amendments of this kind, capable of eliminating specific
sites, would lead to "paralysis," with the waste program unable to move
in any direction.[64] But an amendment by Marriott of Utah to have
transportation requirements "considered" in repository siting was accepted.[65] As Marriott knew, most of the reactors that generate waste are

[60]*Congressional Record* daily ed., November 29, 1982, pp. H8525–H8528.
[61]Ibid., p. H8530.
[62]Before final passage of the bill the Broyhill amendment was again put to a vote, and
it was approved this time 213 to 179. This vote came anticlimactically and without debate
on what was essentially an already settled issue.
[63]*Congressional Record* daily ed., November 29, 1982, pp. H8549–H8551.
[64]Ibid., p. H8540.
[65]Ibid., p. H8538.

located in the eastern half of the country. An aide to one of the Mississippians later confided to me, "While we were trying to shove the repository west, they [the Westerners] were trying to add all the secret code language to shove it east, and they did pretty well at it."

Final House passage of the bill came anticlimactically, on a voice vote. In September Udall had described the House bill as a "delicate fabric of agreements." This fabric would now be tested in the negotiations with the Senate.

House–Senate Negotiations

To reconcile the differences between the House and Senate bills meant accommodating host-state senators who could use the Senate rules to block final action, in view of the lateness of the hour. Of the two bills, the one passed by the House went further in protecting host-state interests. Given the perception that the Senate bill favored the Department of Energy and the industry and the fact that the severe time constraints of the lame-duck session gave the host states and their defenders in Congress a tactical advantage, no waste bill was going to be enacted unless Senator McClure made some concessions.

"The House had done an excellent job in putting us in a box," Keith Glaser, aide to Senator Hart, later told me, not unhappily.[66] "Their timing, deliberate or not, was perfect. They had forced us to negotiate without going to conference." James Curtis of the Nuclear Regulation Subcommittee staff had sized up the situation the same way. "It was clear that we would have to take the House bill, amend it, and send it back," he told me.[67] The Senate principals would have to treat the House bill as the basic legislative vehicle, modifying it through negotiations with other senators and with the House principals, then asking the Senate and the House to accept this prenegotiated product. "We went through a number of drafts," Curtis said. Senators such as Strom Thurmond of South Carolina and Daniel P. Moynihan of New York, who earlier had seen the defunct reprocessing plants in their states as prime prospects for an away-from-reactor facility, now appeared to be satisfied, but those from the potential repository host states remained uneasy.

A Senate draft was sent over to the House on December 7 and discussed with Udall, Ottinger, and other principals a few days later. An agreement seemed possible, but a number of issues remained to be worked out. House and Senate principals met late on the afternoon of

[66]Interview by the author with Glaser, January 6, 1983.
[67]Interview by the author with Curtis, January 6, 1983.

December 14, and by about eight o'clock in the evening they had an agreement and a bill. Senator McClure, a patient and conciliatory bargainer when the situation demands it, had come into the meeting determined to achieve agreement. The bill on the table was basically the House bill, and the issues discussed generally raised no threat to the House consensus, such as it was.

Some of the issues, for all that, were of substantial importance.[68] For example, McClure won agreement that when the federal government took title to spent fuel it would be at the reactor sites, thus relieving the utilities of the responsibility of transporting the fuel to a federal storage or repository facility, with all the possibilities for political hassles and damage claims which that might involve. Senator Gorton got the agreement of the House principals for his provision (already in S.1662) for a tonnage limit on spent fuel or high-level waste at the first repository pending the opening of a second repository.

The only direct challenge to a major House position came when Senator Johnston insisted that the monitored retrievable storage and geologic disposal programs still must move on parallel tracks. The compromise finally agreed to by the House principals split the difference between Johnston's one-year MRS study and Udall's five-year study by specifying a two-and-a-half-year study. The restrictions on the National Environmental Policy Act and licensing reviews were left unchanged.

Last-minute Opposition

Not until December 18, four days after the meeting of the Senate and House principals, did Senator McClure distribute a memorandum describing, none too clearly, what had been decided. So late in the session, McClure needed the support or acquiescence of the host-state senators. Only the three who were members of the sponsoring committees—Johnston of Louisiana and Jackson and Gorton of Washington—had been present at the crucial December 14 meeting. Some senators were feeling cut out of the action on a matter vitally affecting their states. One of these was William Proxmire of Wisconsin, who was threatening to filibuster. In his twenty-five years in the Senate, Proxmire had kept aloof from the Senate establishment or club, with its emphasis on quiet accommodation, and if any senator could mount a credible filibuster threat it was he. A filibuster can be either a grandstanding symbolic political gesture or a lever for the attainment of a limited and achievable goal. In

[68]The results of the negotiations of December 14 were described, after a fashion, in a memorandum of December 18, 1982, from Senator McClure's Energy and Natural Resources Committee to all Senate legislative aides.

the present instance, Proxmire was demanding as his price for not tying up the Senate and killing the waste bill the strong state veto which the House had eliminated by adopting the Broyhill amendment—a veto which would take both houses of Congress to override.

Proxmire had arrived at this states' rights position with strong encouragement from Wisconsin's Radioactive Waste Review Board, an entity established by the Wisconsin legislature in late 1981.[69] Chaired by a powerful member of the legislature and supported by Governor Lee S. Dreyfus (a Republican), this board could speak with the full authority of the state. It had adopted a detailed policy with respect to what a national nuclear waste policy act should contain. Getting a stronger state veto, subject at most to being overridden by a joint resolution of Congress, was the keystone of that policy. Nothing could do more to put the host states on an equal footing with the Department of Energy, or so it was believed. Rightly or wrongly, Wisconsin officials felt that DOE had in the past—in taking steps to study Wisconsin's granite formations—treated the state as anything but an equal partner.

On Saturday, December 18, the showdown was at hand. All week the Senate had been tied up with other business, but now, if McClure could reach an understanding with Proxmire and other host-state senators, the waste bill might be passed during the few days remaining before adjournment. To ratchet up the pressure, Ruth Fleisher and two other Proxmire aides entered the Senate chamber each carrying an armload of nuclear waste discussion material, including reports from the Wisconsin Radioactive Waste Review Board. If a filibuster should be necessary, Proxmire would not be at a loss for words.[70] That evening Charles Traybandt, the McClure committee's chief counsel, asked Fleisher whether Proxmire was still willing to remove his hold on the bill and not filibuster if his state veto amendment was accepted. "I told him, yes," she says.

Late that day Senator Stennis, the Armed Services Committee's ranking Democrat and former chairman, had an Air Force jet dispatched to

---

[69]Interview by the author with Robert Halstead of the Wisconsin state energy division, August 23, 1983.

[70]Interviews by the author with Fleisher on January 6 and August 23, 1983. A week before the final showdown, Ruth Fleisher, Proxmire's aide on the waste issue, had called on Charles Traybandt, the McClure committee's chief counsel. According to Fleisher, Traybandt had earlier accused her of initiating the highly critical *Washington Post* story of July 1982 about McClure's sneak clarifying amendment—an accusation that she has neither confirmed nor denied. (See Ward Sinclair, "Idaho's McClure, Head of Energy Panel, Is a Capitol Hot Potato," *Washington Post*, July 6, 1982.) Now Fleisher was telling Traybandt that either McClure would agree to a stronger state veto or Proxmire would filibuster and kill the bill. Fleisher left confident that McClure would come to terms. Of all the congressional aides who worked on the nuclear waste legislation, Traybandt alone refused to be interviewed, despite the author's repeated requests.

Jackson, Mississippi, to pick up Ronald J. Forsythe, a state nuclear waste specialist and personal emissary of Governor William Winter.[71] Forsythe was briefed on details of the pending legislation by aides to Senators Stennis and Thad Cochran and Congressman Trent Lott, then escorted to a meeting with McClure. Stennis, Cochran, and Senate Majority Leader Howard Baker of Tennessee were also present. Although McClure agreed to no major concessions at this meeting, Forsythe may have helped Senator Proxmire play his hand. Most of that week Forsythe had been at Las Vegas for the Department of Energy's annual waste program review meeting. Forsythe and representatives from other host states had spent hours discussing common complaints about relations with DOE. By giving McClure at first-hand the flavor of these discussions, Forsythe may have helped convince him that feelings in the host states were indeed running deep and that Proxmire's demand for a stronger state voice should not be perceived as an isolated gesture by a maverick senator.

Final agreement for a bill was struck at a meeting that began about noon Sunday, December 19, in the vice-president's office in the Capitol. McClure, Proxmire, and about ten other senators were there. McClure agreed to Proxmire's state veto amendment and made a few minor concessions to other senators.[72]

Meanwhile, Congress Watch and its parent group, Public Citizen, as well as Congressman Markey and his staff, were trying to upset the agreement and kill the bill. They wanted Proxmire to filibuster—although Fleisher told them that Proxmire was sticking to his deal with McClure. Later, Fleisher was to accuse Congress Watch of being absolutist and wanting perfection. "They didn't seem to care about the states and the Great Lakes. Their motivation, frankly, was to cast a blow that would hurt nuclear power," she says.[73] Joan Claybrook of Public Citizen and Congress Watch recalls this encounter only vaguely. Congress Watch was out to kill the bill all right, she says,[74] but for substantive reasons having to do with the legislation itself and not for the purpose of striking at nuclear power. One reason was that the bill did not apply to military as well as commercial waste, she told me, but the other shortcomings she could no longer recall.

As the meeting in the vice-president's office was ending, across the continent in Las Vegas environmental and anti-nuclear activists attending the last session of the Don't Waste America Conference were trying to

[71]Interview by the author on September 29, 1986, with Forsythe, now nuclear waste program manager in the Mississippi Department of Energy and Transportation.

[72]Interview by the author with Ruth Fleisher, August 23, 1983. McClure's formal acceptance of the Proxmire amendment came December 20 on the Senate floor.

[73]Ibid.

[74]Interview by the author with Claybrook, September 11, 1983.

decide on a message to send the Congress about the waste bill. Robert Halstead, a representative of the Wisconsin energy division, who had stayed over from the DOE meeting earlier in the week, got word from Ruth Fleisher that McClure had accepted the Proxmire amendment. He told the Don't Waste America meeting that with the stronger veto now assured, the bill had his state's support. "Our primary concern was to protect the communities who are going to have to eat those sites," Halstead says,[75] and this was essentially his message to the conference. "It was a luxury we couldn't afford to hold the nuclear waste bill hostage to the future of the nuclear industry," he told me. But the suspicion and distrust at the conference were too deep for the activists present to follow Halstead. The consensus of the meeting was that the bill was, on balance, still not acceptable.

Enactment

At noon on Monday, December 20, the waste bill was taken up in the Senate under a unanimous consent agreement, and after fifteen minutes of discussion was passed by voice vote. Senator McClure said, apropos of the Proxmire amendment: "I think we can make it work, and I think, as a matter of fact, it is greater protection for the states."[76] He noted that a "consensus approach" to waste management had been successfully fashioned and that the legislation "carefully balance[d] the constitutional, equitable, legal, and practical interests" of all parties.

It was up to the House to complete action on the bill. With the Ninety-seventh Congress winding down, the members of the House Rules Committee were beginning to scatter. When a quorum was finally assembled, Speaker Tip O'Neill sent one of his aides into the closed-door meeting— a sign that the Speaker wanted to move the bill. It went to the floor on a closed rule, no amendments allowed. Enactment was now a foregone conclusion. The House passed the measure 256 to 32, the nay votes coming from an assorted lot of members, ranging from the anti-nuclear Markey to those still worried about the states having a veto over repositories reserved for military waste.[77]

President Reagan signed the Nuclear Waste Policy Act of 1982 at the White House on January 7, 1983. The signing ceremony was well attended by the members of Congress and staff who had worked on the bill, by nuclear industry leaders and lobbyists, and even by environmental lobbyists Berick and Yeager (who still viewed the bill critically,

[75]Interview by the author with Halstead, August 23, 1983.
[76]*Congressional Record* daily ed., December 20, 1982, p. 15651.
[77] Ibid., pp. 10524–10525.

however). The act was at best a rough blueprint for an effort that would have to continue over more than two decades and that would almost certainly require some major mid-course changes to succeed. The president, however, was sanguine in his comments. "The Act," he said, "provides the long overdue assurance that we now have a safe and effective solution to the nuclear waste problem." He described the act as the culmination of a long legislative effort that "clears the barrier that has stood in the way of developing nuclear power as a vital national energy resource."[78]

What Had Been Achieved

The stated aims of the act represented a clear commitment to the permanent geologic disposal of nuclear waste, to begin before the end of this century. But set against that goal and against the hard-won lessons coming out of the experience in the repository host states, how promising and efficacious did the specific terms of the act appear?

Commissioner John F. Ahearne of the Nuclear Regulatory Commission would allude to some of the problems and uncertainties inherent in the act when the NRC, some months thereafter, issued a preliminary finding of "confidence" that the waste problem would be met. In his view, absent the waste act, the commission could not have found confidence.[79] The question now, he suggested, was could the act carry so heavy a burden? He was worried about continued congressional support because of the act's "ambiguities" and possibly unachievable deadlines, the constituent pressures that were sure to arise, and the act's requirement that Congress "revisit" such major controversial issues as monitored retrievable storage and state objections to repository siting decisions.

The act could be seen in light of the still emerging experience in the host states as an unusual—and in some respects self-contradictory—mix of strong and weak points. A particularly clear strength was the per-kilowatt-hour fee to be levied against the utilities for the new nuclear waste fund. The act's consultation and cooperation provisions, profiting from the ups and downs of Department of Energy experience in New Mexico, could also be a strength, but only if the department and the host states could find a common interest in having the repository siting

---

[78]*Weekly Compilation of Presidential Documents* vol. 19, no. 1 (1983).

[79]Nuclear Regulatory Commission, draft decision on the confidence rulemaking on the storage and disposal of nuclear waste, May 16, 1983, "Additional Views of Commissioner Ahearne."

succeed. The waste fund, whose revenues could be increased as needed by raising the per-kilowatt-hour fee, was a potential source of generous benefits or rewards for the state and locality where the repository would be built.

A striking contradiction in the act was that while potential host states were given the assurance of a rational, participatory siting process, this assurance was then effectively denied, at least in the case of the potential hosts for the first repository. In particular, the act called for guidelines for the evaluation of sites against a variety of disqualifying or potentially favorable and potentially adverse conditions. But the Department of Energy also was directed to notify host states of the presence of "potentially acceptable sites" within ninety days of the act's taking effect, or well before siting guidelines could possibly be adopted.

Congress had in a sense "grandfathered" the sites in the department's existing inventory—sites which the host states had no voice in selecting, and which presented major technical and political problems and conflicts. This glaring contradiction in an act that promised a rational and participatory process made it certain that the Department of Energy's efforts to implement the measure would encounter strong host-state opposition from the start. On the other hand, even if new sites were to be selected and state participation were assured (as in the case of sites for a second repository), host-state resistance would still be intense should there be a perception of sharp conflict, high risk, and little or no benefit. The act might be politically impossible to implement, as was indeed to be the case (see chapter 13). As one might later conclude, Congress had attempted too much, having mandated controversial site screening efforts of wide scope east and west.

But one might also conclude that Congress had attempted too little, having given essentially no attention to the worldwide dilemma of how nations might dispose of growing accumulations of spent fuel without throwing the fuel away. Resolution of this dilemma may lie partly in coupling deep geologic isolation of spent fuel with an option for retrieval if reprocessing and breeders should ever be prudent and make economic sense. In the Nuclear Waste Policy Act Congress actually provided for a retrieval option, but without giving it the emphasis and clarity that seem to be necessary to make the bureaucracy respond effectively.[80] It is to

---

[80]An option for retrieval, for reasons of public health and safety or for resource recovery, is called for in Section 122 of the act. In an interview on September 10, 1986, Andrea Dravo told me that this provision was first written into the act by the House Interior Committee, some of whose members felt that a repository could serve a monitored retrievable storage function for an interim period. Despite its important implications for

the global dimensions of the waste management and spent fuel problems, as seen through the experiences of Europe, Japan, and the international community, that I now turn.

---

repository siting and design, the retrieval provision received little if any mention on the House or Senate floor and is an obscure part of the act. Moreover, although the act can easily be interpreted to call for a practical system of ready retrieval, this has been ignored in Department of Energy and Nuclear Regulatory Commission regulations. See Department of Energy, *Waste Management Systems Requirements and Descriptions (SRD)*, DOE/RW-0063 (Washington, D.C., January 1986) especially p. 49; Nuclear Regulatory Commission, "Disposal of High-Level Radioactive Wastes in Geologic Repositories Technical Criteria," 10 CFR, pt. 60, *Federal Register* vol. 48, no. 120 (June 21, 1983) pp. 28197, 28198.

Part 3
Europe, Japan, and the International Waste Problem

Introduction to Part 3

Like the Americans, the Europeans and Japanese launched nuclear de-velopment on a commercial scale without the means to dispose of the radioactive waste to be generated. Such means are still lacking today. In certain nations, repository siting faces especially severe geologic and political constraints. Japan, noted for its earthquakes, volcanic activity, and high population density, is perhaps the clearest case in point. The United Kingdom, as a small, wet, densely populated island nation that has made serious mistakes in its past handling of nuclear wastes, also has proved to be an unsympathetic political environment for repository projects. Then there are such small nations as Austria, Belgium, the Netherlands, and Switzerland: they face a variety of constraints, geologic or political (if not both), and for any one of them to invest hundreds of millions of dollars in a repository for the waste from one to fewer than a half-dozen reactors makes little economic sense.

To see how Europe's leaders in nuclear power and technology have fared and are faring with their nuclear fuel cycle and radioactive waste problem, one should turn to France, Germany, the United Kingdom, and Sweden. France has by far the largest nuclear program of the four, but each of them has been importantly committed to fission energy.[1] The Swedish program is the smallest, but as measured in generating capacity it is only slightly behind the British program, and on a per capita basis Sweden's commitment to nuclear power has been greater than that of any country in the world. Sweden is of interest also because of its policy—initiated by a somewhat ambiguous referendum decision—to phase out nuclear power entirely in the next century.

[1]Total nuclear generating capacity in 1985 for these countries follows (with the capacity expected in the year 2000 shown in parentheses): France, 35.6 gigawatts (61.2); Germany, 16.4 (29.3); the United Kingdom, 10 (13.6); Sweden, 8.4 (9.4). Pacific Northwest Laboratory, *International Nuclear Fuel Cycle Fact Book* (Richland, Wash., 1985).

France, the United Kingdom, and Germany stand out as well because of their past and present commitments to commercial reprocessing of spent fuel and, sooner or later, reuse of the separated plutonium. Sweden, by contrast, stands out because of its firm commitment to disposing of most of its spent fuel as unreprocessed waste in a geologic repository. The influence of fuel cycle choice on the waste strategy in each country has been important. Although officially committed to reprocessing, the Germans have not said no to direct geologic disposal of spent fuel as emphatically as the French; the Germans have left the door ajar.

Each of the four countries represents a distinctive cultural, institutional, and political environment. Furthermore, as they have attempted to come to grips with questions about plutonium and radioactive wastes, circumstances and events have led each of them to follow its own, quite different course, often for reasons not readily explained. Why, for instance, have the United Kingdom's political leaders bowed to local protests and abandoned field studies for the siting of high-level waste repositories, when French leaders have in most cases pushed through the siting of nuclear projects even when faced with spectacularly large demonstrations and unanimous opposition by local officials? Institutional differences can be pointed out, but the outside observer may in the end be reduced to saying, "The French are the French, and the British are the British."

But a look at the British experience from the standpoint of the nuclear imperative to contain radioactivity may shed light on why British public opposition to the siting of waste facilities is as widespread and apparently unremitting as it is. In Britain the policy was established years ago to release to the environment low-level reprocessing wastes which contain biologically active species such as cesium and plutonium, with half-lives in the intermediate to very long-lived range. Whether of any consequence to the public health and safety or not, this practice, long in controversy and now to be reduced to very low levels, is one for which the British nuclear industry is paying a significant political penalty. Yet, as will be noted, the British practices never defied or flouted the prevailing wisdom of recognized national and international scientific bodies concerning radiation protection.

Successful national programs in repository siting and development now appear possible in one or more of the four European countries. Success would make more promising any subsequent efforts at establishing international programs. But as the French and German experiences suggest, the present drift toward use of plutonium fuel in many European light water reactors poses genuine and not easily calculated risks for the public, and shows that a nuclear industry which has been ahead of itself all along is about to take yet another premature step.

7
The United Kingdom: Problems of Containment

The United Kingdom's nuclear waste problem can be best examined by looking at British Nuclear Fuels Limited's reprocessing center on the Irish Sea not far south of the Firth of Solway, the broad estuary that separates northwest England from the south of Scotland. All of the spent fuel from British reactors—and hence nearly all of the radioactivity—fetches up at this center. Not only do several major kinds of waste arise here, but the greater part of the collective dose of radiation from commercial nuclear power to which the British public has been exposed has come from here.

The River Calder flows through the center, with the Windscale fuel reprocessing facility just to the north of the river and the Calder Hall power reactors just to the south. Construction of nuclear facilities at this site began in the early postwar years with the original plutonium-production reactors, or "piles." In 1956 the first of the Calder Hall reactors came on line and produced the first nuclear power for the commercial grid. From that time on the center assumed a key role not only in Britain's nuclear weapons program, but also in its emerging nuclear energy program, and especially in the reprocessing of fuel from magnox reactors.[1]

---

[1]Today there are eleven magnox reactors. This reactor takes its name from its magnesium-alloy-clad natural uranium fuel. Without reprocessing and the recycling of the remaining fissile U-235 isotopes, the magnox would be a grossly inefficient uranium burner because of the fuel's low burnup or irradiation. Even with uranium recycling as part of the magnox program, this reactor, with its low power density (or large size in relation to the power produced), is considered obsolete. Britain's second-generation reactor is the advanced gas-cooled reactor (AGR), of which there are eight; the burnup of the AGR's slightly enriched uranium oxide fuel is greater than that of the magnox fuel, but much less than that of light water reactor fuel. The Central Electric Generating Board proposes to build Britain's first light water reactor in Suffolk at Sizewell. In the early fall of 1986 the fate of the Sizewell project, which is controversial on economic and other grounds, was still undecided, but the report from a public inquiry was expected at any time.

Until 1981 the overall complex was known in England and the rest of the world by only one name: Windscale. But in March of that year British Nuclear Fuels Limited (BNFL) renamed the site Sellafield, for a village nearby. Although BNFL officials say this interpretation is mistaken, the change was widely viewed as being inspired by a desire to make a clean start at Sellafield.[2] For over the previous twenty-five years Windscale had acquired a troubled and sullied reputation—a place known for bad news.

Windscale's bad news began when a plutonium-production reactor caught fire in 1957. An estimated 20,000 curies of iodine-131 escaped up the stack. This was, and is, the largest accidental release of radioactive iodine from a reactor to have occurred in any of the democratic industrial societies.[3] It led to the temporary banning of milk production by dairy farmers in an area of some 300 square miles and to pouring 2 million liters of milk into rivers and the sea.[4] From the early 1970s into the 1980s there were repeated embarrassments because of highly publicized accidental discharges of radioactivity from the reprocessing plant. But in addition there was growing public concern that routine, planned releases of liquid low-level waste to the Irish Sea might be putting public health and safety at unacceptable risk, even though at no point did they exceed authorized limits, which had been set in accordance with recommendations of the International Commission on Radiological Protection (ICRP) and Britain's National Radiological Protection Board (NRPB).[5] The routine releases to the Irish Sea have been a major source of political trouble for Windscale and the nuclear industry, and I shall come back to them (see Liquid Waste Discharges to the Irish Sea, below). Suffice it to say here that however valid the nuclear industry's claim that the

[2]According to Donald Avery, formerly deputy managing director of BNFL, "the overall site name of Sellafield was not new, the locals had never used any other, and the two units were, and are, called Windscale and Calder Works. We can perhaps be judged naive for not anticipating the interpretation which was put upon the change but we cannot properly be accused of an intent to deceive." Letter to the author, April 3, 1985.

[3]The release at Chernobyl was much larger, and the reported releases of radioactivity from the Soviet Union's military plutonium production center at Kyshtym in 1958 appear to have been vastly more significant radiologically than the release at Windscale.

[4]Walter C. Patterson, *Nuclear Power* (Baltimore, Penguin Books, 1976) pp. 162–166. According to *Nucleonics Week* (February 24, 1983), a report by the National Radiological Protection Board concluded on the basis of "theoretical calculations" that the iodine releases from the Windscale fire of 1957 might cause as many as 260 thyroid cancers over a forty-year period, with 13 of them fatal.

[5]None of the accidental discharges of radioactivity associated with fuel reprocessing at Windscale have resulted in significant exposure of the public to radiation, according to British authorities. Radioactive Waste Management Advisory Committee (RWMAC), *First Annual Report* (London, Her Majesty's Stationery Office, May 1980) pp. 13–15; interview by the author, September 30, 1980, with John Dunster, now director of the NRPB.

public health has not suffered from the releases, criticism of the industry has nonetheless been intense. The evidence is that the nuclear enterprise in Britain has been poorly served politically by its once strongly held—but now increasingly constrained—philosophy that releases of radioactivity to the environment can have a large place in waste management so long as authorized limits and ICRP recommendations are not exceeded. Indeed, the record shows that the complexities, the controversy, and the fears associated with radiation and things nuclear have created a situation in which official discharge limits and the recommendations of the NRPB and ICRP have become suspect and could lose their political authority.

The accidental releases that have occurred at Sellafield bear witness to the demanding nature of reprocessing as a technology that is both politically sensitive and technically complex. For example, in 1972 some fuel only recently removed from a reactor, and hence not aged long enough for the shorter-lived fission products to decay, was mixed by mistake with properly aged fuel and was reprocessed; as a result, the total quarterly emissions of the short-lived iodine-131 isotope shot up thirty-nine times above normal.[6] This led to official remonstrances and corrective measures. Yet nine years later, in 1981, another accidental release of iodine was triggered in the same way,[7] again leading to official remonstrances and promises of new precautions in the future.

Two major leaks of highly radioactive liquids into the soil came to light in the 1970s. Each had gone on for some years before being detected, and each was played up in embarrassing headlines. One of these had to do with storage of the magnesium alloy cladding that is stripped from the magnox reactor fuel. Because the cladding may catch fire spontaneously, it is stored in water-filled concrete silos, where it corrodes rapidly and forms a radioactive sludge that eventually will have to be retrieved and prepared for off-site disposal. The original silo is believed to have developed a leak in 1972, possibly because chemical reactions caused high temperatures that cracked the concrete. Although first detected in 1976, the leak could not readily be stopped, and in 1980 it was still going on at the rate of about 185 gallons per day. By then an estimated 50,000 curies of radioactivity (mostly cesium and strontium) had escaped, with the greater part of the radionuclides believed to be bound to clay soil particles near the silo. How to stop the

---

[6]*The Windscale Inquiry*, report to the Secretary of State for the Environment by the honorable Mr. Justice Parker, vol. 1 (London, Her Majesty's Stationery Office, January 1978) p. 46.

[7]Jim Trotter, "BNFL Resumes Reprocessing Plant Operation After Iodine Release," *Nucleonics Week*, October 15, 1981; Jim Trotter, "BNFL and CEGB Were Chastised for Shortcomings," *Nucleonics Week*, December 24, 1981.

leak and recover, treat, and permanently dispose of both the radioactive sludge from the silo and the heavily contaminated soil from around it was a problem for which there was no solution in sight.[8]

In the other big leak, some 100,000 curies of high-level waste escaped into the soil beneath an all-but-forgotten waste sampling building, a small, concrete, boxlike structure. The last sample was collected in 1958, but because of faulty equipment and careless management some waste continued to be diverted to this building, where it eventually overflowed the sampling tank and the sump beneath. Still worse, a sump emptying line seems to have been mistaken for a sample line and cut and capped, and the pointer on the sump-level gauge was evidently on its second circuit around the dial, giving a falsely reassuring reading. Waste was found to have been escaping into the ground through breaks in the building's stainless steel liner when the leak was discovered and stopped in March 1979.[9]

The sampling building, known as B701, was one of about 600 buildings on the site, some new and modern, others obsolescent but still in use, and still others abandoned, like B701 itself.[10] At a conventional industrial site the removal of abandoned structures poses no special problem. But at a nuclear fuel reprocessing plant removal of such structures represents a costly and complex decommissioning problem when they are heavily contaminated with radioactivity.

In terms of production goals, reprocessing at Windscale has in the main been successful, with 1,200 tons or more of the magnox natural uranium fuel reprocessed in most years. Both the original reprocessing plant and its successor were built to "no maintenance" standards whereby failure-prone equipment such as pumps and valves were put outside the process cells so that radiation workers should never have to enter them. But there have been setbacks at Windscale, the worst being the failure of the venture into the reprocessing of oxide fuels—a greater challenge than the magnox fuels because of their much higher burnups. After the original reprocessing plant was shut down in 1965 it was equipped with a new oxide fuel receiving unit, then reopened in 1969 to receive fuel from British advanced gas-cooled reactors (AGRs) and foreign light

[8][R. Gausden], Chief Inspector of Nuclear Installations, Health and Safety Executive, "Report on the Silo Leak at Windscale" (1980).

[9]Health and Safety Executive, "The Leakage of Radioactive Liquor into the Ground, British Nuclear Fuels Limited, Windscale, 15 March 1979" (July 31, 1980); David Fishlock, "Windscale Safety Rules Neglected, Management Admits," *London Financial Times*, August 1, 1980.

[10]Fishlock, "Windscale Safety Rules Neglected"; David Fishlock, "The Building They Forget at Windscale," *London Financial Times*, August 1, 1980; visit by the author to Windscale, October 1980.

water reactors. Although its nominal capacity was 100 tons a year, the plant processed only 90 tons in over three years, before a ruthenium-blowback incident in 1973 caused the plant to be shut down again, apparently for good. An unexpected chemical reaction inside a rotor-equipped feeder vessel used to circulate process liquor apparently forced ruthenium-contaminated gases along the rotor's drive shaft, which passed through the heavily shielded wall separating the hot cell from a working area. The ruthenium then spread quickly throughout the plant by the air circulation system.[11] Thirty-four workers were exposed, though not severely.

Undaunted, British Nuclear Fuels in late 1974 announced plans for a proposed thermal oxide reprocessing plant (THORP) as Britain's first major entry into the international commercial reprocessing market. Though challenged by many of the participants in the official Windscale inquiry of 1977, these plans were approved by the government in 1978,[12] and construction was to start in the early 1980s. Therefore, if all goes as planned, by the mid 1990s Sellafield would be the site of two major reprocessing plants: the THORP facility (see chapter 12) and the plant for reprocessing fuel from the magnox reactors (until those reactors are decommissioned sometime around the turn of the century).

But a necessary condition for a successful expansion of reprocessing at Sellafield is a program of waste management that is acceptable to regulatory authorities and the British public. To satisfy this condition with respect to each of Sellafield's principal waste streams is a considerable technical and political challenge. These waste streams include: the high-level waste that BNFL must solidify for geologic disposal off site or for long-term storage in surface or near-surface vaults on site; the relatively voluminous intermediate-level and plutonium-contaminated wastes that must eventually go to a repository off site, because to allow these wastes to accumulate for long at the reprocessing center would impose substantial storage costs and, with the containers of waste proliferating by the thousands, political burdens as well; the solid low-level waste that will probably all have to go to off-site burial because of opposition to sea dumping (a permanent ban on sea dumping, which

[11]Nuclear Installations Inspectorate, "Report by the Chief Inspector of Nuclear Installations on the Incident in Building B204 at the Windscale Works of British Nuclear Fuels Limited on 26 September 1973," CMND 5703 (London, Her Majesty's Stationery Office, July 1974) p. 8.
[12]*The Windscale Inquiry*, p. 10. The report by the presiding officer for the Windscale inquiry, Mr. Justice Parker, was presented to the secretary of state for the environment in January 1978 and debated in the House of Commons on March 22, 1978. Favored by both the Labour government and the Conservative opposition, the THORP project was cleared for construction by the House of Commons on May 15, 1978, by a vote of 224 to 80.

now seems likely, would make it essential to find a place for a new burial facility); and the liquid low-level waste now still being discharged to the Irish Sea, most of which must sooner or later be retained, reduced to solid form, and disposed of off site. Although normally discharged in a dilute form, the liquid low-level waste has cumulatively contained tens of thousands of curies a year, including hundreds of curies of such alpha-emitters as plutonium and americium. Two political traumas in the fall of 1983 were to contribute to growing pressure on British Nuclear Fuels to go well beyond its then-existing plans for dealing with these discharges.

High-level Waste: Finessing the Geologic Disposal Issue

Just as in the United States, nuclear technology in Britain was allowed to get ahead of itself, with reactors and reprocessing plants being built and operated in the absence of any plans for or means of permanently getting rid of the waste. In one respect, however, the British improved on the American record. Britain's high-level waste storage tanks were constructed of corrosion-resistant stainless steel. This allowed the waste to be left in its acidic state, which meant that there would be less of it and that it would be easier to vitrify than the U.S. military waste that has been increased in bulk and complexity by the addition of a neutralizing agent. But the British Atomic Energy Authority (AEA) made the same mistake as the Atomic Energy Commission by not carrying waste solidification from research and development to commercial-scale operation—which only the French had the foresight to do.

In the matter of geologic disposal, the Atomic Energy Authority actually had done little or nothing when, in 1974, Britain's nuclear waste problems first came under the scrutiny of the Royal Commission on Environmental Pollution. The commission's report, issued in September 1976, decried the lack of progress, and said that with respect to the problem of ultimate disposal the United Kingdom appeared to be "conspicuously backward."[13]

To the nuclear industry's alarm, the commission recommended that "there should be no commitment to a large programme of nuclear fission power until it has been demonstrated beyond reasonable doubt that a

---

[13]Royal Commission on Environmental Pollution, *Sixth Report—Nuclear Power and the Environment* (London, Her Majesty's Stationery Office, September 1976) p. 154, para. 406 (hereafter cited as the Flowers report, after Sir Brian (now Lord) Flowers, commission chair at that time).

method exists to ensure [these wastes'] safe containment... for the indefinite future."[14] The commission specifically rejected the concept of indefinite storage of solidified high-level waste in surface vaults—an option not clearly distinguishable from disposal after a minimum of fifty years in storage, which the Atomic Energy Authority, British Nuclear Fuels, and eventually the government itself would accept. The commission argued that "any storage system implies some degree of maintenance and could be regarded as an imposition on our descendants who would have received no direct benefit from the power generated.... In addition, the storage of such immense quantities of radioactivity presents dangers from accidents or from malefactors or acts of war. Engineered surface storage facilities might also be vulnerable to major climatic changes. They may serve adequately for a limited time, but they may also by their mere existence direct attention from the need for a permanent and final resting place for the wastes."[15]

The Atomic Energy Authority was being criticized by one of its own, for the commission's chairman, Sir Brian (now Lord) Flowers, Rector of the Imperial College of Science and Technology, was a former head of theoretical physics at the authority's Harwell research establishment. Flowers and his colleagues on the commission represented a prestigious slice of Britain's scientific and technological elite. The government's response to the commission's report was on the whole positive and accepting. In a white paper issued in response to the Flowers report, the government promised to "ensure that waste management problems are dealt with before any large nuclear programme is undertaken."[16]

The Atomic Energy Authority, which saw what might be coming when the Flowers commission first began looking over its shoulder back in 1974, had commissioned the Institute of Geological Sciences in 1975 to investigate the possibilities of geologic disposal.[17] As it turned out, only in the isolated and remote county of Caithness in the extreme northeast of Scotland did the authority find local officials willing to grant the planning permission necessary for borehole drilling. Caithness was familiar with nuclear activities because it was in this far corner of the United Kingdom that the AEA, as a safety precaution, had chosen years before to build its experimental breeder reactor research facility.[18] But

[14]Flowers report, p. 202, para. 533.

[15]Ibid., p. 148, para. 390.

[16]*Nuclear Power and the Environment*, CMND. 6820 (London, Her Majesty's Stationery Office, May 1977) p. 7.

[17]Flowers report, p. 149, para. 391.

[18]John Hill, "Nuclear Power's Crisis of Confidence," *New Scientist*, November 19, 1981, p. 517.

requests for planning permission, which is required under the Town and Country Planning Act, were emphatically denied in southwest Scotland and in Northumberland, a county in the northeast corner of England below the Scottish border. The prospect elsewhere in England and in Wales was for protest and more denials. Under British law, denials can be overturned by the responsible minister in London, but would the government give enough priority to the geologic investigations to proceed with them in the face of all the political heat?

One site where the Institute of Geological Sciences had plans for extensive drilling was a modest eminence of 2,270 feet called Mullwharchar Hill, the most visible manifestation of a large granite mass in the Galloway Hills of southwest Scotland. Mullwharchar is part of a wild landscape characterized by deep glacial lakes and heather-covered moors attractive to hillwalkers and other summer visitors. Largest of the lakes is Loch Doon, from which the River Doon flows through the Doon Valley, a sheep-farming and coal-mining area that has been down on its luck, with much joblessness in Dalmellington, the biggest town in the upper valley. Dalmellington is the home of Dr. Mir Sadig Ali, a Pakistani-born general practitioner and respected citizen, and with his wife—a native Scot—a leader of the opposition to proposed borehole drilling in the area. In late 1976 reports in the local press confirmed rumors that the Atomic Energy Authority proposed to investigate the suitability of the Mullwharchar Hill granite for a nuclear waste repository. Local citizens learned more about the project at a public meeting in January 1977, which was addressed by Frank S. Feates, then a staff scientist at Harwell. As Mrs. Ali remembers it, Feates described Mullwharchar Hill as not merely a site for geologic research—which is how the government was later to characterize the proposed test drilling—but as a potential repository site. "We were not satisfied in any way with any of the answers given," Mrs. Ali told me.[19]

By the time of the meeting with Feates the Alis already had organized the Campaign Opposing Nuclear Dumping, or COND. One COND rally attracted 4,000 people even though it was held in bad weather and in an out-of-the-way place near Mullwharchar Hill. Their principal worry, Mrs. Ali told me, was the possibility that repository leaks, by contaminating groundwater, might spread radioactivity through much of Doon Valley. COND itself was merely "antidumping" rather than anti-nuclear, but it formed a close alliance with the Scottish Campaign to Resist the Atomic Menace, an Edinburgh-based coalition opposed to both civil and

---

[19]The Alis were interviewed by the author in October 1980; remarks attributed to them in this chapter are from that interview.

military nuclear activities. According to Dr. Ali, because of the Mull-wharchar project many people never previously concerned about nuclear power were becoming anti-nuclear.

Sheep farmers in particular felt threatened. Douglas Baird, who rents forest commission land for his flocks, was a spokesman for the farmers of the Doon Valley. His attitude was akin to that of farmers in west Texas who have worried about their farm products getting a bad name. "Who would like to buy animals from a radioactive area?" Baird asked when I visited his cottage by Loch Doon in the autumn of 1980. "People have been affected by it [radiation] at Windscale and various places," he went on to say; "it doesn't take very much to cause a lot of trouble."[20]

Faced with opposition to proposals for borehole drilling, the government was insisting that geologic disposal was only one of three options being considered for ultimate disposal of high-level waste. The other two were for disposal of waste either on or beneath the seabed. One option might be tested on a small scale during the 1990s, the government said, but no decision on how or whether to proceed with actual disposal would be taken before the next century.[21]

Formal proceedings under the Town and Country Planning Act got under way in Scotland in January 1978. The Ayrshire district council, as expected, denied permission to plan the proposed borehole drilling; the secretary of state for Scotland would have to decide whether to order a public inquiry and then, from the record of that inquiry, rule on the matter himself. Although the Labour party was in power at the time, in the spring elections of 1979 the Conservatives defeated Labour, and the secretary of state for Scotland in Prime Minister Margaret Thatcher's new government turned out to be none other than George Younger, a Conservative member from Ayr—only fifteen miles from Dalmellington—and a vocal critic of the Mullwharchar project who had once led a protest march through the streets of Ayr. Younger was now responsible for that project's fate.

[20]The impression of many British citizens that the radiation hazards at Windscale have had severe health consequences is contradicted by J. A. Preece, director of information services at BNFL, who has called the firm's safety record "second to none." But as Preece has told the author, he does not mean to imply "absolute safety, which clearly cannot be guaranteed and probably not achieved" (personal communication, October 3, 1985). Under the Nuclear Installations Act of 1965 BNFL is liable for damages if "on the balance of probabilities" a death is attributable to radiation exposure. The *London Times* reported on September 17, 1986, that BNFL had paid more than £600,000 in compensation to families of cancer victims pursuant to an agreement with the four unions at Sellafield. The awards did not represent acceptance of liability, according to BNFL.

[21]Press notice and statement by the Scottish Development Department on behalf of the secretary of state for Scotland, issued February 1, 1980.

Younger called for the public inquiry, which began in February 1980 and continued over a period of some six weeks. Opposition to the project from people within the region was massive, and there was much skepticism about the government's attempt to show that its borehole drilling program was a broad research effort by filing drilling applications in several other places, including three in the south of England. As one official in the Department of the Environment put it to me, "This just tends to spread the misery and opposition. I don't think it will help."[22]

In the end, the government would choose to abandon its plans for the borehole investigations. On December 16, 1981, Tom King, Minister for Local Government and Environmental Services, announced to the House of Commons that the investigations no longer had "any present priority." The new priority was to be "ensuring the safe and acceptable storage of [solidified high-level] waste." In support of this new policy King cited recommendations made the previous May by his department's Radioactive Waste Management Advisory Committee (RWMAC), a group made up of independent scientists and of technical managers from the nuclear industry and trade unions.[23] But the decision was in fact a policy triumph for British Nuclear Fuels and the Atomic Energy Authority.

Donald Avery, BNFL's deputy managing director and a member of the advisory committee, had set out for me earlier the rationale for such a change of policy. The critical need, in his view, was not to demonstrate ultimate disposal, as the Flowers commission had argued, but to solidify the liquid high-level waste that was in tank storage at Windscale by using the vitrification technology developed by the French at Marcoule. The waste would then be stored for fifty years or longer in engineered structures, built either on the surface or just beneath the surface. "The

---

[22]Department of the Environment press notice of January 17, 1980, identifying additional areas for investigation by the Institute of Geological Sciences—three in the south of England and one in Wales; interview by the author with Dinah Nichols, assistant secretary of the Department of the Environment, September 30, 1980.

[23]The advisory committee wrote: "We think serious consideration should be given to the possibility that containment in an engineered storage system might be the best way to deal with solidified high-level waste for decades or even centuries. . . .storage followed by disposal would seem to be [best]. . . .our present view is that the transition is unlikely to take place until at least 50 years after the wastes have been vitrified. . . .the ability to store for long periods means that there is no immediate requirement to select a disposal route." Radioactive Waste Management Advisory Committee (RWMAC), *Second Annual Report* (London, Her Majesty's Stationery Office, May 1981) pp. 21–23. This advice, strongly influenced by industry people on the committee, gave the government the technical rationale it needed for its decision. RWMAC was saying also (but none too forcefully) that continued field research was needed to support geologic disposal in the future. See RWMAC, *Third Annual Report* (May 1982) para. 2.2 through para. 2.4.

longer you store, the less the heat output and the easier, or more relaxed, are the criteria for the disposal site," Avery told me.[24]

It was by no means obvious, he said, that the public would be satisfied either with a policy of immediate geologic disposal or with one of long-term storage. "At the one extreme," he added, "you have the argument: if you can store these things on the surface and look after them, why bother about disposal? ... At the other extreme is the argument that, if they are so dangerous you can't dispose of them, why produce them at all?" In his opinion, the best course was to continue gathering information about storage and disposal alternatives, but to leave the choice to a later generation. "We should not commit ourselves now to a particular course which leaves the poor wretches fifty years from now with no option," he said. Avery acknowledged that this was not yet the policy of the government, which was still formally committed to the Flowers commission philosophy that there should be no major expansion of nuclear power until safe disposal had been demonstrated. "What we have been trying to do," Avery said, "is to persuade the government that Brian Flowers was wrong."[25]

John Dunster, then deputy director of the Health and Safety Executive, which regulates British Nuclear Fuels through the Nuclear Installations Inspectorate, believed that long-term surface storage was quite acceptable from the technical, health, and safety standpoints. While it was less clear to him that such storage would be acceptable politically, Dunster felt that the political problems of siting geologic repositories made this disposal option distinctly uninviting. "As our planning laws go at the moment," he told me, geologic disposal "is feasible because you have a local inquiry, you listen to local objections, and if the national benefits override the objections, then the minister decides. But he will probably have to have troops defending his [site exploration and development crews]... I'm not sure the political will is going to be there."[26]

Lord Flowers, as much as he favored getting on with geologic disposal, also was conceding that the problem of borehole drilling might be politically intractable. The proposed drilling, he told me before the government abandoned this program, "arouses great local opposition, yet it cannot realistically be put in the context of great national priorities because in itself it's a very small thing."[27]

[24]Interview by the author with Donald Avery, October 1980.

[25]Avery says, however, that he and the advisory committee as a whole regretted the government decision to abandon the program of geologic field investigations looking to the eventual siting of repositories. Letter from Avery to the author, April 3, 1985.

[26]Interview with Dunster, September 30, 1980.

[27]Interview by the author with Lord Flowers, October 1, 1980.

Among those opposed to borehole drilling in their areas, a number favored extended surface storage. "If we have a limited nuclear program we can keep the waste under surveillance," George Foulkes, a Labour M.P. from Ayr, remarked to me. "What worries me is to have the waste underground, out of sight, and then, a hundred years from now, having it come back, maybe in our water supplies." Gibson MacDonald, chair of the Conservative party in Ayr and a leader of the opposition to the Mullwharchar Hill project, spoke similarly, emphasizing the importance of being able to retrieve the waste should anything go wrong.[28]

The beauty of the government's new policy from a political standpoint was that the Atomic Energy Authority not only would have no need to find and develop repository sites for a long time to come, it probably would not have to bother with finding a new site for engineered storage either. Donald Avery of BNFL told me that there is room enough for such storage at Windscale. Clearly, the convenient arrangement is for the storage facility to be adjacent to the waste vitrification plant, which is how the French have arranged things at Marcoule. And although British Nuclear Fuels is not saying as much, Cumbria County, which has lived with the waste in its more hazardous liquid form for three decades, would be unlikely to object strenuously to on-site storage of solidified waste.

In sum, British Nuclear Fuels and the government appeared to be politically in the clear for the moment on the high-level waste issue. But this did not mean that the geologic disposal issue would never reassert itself. Two problems remained. The longer term problem was that, absent a means of permanent disposal, the nuclear industry would almost inevitably be dogged (as the Flowers report had indicated) by the perpetual custodial care issue, particularly if it appeared ready to embark on a major expansion. The shorter term problem was that while the need for geologic disposal of high-level waste had been politically finessed for the moment, the need for a repository for intermediate-level and plutonium-contaminated waste posed political quandaries that could not be similarly avoided.

Intermediate-and Low-level Waste

British Nuclear Fuels' plans for disposal of waste in the intermediate- and low-level categories were to be beset by severe and worsening political difficulties. One major setback was the temporary and possibly permanent loss of deep ocean dumping. In the fall of 1980 there was

---

[28]Foulkes and MacDonald were interviewed by the author in the fall of 1980.

still hope that this avenue of disposal would be increasingly available. "One of the most important things for us in this country is to keep open, and if possible increase, the use of the deep oceans for dumping low- and intermediate-level wastes, including transuranics," Donald Avery of BNFL told me then. "When you've got a small, highly populated island, you don't have to go through much science to come to that conclusion. We see sea dumping as really very central to a satisfactory waste disposal strategy."

But the Flowers commission had already warned that hopes for expanded use of ocean dumping might well be illusory. The commission noted that in the 1990s British Nuclear Fuels hoped to be permitted to dump each year some 2 million curies of beta-emitters (such as strontium and cesium) and 24,000 curies of alpha-emitters (such as plutonium and americium). These levels would represent roughly a hundredfold increase over the dumping by Britain in the mid 1970s. Doubting that this would be acceptable, the commission saw a need for a suitable national disposal facility on land.[29] That this was a correct and farsighted judgment was to be borne out by the resolution adopted in February 1983 by member nations of the London Dumping Convention, which called for a moratorium on dumping pending completion of international reviews of the practice.[30] Following the adoption of that nonbinding resolution, and after demonstrations against dumping by the environmental group Greenpeace, protests in Spain, and the announcement of a boycott by British transport and seamen's unions, the British government cancelled the dumping planned for 1983 (dumping has not since been resumed).

British Nuclear Fuels was now in an increasing bind with respect to both low- and intermediate-level wastes. The shallow-burial facility at Drigg, a site on the Cumbrian coast not far south of Sellafield, would be available for Sellafield's low-level waste through the turn of the century. But for much of Sellafield's intermediate waste there was no existing means of disposal at all. Its radioactivity was too concentrated or long-lived to meet existing national and international standards for either

[29]Flowers report, p. 142, para. 372. By comparison, in 1980 Britain dumped waste containing 106,000 curies of beta/gamma radioactivity and about 1,800 curies of alpha radioactivity. The waste was contained in drums together weighing on the order of 2,500 tons. Department of the Environment press release, July 3, 1980.

[30]The international treaty formally known as the London Convention on the Prevention of Marine Pollution by Dumping of Waste and Other Matter was adopted in 1972 and entered into force in 1975. The convention prohibited dumping of high-level waste; dumping of low-level waste was to be regulated under standards adopted by the International Atomic Energy Agency. See Daniel P. Finn, "Ocean Disposal of Radioactive Wastes: The Obligation of International Cooperation to Protect the Marine Environment," *Virginia Journal of International Law* vol. 21, no. 4 (summer 1981) pp. 621–690.

shallow burial or ocean dumping. Two new disposal facilities were needed: a shallow-burial facility to extend the life of the Drigg site by accepting low-level waste from the reactor stations and other sources; and a geologic repository of one kind or another to receive the intermediate waste. In principle, the intermediate waste could be allowed to accumulate in storage on site; but in practice, as Donald Avery put it to me, "this would be pushing it some" because not only are large amounts of intermediate waste generated, but much of the waste—in particular the sludges and ion-exchange resins—is of a kind that cannot be compacted.[31]

In October 1983 the secretary of state for the environment, Patrick Jenkin, announced that sites for two disposal facilities had been tentatively selected by the Nuclear Industry Radioactive Waste Executive (NIREX). NIREX, a coordinating entity made up of waste generators such as BNFL and the Central Electric Generating Board, had identified one site on land (owned by the board) at Elstow in Bedfordshire, some forty miles north of London. At this site low-level waste and non-alpha-contaminated intermediate-level waste would be buried in trenches excavated in a thick clay layer. NIREX also had identified a closed anhydrite mine in the north of England as a candidate site for geologic disposal of long-lived or alpha-contaminated intermediate-level waste. This site was at Billingham, an industrial town just inland from the North Sea and some ninety miles across the United Kingdom's narrow northern waist from Sellafield, on the opposite coast. The mine, owned by Imperial Chemical Industries, is at a depth of about 750 feet and extends laterally over about two square miles.[32]

According to the government's announcement, final selection and development of the sites would depend on findings from geologic studies and on the outcome of requests by NIREX for planning permission. Local authorities would first be asked to approve plans for any exploratory field studies (such as borehole testing) that required permission, then later to approve plans for construction of the disposal facilities themselves. Secretary of State Jenkin promised that at each stage formal public inquiries would be held.

The timing of this announcement hardly could have been worse because, as I shall describe, there would soon be news of another spill of radioactivity at Sellafield and of a cancer scare at a village near the reprocessing center. The criticism heard in the House of Commons from November 1983 through the spring of 1984 of the tentative selection

[31]Interview with Avery, October 1980.
[32]U.K. Atomic Energy Authority. "British Waste Sites Announced," *Atom*, December 1983, pp. 279–280.

of the Elstow and Billingham sites was often mixed with criticism of waste management at Sellafield. Conservative members of Parliament loyal to the Thatcher government, as well as Labour M.P.s, were speaking out. A Conservative from Darlington, not far from Billingham, spoke of a "continuing erosion of public confidence" resulting from reports of problems at Sellafield; Frank Cook, a Labour M.P. from the Billingham area and a leader of the opposition to the anhydrite mine site, said that "anxiety within the electorate" arose because events had shown the "bland assurances" of the past to be ill founded; another Conservative M.P. warned that if waste from Sellafield were sent to the Billingham mine there would be "great distress in the northeast of England."[33] M.P.s from the affected areas were bothered that, inasmuch as NIREX had not indicated alternative sites, it was not clear how or why these two particular sites were chosen. It was suggested that the choices could have been more political than technical, that lacking alternative sites NIREX would have nothing to fall back on should either be rejected, and that NIREX would therefore go into the investigation of these sites necessarily prejudiced in their favor.[34]

An especially persistent criticism was that nuclear waste should not be disposed of in areas so highly developed and densely populated as Elstow and Billingham. Frank Cook hammered hard at NIREX for its "irresponsible" and "preposterous" proposal to put waste in a mine found in an urban region of almost 750,000 people and situated almost directly beneath a community of 35,000 inhabitants.[35] Another persistent theme was that Elstow and Billingham already had industrial wastes of their own, and that to burden them with nuclear waste from other parts of the United Kingdom was wrong and unfair. Cook described the attitude of his constituents in Cleveland County, where the Billingham mine is located, in this way: "They already tolerate more than 14.5 percent of the nation's registrable hazardous locations, one in seven of Britain's gigantic dangerous dustbins. They do not like that, but they tolerate it. They do so because the hazardous chemicals and the like were produced in the area. Their production provided employment and a degree of prosperity, however limited, for the whole community. This radwaste is unwanted and it is quite alien." According to Cook, his constituents were also disturbed because the NIREX proposal could

---

[33]Excerpts from discussions in the House of Commons on December 20, 1983 (comments of Michael Fallon), February 14, 1984 (comments of Frank Cook), and December 5, 1983 (comments of Richard Holt) reprinted in "In Parliament," *Atom*, February 1984 and April 1984.

[34]See excerpts from discussions in the House of Commons on October 25, 1983, and May 3, 1984, reprinted in "In Parliament," *Atom*, January 1984 and July 1984.

[35]Discussion in the House of Commons, May 3, 1984, "In Parliament," *Atom*, July 1984.

have "grievously adverse effects on the regional economy.... We have evidence of the serious effect on land values, house purchase, industrial development and job provision, all directly attributable to this nonsensical idea."[36]

Public opposition in the Elstow and Billingham areas appeared massive. For instance, in May 1984 a petition signed by some 85,000 Cleveland County residents protesting consideration of the Billingham mine for a repository was presented to Prime Minister Thatcher. Imperial Chemical Industries, originally willing to allow its mine to be used for a repository, was now bowing to the fervent local opposition by declaring itself opposed to the project.[37] The upshot was that in January 1985 the secretary of state for the environment announced that consideration of the Billingham site was being abandoned and that NIREX had been instructed to come up with a list of several alternative sites. He said that Elstow remained a possibility for a low-level waste site, but that other sites would be considered as well.[38]

In February 1986 the government announced that three additional candidate sites had been chosen, all on land owned by the Central Electric Generation Board or the ministry of defense in the eastern counties of Essex, Lincolnshire, and Humberside. Strong local opposition immediately developed, and the county councils in Lincolnshire and Humberside formed a coalition with the council in Bedfordshire (host to the Elstow site) to lobby against the siting investigations.[39]

In a change of tactics, the government had Parliament approve a special development order to permit the necessary field studies instead of seeking planning permission from the county councils. This meant that the public inquiry would not come until the field studies were completed and the Nuclear Industry Radioactive Waste Executive had prepared an environmental statement comparing the sites and explaining its choice of one of them for the disposal facility. The studies were expected to take up to eighteen months, and no public inquiry was contemplated until late 1988 or early 1989[40]—and therefore not until after the next British election, which would have to come by the summer of 1988.

[36]Ibid.

[37]Ibid.

[38]Stephanie Cooke, "The British Government Is Dropping the Controversial Billingham Site," *Nucleonics Week*, January 31, 1985.

[39]Interview by the author on September 27, 1986, with Andrew Blowers, member of the Bedfordshire County Council and geographer at the Open University, Milton Keynes.

[40]Secretary of State for the Environment, *The Government's Response to the Environment Committee's Report*, CMND. 9852 (London, Her Majesty's Stationery Office, July 1986) (hereafter cited as *Government Response to Environment Committee Report*).

In its response to the continuing criticism of the waste management effort, the government also announced in 1986 that the waste that could be received at the near-surface disposal facility would be tightly restricted—far more restricted than are similar facilities in France or the United States. Intermediate-level wastes considered short-lived because they contain radioactive species having half-lives of 30 years or less had originally been intended for near-surface disposal along with low-level waste, but now the government had decided that such waste should be stored until a deep repository became available.[41] Commenting on this change, the nuclear industry said that while deep disposal of waste of this kind was not technically required, it found such "Rolls Royce solutions" acceptable for the sake of gaining public confidence.[42]

Exploratory drilling at the candidate sites finally began in September 1986, after a delay of several weeks during which protesters blocked access by work crews.[43] If the government shows the political fortitude to stay the course, NIREX may have a low-level waste site by the early 1990s.

But an answer to the industry's pressing need for a deep geologic repository for intermediate-level waste lies still further ahead. NIREX has tried to recover from the setback at Billingham, which technically was a promising site, by seeking the help of the British Geological Survey (BGS). The survey was asked to consider not only land sites, but also sites beneath the seabed with access either from the coast (as in Sweden; see chapter 9) or from platforms similar to those used in North Sea oil fields. BGS is also considering sites on small offshore islands, of which there are reported to be more than one hundred that might be suitable.[44]

Liquid Waste Discharges to the Irish Sea

Three equally remarkable things can be said about liquid waste discharges to the Irish Sea. First, Windscale has been the single greatest continuing point source of radioactive pollution to the world's seas and oceans.[45] Second, as far as is known, at no time have these discharges

[41]*Government Response to the Environment Committee Report*, p. 4, paras. 19–20.

[42]"Radioactive Waste," the nuclear industry's response to the Environment Committee's Report (July 1986) p. 5.

[43]Trudi McIntosh, "Blockade Ban Won by NIREX," *London Times*, September 12, 1986.

[44]Neil Chapman and Tim McEwen, "Geological Solutions for Nuclear Waste," *New Scientist*, August 28, 1986, pp. 36–40.

[45]House of Commons, Environment Committee, *Radioactive Waste* (London, Her Majesty's Stationery Office, March 1986) chap. 6, "Discharges"; Vaughan T. Bowen, formerly of the Woods Hole Oceanographic Institution and specialist on radioactivity in the marine environment, interviewed by the author, February 6, 1985. In a letter to the author, June

exceeded authorized limits, which have been set by the British government in the light of maximum dose criteria adopted by the International
Commission on Radiological Protection.[46] Third, public disquiet about
the Windscale discharges—disquiet not only in the United Kingdom,
but in other countries as well—is now leading the British government
to require discharge reductions going beyond those already achieved or
planned.

As recently as 1978, the collective dose commitment to the public
from the Windscale discharges to sea and air (the discharge to the sea
being overwhelmingly predominant) was twelve times greater than that
from all the rest of Britain's nuclear facilities together. By the early
1980s the Windscale discharges had been substantially reduced, but
they still accounted for the greater part of the collective dose commitment from British nuclear installations.[47] For most Britons, however, that
dose would probably be but a fraction of the dose received from medical
and dental X rays.

Cesium from Sellafield shows up not only in the waters of the Irish
Sea and in the flesh of fish caught there, but in the North Sea and in the
fish from its waters. Indeed, the destination of much of the cesium that
is carried by currents around Scotland into the North Sea is the Arctic
Ocean, where cesium serves as an important tracer in oceanographic
studies.[48] Calculations made by the British fisheries laboratory in the

---

11, 1985, Bowen points out that in the 1960s the discharges of radioactivity from plutonium-production reactors at the Hanford works to the Pacific via the Columbia River, as
measured in total curies, were greater in some years than any of the annual discharges
since recorded at Windscale. But the radioactivity from Hanford was largely from relatively
short-lived activation products, with only trace amounts present of such fission products
as cesium-137 and strontium-90 and of actinides such as plutonium-239 that may pose
biological hazards for hundreds or thousands of years. According to Bowen, the plume of
radioactivity extending from the mouth of the Columbia could be followed only a relatively
short distance before it disappeared. By contrast, he says, "the Windscale stream [or plume]
contains mostly nuclides of intermediate half-life range, or one to tens of years, with a
significant amount of really long-lived things like plutonium. Traces from Windscale will
be readily measured for centuries over [the plume's] whole enormous trajectory."

[46]Radioactive Waste Management Advisory Committee, *First Annual Report*, pp. 14–
15.

[47]Independent Advisory Group [to the Minister of Health], *Investigation of the Possible
Increased Incidence of Cancer in West Cumbria* (London, Her Majesty's Stationery Office,
1984) p. 9, fig. 1 (hereafter cited as the Black report, after the group's chairman, Sir
Douglas Black). The figure cited shows that in 1978 the collective dose commitment from
discharges to the sea and air at Sellafield was 13,700 man-rems, of which all but 350 man-
rems was to the sea. By 1982 the annual dose commitment from the liquid discharges had
been reduced to 7,300 man-rems.

[48]Hugh D. Livingston, Vaughan T. Bowen, and Stuart L. Kupferman, "Radionuclides from
Windscale Discharges I: Nonequilibrium Tracer Experiments in High-latitude Oceanogra-
phy," *Journal of Marine Research* vol. 40, no. 1 (1982).

mid 1970s from cesium levels in the fish catch showed a collective annual national dose of 8,300 man-rems. For all nations fishing the Irish Sea and the North Sea the total dose was 14,000 man-rems.[49] If continued over a generation, a collective dose of this magnitude would lead to from 54 to 350 radiation-induced cancers (depending on the risk coefficients used), about half of them fatal.[50]

Windscale's discharge limit for beta-emitters (excluding tritium[51]) was established in the 1950s at 300,000 curies per year, or 75,000 curies for any consecutive three-month period. This limit was to remain unchanged over the next three decades.[52] The peak beta release for any year was 245,000 curies in 1975. The limit for alpha-emitters, first set at 1,800 curies a year, was raised to 6,000 curies in 1970 to accommodate the rising discharges and planned growth of the nuclear program. The peak year for alpha releases was 1973, when 4,900 curies were discharged, mostly in the form of plutonium. Much of the plutonium, instead of being flushed from the Irish Sea like the cesium, precipitates out into bottom sediments, from which it may be resuspended by currents or wave action and remobilized.

By 1983 the annual Sellafield discharges—made through three pipelines extending two and a half kilometers off shore—were down to 67,000 curies of beta-emitters such as cesium and ruthenium and to 380 curies of alpha-emitters such as plutonium and americium. They were equivalent to only about 27 percent and 8 percent, respectively, of the peak beta and alpha discharges of the 1970s, and were far below the authorized limits.[53] Nonetheless, they were high by comparison with

[49]Roy E. Ellis, professor of medical physics at the University of Leeds, statement or "proof of evidence" prepared for the Windscale inquiry, 1977, reprinted in Town and Country Planning Association, *Planning and Plutonium*, Evidence of the Town and Country Planning Association to the Public Inquiry into an Oxide Reprocessing Plant at Windscale (London, 1978) pp. 62–63.

[50]The coefficient (or measure) of risk for radiation-induced cancers varies with the model used. Low and high estimates of about 27 to 175 cancer deaths over a generation from a collective dose of 14,000 man-rems can be derived from models used in National Academy of Sciences, Committee on the Biological Effects of Ionizing Radiation, *The Effects on Populations of Exposure to Low Levels of Ionizing Radiation: 1980* (Washington, D.C., National Academy Press, 1980) tables V-16, V-17, V-19, and V-20, pp. 203–207. With cancer deaths in the above range, the total incidence of cancer cases, fatal and nonfatal, would be about 54 to 350, or twice the cancer mortalities. For a study indicating that total cancer incidence is about twice cancer mortalities, see National Cancer Institute, *Surveillance Epidemiology and End Results: Incidence and Mortality Data, 1973–77,* Monograph 57, NIH Publication no. 81-2330 (Washington, D.C., 1981).

[51]Tritium, or radioactive hydrogen (H-3), undergoes an immense dilution by ordinary stable hydrogen when discharged to the sea.

[52]Black report, figs. 3.5 and 3.6, pp. 47–48.

[53]Radioactive Waste Management Advisory Committee, *Fifth Annual Report* (London, Her Majesty's Stationery Office, June 1984) table 2, p. 19.

the levels achieved elsewhere. For example, at the French plant at Marcoule, which is a sizable operation even though only about a third as much magnox fuel is reprocessed there as at Sellafield, the 1983 liquid discharges contained 1,700 curies beta and 1.5 curies alpha.[54] Discharges at Marcoule have been kept low by making greater use of evaporators and various filtration and precipitation methods than has been the practice at Windscale.

At Windscale some steps have been taken to reduce the radioactivity of the wastes discharged to the sea. For instance, some of the waste has been retained in holding tanks long enough for the short-lived fission products to decay. But the Windscale works' present so-called effluent treatment plant goes by a misnomer. As the British Department of the Environment has acknowledged, this is a facility where "little treatment takes place."[55] The magnox fuel cooling ponds have been no less important than the reprocessing plant itself as a source of radioactivity discharged to the sea. Magnox fuel cannot be kept long in such ponds without undergoing severe corrosion. The cladding on fuel in pond storage begins to fail, cesium leaks into the pond water, and the radioactivity in the liquid waste effluents rises precipitously. Reductions following the peak cesium discharges of the mid 1970s were achieved by placing resin bed filters in the cooling ponds,[56] pending start-up of a major new site ion exchange plant (a unit that finally came on line in 1985).

The Radioactive Substances Act of 1960, under which Windscale discharges are regulated, provides that for any member of the general public the maximum annual exposure from nuclear installations shall not exceed 0.5 rem for the whole body (other dose limits are organ-specific and vary according to the biological potency of the specific radionuclide); that the collective dose commitment for the entire national population shall not exceed an average of 1.0 rem per person over a period of thirty years; that these dose levels, to be observed regardless of cost, shall be outer limits; and that all that is reasonably practicable will be done—but with cost, convenience, and national needs taken into account—to keep the actual doses far below those maximums.[57]

---

[54]Ibid.

[55]The effluent treatment plant's purpose is to receive low-level waste streams from various parts of the Sellafield site and neutralize their acidity before discharge to the sea. This facility is discussed in the Department of the Environment report, "An Incident Leading to Contamination of the Beaches Near to the British Nuclear Fuels Limited Windscale and Calder Works, Sellafield, November 1983" (January 1984) annex 2.

[56]Radioactive Waste Management Advisory Committee, *Fourth Annual Report* (London, Her Majesty's Stationery Office, June 1983) p. 21, para. 4.6.

[57]Flowers report, p. 90, para. 213.

The last requirement reflects the International Commission on Radio-logical Protection (ICRP) principle of ALARA—"as low as reasonably achievable."

In arriving at the release limits for the discharges to the Irish Sea, the British authorities have sought to ensure that the radiation dose received by the most exposed members of the public—the critical group—is not more than a modest fraction of the ICRP recommended limit.[58] To do this in a defensible manner requires correct judgments as to the potential contribution of a number of different radionuclides to the total dose; an understanding of the environmental pathways followed by each of the more troublesome radionuclides; and correct identification of the critical group and a good knowledge of its habits.

Setting release limits has been a complex scientific undertaking, re-plete with uncertainties. Since the early 1970s the principal critical group has been a small fishing community where people eat heartily of the local catch. Exposure from ingesting fish and shellfish contaminated with radioactivity rose during the mid 1970s, then declined as discharges of radioactivity were reduced. But 1981 and 1982 reports by the Min-istry of Agriculture, Fisheries and Food (MAFF) showed that the critical group's appetite for molluscs was much greater than previously believed. At the same time, a study by the National Radiological Protection Board indicated that the group's uptake of plutonium through the human gut was five times greater than previously understood.[59] Together these findings meant that estimates of the contribution of plutonium to the overall dose was about fifteen times greater than previously believed. As a result, new estimates for the total dose went up from 24 percent of the ICRP recommended limit to 69 percent, or about seven times higher than considered desirable by the Department of the Environment's Ra-dioactive Waste Management Advisory Committee.[60]

Not until 1985 were the full reductions in the Sellafield discharges, promised by British Nuclear Fuels at the time of the Windscale inquiry,

[58]The setting of discharge limits for liquid effluents from nuclear installations is a joint responsibility of the secretary of state for the environment and the minister of agriculture, fisheries and food.

[59]Ministry of Agriculture, Fisheries and Food Fisheries Research Laboratory, "Radioac-tivity in Surface and Coastal Waters of the British Isles, 1980" (Lowestoft, 1982); Ministry of Agriculture, Fisheries and Food Directorate of Fisheries Research, "Aquatic Environment Monitoring Report Number 8[,] Radioactivity in Surface and Coastal Waters of the British Isles, 1981" (Lowestoft, 1982); J. D. Harrison, "Gut Uptake Factors for Plutonium, Amer-icium, and Curium," NRPB-R129 (National Radiological Protection Board, January 1982). See also National Radiological Protection Board report NRPB-GS3 (January 1983).

[60]The committee recommended that exposure to the critical group be no greater than 10 percent of the ICRP recommended limit. Radioactive Waste Management Advisory Committee, *Fifth Annual Report*, p. 20.

finally to be accomplished. The new Site Ion Exchange Plant (SIXEP) removes most of the cesium from the cooling-pond water before it is released to the sea. Also, a new salt evaporator reduces the residual radioactivity of plant process liquids. But SIXEP and the salt evaporator cannot by themselves reduce releases to the very low levels demanded by BNFL's critics. Without further treatment of its wastes, the magnox plant will still be releasing annually some 200 curies of plutonium and other alpha-emitters and more than 32,000 curies of cesium and other beta-emitters.[61] (At BNFL's new THORP facility, on the other hand, little radioactivity is to be discharged to the sea other than in the form of tritium.)

That the continuing discharge to the Irish Sea was becoming an intolerable embarrassment for British Nuclear Fuels was made evident by two political traumas in the fall of 1983. The first grew out of the showing of a documentary entitled "Windscale—the Nuclear Laundry" by Yorkshire Television. The second trauma resulted from an accidental discharge of radioactivity which caused contaminated flotsam to wash up along twenty miles of beach.

The Yorkshire Television program reported that since 1952 when the Windscale works began operating, twenty-five young people had died of cancer or had been diagnosed as having cancer in the five coastal parishes to the south of Sellafield.[62] Moreover, eleven of these cancer victims lived in Seascale, a village only a mile to the south of the reprocessing plant, and seven of them had leukemia, a disease for which radiation is one of few established environmental causes.[63] The Yorkshire Television investigative team claimed that the incidence of leukemia among children in Seascale under ten years old was ten times the national incidence. British Nuclear Fuels attacked the reporting methods used and called the show's findings "unvalidated." But the government was enough impressed to order an immediate investigation by a special independent advisory group headed by Sir Douglas Black, a former president of the Royal College of Physicians.[64]

The beach contamination incident provoked an even stronger response. Investigations were undertaken by the Department of the Environment, the Nuclear Installations Inspectorate, the Ministry of Agriculture,

[61]Ibid., p. 19, table 2.

[62]Black report, p. 11, paras. 2.3 to 2.6.

[63]The Black report, p. 34, para. 2.46, refers to radiation as the only established environmental cause of leukemia. But Merril Eisenbud, professor emeritus at the Institute of Environmental Medicine, New York University Medical Center, has said that benzene also causes leukemia, and that viruses are a possible but not yet established cause. Personal communication to the author, May 29, 1985.

[64]Jim Trotter, "Television Report for Cancer Incidence Prompts Inquiry into BNFL Windscale Works," *Nuclear Fuel*, November 7, 1983.

Fisheries and Food, and the Director of Public Prosecutions. The latter brought charges against BNFL, which ultimately led to fines totaling £10,000 being imposed by a Crown court.[65]

The reports issued by the Department of the Environment and the nuclear inspectorate indicated that the beach contamination incident was attributable to inadequate equipment, lax management, and bad judgment. The reprocessing plant had been shut down on October 30, 1983, for routine maintenance. The system was to be flushed out to reduce radioactivity levels, with the resulting waste going to a number of highly radioactive wash tanks. The problem first arose following a change of shifts on November 10, when workers on the new shift assumed that one of the tanks contained only low-level waste, whereas in fact it also contained highly active solvent and crud layers; these materials—now believed to have contained some 4,500 curies, mostly ruthenium—were transferred along with the low-level waste to one of the sea tanks from which discharges are made to the Irish Sea. The mistake was discovered before any discharge actually occurred, but much of the high-level waste was discharged anyway, at different times and in different ways over a period of almost a week. Some of it was discharged deliberately in the belief that this was the best alternative, but most of the discharge was inadvertent, through poor information and misjudgment.[66]

The problem of coping with the original mistake was made more difficult by a number of aggravating circumstances: rising radiation levels on piping and in working areas; an inadequate system of tanks, pipework, and valves, including an emergency return line too small for the job at hand; inadequate arrangements and instrumentation for waste sampling and monitoring; and flawed and insufficient management procedures and record keeping.[67] At a meeting a few weeks after the incident, 100 British Nuclear Fuels process workers voted no confidence in the BNFL management.[68]

---

[65]The seven-week trial, conducted by the Carlisle Crown Court, took place in June and July 1985. BNFL pleaded guilty to keeping inadequate records in part of its operation. It was convicted on three other counts, the most important of which was failing to keep radiation exposures from plant discharges as low as reasonably achievable. The court rejected a charge of "casual and haphazard" management, and found that the accidental discharge had done no harm to public health. British Nuclear Fuels, "BNFL court case— Day 33 (23 July 1985)," press notice.

[66]Department of the Environment, "An Incident Leading to Contamination of the Beaches"; Nuclear Installations Inspectorate of the Health and Safety Executive, "The Contamination on the Beach Incident at British Nuclear Fuels Limited, Sellafield, November 1983" (February 1984).

[67]Department of the Environment, "An Incident Leading to Contamination of the Beaches"; Nuclear Installations Inspectorate, "Contamination of the Beach Incident."

[68]"On the Beach," *Nuclear Engineering International*, January 1984, pp. 14, 15.

A particularly embarrassing aspect of the incident for British Nuclear Fuels was that the first word to the public that something was amiss came from Greenpeace activists who were attempting to plug up the discharge pipes. "I was sitting in an inflatable Zodiac by one of the submerged pipes, when suddenly this big gob of oil came out," Steve McAllister, a Greenpeace leader, later said. "Our geiger counter went crazy; we thought the damn thing had broken. Later the British government announced they'd had a high-level radioactive spill, while we were sitting right there." Meanwhile, BNFL was pressing charges against Greenpeace for interfering with its discharge pipes. Greenpeace was fined $75,000, but news of the accidental discharges brought Greenpeace public sympathy and enough contributions to pay off the fine (which the court ultimately reduced).[69]

Besides putting British Nuclear Fuels in a poor light, the beach contamination incident made it clear that the authorized discharge limits should be tightened, for public relations reasons if for no other. The radioactivity released, concentrated in an oily solvent slick, had led to so much contaminated seaweed and flotsam being washed up that the government felt it necessary to keep much of the Cumbrian seashore posted against unneccessary public use for months. Yet the discharge limits actually had not been exceeded.

Members of Parliament from West Cumbria and southwest Scotland reacted to the new evidence of problems at Windscale with outrage and concern. In responding, the minister for the environment, Patrick Jenkin, commented that the risk to individuals even from such unfortunate happenings as the beach contamination incident was minor compared to the dangers faced by miners and North Sea divers in obtaining fossil fuels. He also noted that just the week before the Copeland borough council had decided to grant planning permission for the THORP project, which would provide some 3,000 jobs for construction workers and later about 1,000 jobs for radiation workers.[70]

But for D. N. Campbell-Savours, Labour M.P. from West Cumbria, the question was how the community could keep the nuclear industry without fear that other industries—in particular tourism and commercial fishing—would suffer and without fear that people's health would be put at risk.[71] Campbell-Savours pointed out that immediately following

[69]Dick Russell, "Greenpeace: The Hippie Navy Gets Organized," *The Amicus Journal*, September 1984. A Zodiac is an inflatable boat.

[70]See discussion in the House of Commons on December 20, 1983 (comments of George Foulkes, John Cunningham, and Patrick Jenkin), and December 13, 1983 (comments of Robert Rhodes James) reprinted in "In Parliament," *Atom*, February 1984.

[71]Discussion by D. N. Campbell-Savours, M.P. from Workington, in the House of Commons, December 20, 1983, reported in *Atom*, February 1984. Material on Campbell-Savours in the following pages is from discussions in the House on February 14 and April 25, 1984, reported in *Atom*, April 1984 and July 1984.

the beach contamination incident, some local fishermen had trouble marketing their catch, even though the Ministry of Agriculture, Fisheries and Food found no evidence that fish had been affected by the accidental discharge. He and another M.P. later expressed their lack of confidence in ICRP guidelines and dose limitation criteria, Campbell-Savours saying that these are "surrounded by unending controversy" and are "under increasing attack by scientists throughout the world."[72]

Campbell-Savours called on the government to drop its adherence to ICRP's ALARA principle, with its concern for economic impact, in favor of ALATA—"as low as technically achievable," regardless of cost. ALARA, as he saw it, had been used as much to justify keeping discharge limits up as to bring them down. He said that ALATA was supported by Ireland (the foreign state most directly affected by the contamination of the Irish Sea); the five Nordic nations of Iceland, Norway, Sweden, Finland, and Denmark; West Germany; and (as of that time, April 1984) by 274 members of the House of Commons, or better than 40 percent of the total membership. In addition, according to Campbell-Savours, the European Community and the British and Irish trade unions had all spoken out or adopted resolutions against the Sellafield discharges. In his view, whatever the findings of Sir Douglas Black and his group from their Seascale leukemia study, there would remain a "perceived risk" disturbing to Cumbria and its peace of mind.

The Black report was issued several months later. British Nuclear Fuels and the British nuclear industry needed a clear, unambiguous statement absolving Sellafield and the Windscale works of responsibility for the apparently unusual incidence of childhood leukemia at Seascale. But the very nature of the statistical problem being addressed meant that no such statement was possible, and the Black group could at best

[72]On the first of these points he was certainly correct; on the second he overstated the case. The prestige and authority of the ICRP among national radiation protection boards, regulatory agencies, and radiation biologists, health physicists, and epidemiologists appear to remain high. Its recommended dose limits, however, are by no means free of professional controversy. For instance, in 1977 at the Windscale inquiry on the THORP project, Edward P. Radford, then a University of Pittsburgh epidemiologist and chair of the National Academy of Sciences' Committee on the Biological Effects of Ionizing Radiation, contended that the ICRP limits were too permissive. In his view, the whole-body dose limit of 500 millirems a year for nonoccupational exposure should be reduced to 25 millirems, and the maximum permissible concentration for plutonium and americium should be reduced by a factor of 200. *Windscale Inquiry* vol. 1, p. 49, para. 10.48. According to Merril Eisenbud of the New York Medical Center's Institute of Environmental Medicine, Radford's risk coefficients are not greatly different from the ICRP's. Eisenbud suggests that the argument between Radford and the ICRP has had principally to do with how far the ICRP should go in defining acceptable levels of risk. The ICRP principle of ALARA itself calls for reducing exposure below the recommended dose limits, subject, however, to a weighing of other considerations, notably costs. Personal communication to the author, May 29, 1985.

offer a "qualified reassurance" that no evidence had been found of a general health risk for children or adults living near Sellafield.[73]

The group was forthright in acknowledging the many uncertainties that had to be taken into account. The report noted that while the incidence of leukemia in Seascale and the surrounding rural district was well above the average for Cumbria and the nation as a whole, this could mean either a great deal or nothing at all. "It is in the nature of an average that a proportion of its components fall above it," the report said, "so it becomes a matter of judgment at what level a raised incidence of leukemia becomes significant."[74] According to calculations by the National Radiological Protection Board the total radiation exposure, from natural and artificial sources, of young people under twenty years of age in Seascale would have resulted in only slightly more than "half a death" (in statistical terms) from leukemia.[75] But the fact was that among Seascale's eleven young cancer victims, there were four leukemia deaths. For the NRPB, using widely accepted assumptions about the dose-response relationship, to have found that these deaths were attributable to Sellafield discharges would have required discharges at least forty times greater than those reported; it would also have required that the actual dosage received by the public be correspondingly greater than previously estimated.[76]

As Black and his group acknowledged, the circumstance that most set Seascale and its environs apart from other places in the nation was the presence of fuel reprocessing close by at Sellafield, an operation unique in scale and complexity and in the amount of radioactivity discharged. Time and again the group came back to the question of whether the full extent of the discharges was known. Over the history of the facility there had been fourteen known accidental releases of radioactivity, of which only the Windscale reactor fire of 1957 was believed to have resulted in significant exposure to the public. But the deficiencies in plant management brought to light by the November 1983 beach contamination incident made it only natural for the Black group to wonder whether some major releases could have gone undetected, even though British Nuclear Fuels and the responsible regulatory agencies assured the group that this was improbable.[77]

On December 18, 1984, Secretary of State for the Environment Jenkin informed the House of Commons that the government intended to

[73]Black report, p. 93, para. 6.13.

[74]Ibid., p. 89, para. 5.17.

[75]Ibid., p. 70, para. 4.48.

[76]Ibid., p. 81, para. 4.88.

[77]See the Black report, p. 77, para. 4.71; p. 78, para. 4.75; p. 89, para. 5.15; and p. 93, para. 6.13.

reduce the Sellafield alpha discharges to less than 20 curies a year, and beta and gamma discharges to about 8,000 curies. The first would be equal to and the latter lower than the discharges from La Hague—"the only plant," the minister said, "in any way comparable" to Windscale. If this goal still fell short of the as-low-as-technically-achievable standard espoused by some Sellafield critics, it nonetheless represented a major step in that direction. However, the Enhanced Alpha Recovery Plant (EARP) that is to be built will increase the generation of alpha-contaminated solid waste for which a deep repository is needed off site by the equivalent of almost 3,000 drums a year.[78]

Whatever is done about Sellafield's present and future discharges to the Irish Sea, the effects of some of the past discharges will persist. In particular, the plutonium that is being constantly resuspended and remobilized from bottom sediments may be an almost infinitely renewable source of public uneasiness and controversy.

A New Emphasis on Containment

Sellafield's waste problem was central to the recommendation made by the House of Commons Environment Committee in March 1986, which called on the government to consider abandoning reprocessing, and in particular the reprocessing of oxide fuels at THORP, now under construction at Sellafield.[79] Chaired by Sir Hugh Rossi, a prominent Conservative M.P., this committee had made the radioactive waste problem the subject of its first report, in which Sellafield figured large because "of all the methods of disposal of radioactive waste, discharge direct to the environment causes the greatest anxiety ... [and] Sellafield is the largest recorded source of radioactive discharge in the world."[80]

The committee noted that a stated objective of the government was that all practices giving rise to radioactive wastes "must be justified ... in terms of [the resulting] overall benefit." The critical question, then,

---

[78]This alpha recovery plant, to cost £150 million, will give rise to 600 cubic meters of intermediate-level waste annually. Donald Avery, "Liquid Effluent Discharges from the Sellafield Site," BNFL memorandum, December 18, 1984, prepared for the Sizewell B Power Station Public Inquiry.

[79]House of Commons, Environment Committee, *Radioactive Waste* vol. 1, paras. 205–216. The committee also recommended that an analysis be made to determine if there was no alternative to reprocessing the fuel from Britain's obsolescent magnox reactors, which are to be phased out over the next few decades. But the committee said that the highly corrosion-prone magnox fuel was unlikely to be suitable for permanent disposal.

[80]Ibid., para. 116.

was "whether the waste and radiation problems associated with reprocessing outweigh its economic, commercial, and long-term benefits."
The committee examined the various arguments made for reprocessing
at Sellafield and found all of them wanting, at least for the present.[81] In
particular, the economic prospects of the THORP project were called
into question, the committee noting that the prices charged reprocessing
clients had risen sharply—a 30 percent increase having been announced
in 1983—even as the price of uranium was falling steeply. "BNFL's
commercial success in reprocessing in the future depends upon taking
on large amounts of foreign reprocessing for which charges bearing no
relation to the world price of uranium can be made," the committee
said. It recommended that THORP be abandoned unless there was no
way to keep the resulting financial penalties and job losses from being
unbearably high.[82]

In its reply to the environment committee, the Thatcher government
insisted that the financial costs and job losses would indeed be unacceptable. The government said it was taking the long view of Britain's
energy security needs, and argued that to abandon "the experience and
expertise gained from several decades of safe and successful development of the nuclear fuel cycle" would be wrong.[83] The government also
insisted that progress was being made on a broad front in the management and containment of radioactive waste. The declining discharges at
Sellafield were noted, as were the efforts to find sites for near-surface or
deep disposal of low-level and intermediate-level wastes.

This reply by the government was predictable. There was never really
any possibility that Prime Minister Thatcher and her cabinet would
abandon THORP at a point when £600 million had already been spent
on it or committed to it. The government was betting, and assuming,
that management of waste from THORP would result in much more
stringent containment than had been applied to waste generated at
Sellafield previously.[84]

A practice of "dilute and disperse" was still defended for certain wastes
(such as tritium, which would continue to be discharged to the Irish
Sea),[85] but the trend clearly seems to be toward more stringent containment. This is evident in the government's decision to commit all but

[81]Ibid., paras. 116, 177–205.
[82]Ibid., paras. 198, 200, 216.
[83]*Government Response to Environment Committee Report*, pp. 16–18, para. 75 through
para. 82.
[84]Ibid.
[85]Ibid., p. 8, para. 37 through para. 40.

truly low-level waste to geologic disposal. Furthermore, continued conflict and controversy over the siting of the near-surface facilities for low-level waste in or near populated areas might eventually lead to all solid wastes being put into geologic repositories on small coastal islands or beneath the seabed.

The approach of the British general election that must be held during 1987 or the first half of 1988 is likely to accentuate the trend toward more stringent and complete containment of radioactive wastes. The Labour party is deeply divided over nuclear power.[86] Because of the large number of jobs at stake, Labour, should it gain power, is unlikely to follow the lead of those important factions within its ranks that want an early end to the nuclear industry. But Labour, in power or out, is likely to increase the political pressure on the industry to observe the Rolls Royce standards of containment favored by the House of Commons Environment Committee. The Conservative party will almost certainly do the same. Ironically, it includes several functionaries who must stand for reelection in constituencies where the Conservative government has selected candidate low-level waste sites.[87] The party will be feeling increasing pressure from within its own ranks as well as from Labour and the general public to follow a course in radioactive waste management calculated to minimize conflict and increase containment.

[86]Philip Webster, "Scargill Defeated in Bid to Phase Out Nuclear Power Within Five Years," *London Times*, September 29, 1986.

[87]Trudi McIntosh, "Blockade Ban Won by NIREX"; Pearce Wright, "Exploration to Start at Three Nuclear Waste Disposal Sites," *London Times*, August 13, 1986.

8
Germany: Wastes, Fuel Cycle Choice, and Politics

The Gorleben Salt Dome Project

The geologic investigation of the salt dome at the village of Gorleben in Lower Saxony is the most advanced effort in the world to establish a repository for high-level waste from commercial nuclear power facilities. The Waste Isolation Pilot Plant project in New Mexico is further along, but that repository will be for transuranic waste from nuclear weapons programs. The Germans are gambling boldly, even a bit imprudently, by investigating only one kind of rock (dome salt) and a single site (the Gorleben dome) without identifying back-up sites. If the Gorleben dome is confirmed as a geologically acceptable site, construction of a repository could begin by the mid 1990s, probably putting the Germans ahead of every other country in establishing such a facility. But a repository at this salt site, or at any salt site, will almost certainly not allow retrievable storage of spent fuel (see chapter 5). This could be important in the light of controversies currently raging in Germany over fuel reprocessing—controversies which may yet lead to reprocessing's abandonment, long deferral, or severely constrained development.

When the Gorleben project was conceived some ten years ago, it was to be much more than a nuclear waste repository. But much has changed since then. There may be no better way into the political thicket of German nuclear fuel cycle policies and prospects than to start with an account of what has happened at Gorleben.

The Gorleben salt dome is at an elbow of the Elbe River, which forms the frontier between the Federal Republic of Germany (FRG, or West Germany) and the Democratic German Republic (DDR, or East Germany), and extends beneath the Elbe several miles into East Germany. This part of Lower Saxony is devoted principally to agriculture and forestry, and is about as far removed from the mainstream of the highly industrialized West German society as it is possible to get. From Hannover, the capital of Lower Saxony, Gorleben is about 140 kilometers to the northeast, across the slightly rolling North German Plain where the traveler passes broad sugar beet fields, pastures with black and white Holstein cattle, scattered woodlots of pine, oak, and birch, and occasional villages with houses and barns of red brick and tile. Beyond the town of Dannenberg one sees on the far side of the Elbe the closely spaced guard towers of the East German border police. A bit further on is Gorleben, a village typical of the region, and then in a large field bordered by pine woods is borehole site no. 1004, where in the spring of 1980 one of the most remarkable protests against nuclear power yet seen took place.

How did the Gorleben project begin and how did it become an important symbol of the conflict over nuclear energy in Germany? Originally, the repository was to be part of a completely integrated nuclear fuel cycle center proposed for the Gorleben site. The center, referred to by the Germans as the *Entsorgung-Zentrum*, was to include facilities to receive, store, and reprocess spent fuel at the rate of 1,400 metric tons a year; fabricate mixed oxide fuel from the plutonium recovered; treat the wastes from the reprocessing and fuel fabrication plants; and then dispose of the treated wastes in a repository to be built in the salt dome underlying the complex. The facility was to service the equivalent of about forty-five large reactors, including all of the twenty-odd reactors then operating or under construction in Germany, plus that many more.[1]

An integrated fuel cycle center of this kind would offer, in principle, two important advantages. First, from the standpoint of security, the plutonium separated from the fuel could be transported in a form that could not be used directly in bombs, if stolen. After reprocessing, the plutonium would be converted from a nitrate (a liquid form) to an oxide (a solid) and combined with uranium oxide into a light water reactor fuel that would have too low a fissile content to serve as a nuclear explosive. Second, from the standpoint of nuclear waste disposal, transport of the high-level, transuranic, and low-level wastes would be unnecessary because they could be committed to the repository beneath the center.

Under the German Atomic Energy Act, DWK[2]—the nuclear fuel cycle company created by twelve electric utilities—would apply to the state government of Lower Saxony for a license to build the reprocessing and allied surface facilities. The Physikalisch-Technische Bundesanstalt

---

[1]Carsten Salander, "The Concept of the German Electric Power Industry for the Disposal of Spent Fuel from Nuclear Power Plants," *Kerntechnik*, 20 Jg. 1978, Heft 5, S.229–237.
[2]Deutsche Gesellschaft für Wiederaufarbeitung von Kernbrennstoffen mbH, Hannover.

(PTB), a German technical agency, would apply for permission for the site investigation and development work necessary to build the waste repository. As a principal determinant in the siting of the center was to have a repository site, a suitable geologic formation had to be identified. Some 200 salt domes are found in northern Germany, most of them in Lower Saxony, and at the outset the Gorleben dome did not figure in plans for the *Entsorgung-Zentrum*. The federal government in Bonn initially proposed to the government of Lower Saxony three alternative sites for the center, intending to investigate all three and to choose the one most suitable. For security reasons, the Gorleben dome and other domes lying within about twenty kilometers of the East German frontier had been deliberately excluded from consideration.[3]

Announcement of the three potential sites in early 1976 by officials of Lower Saxony immediately triggered controversy and local resistance. For example, in late 1976 local farmers mounted an extraordinary demonstration at the site which had been designated north of Lingen, near the Dutch border; about 800 showed up on their tractors to call the lie to any suggestion that the opposition was the work of outside radical elements.[4] As a result Prime Minister Ernst Albrecht of Lower Saxony chose to reject all three of the sites proposed by Bonn; after consulting an expert group within his government, he decided instead that DWK should apply to build the fuel cycle center at Gorleben.[5] Albrecht apparently felt that the proposed center would be seen by most people there as a way to turn the depressed local economy around and to begin gaining rather than losing population. Meanwhile, the Federal Ministry of Research and Technology also had concluded that the Gorleben dome appeared suitable, and German defense officials did not, after all, object to having a fuel cycle center near the East German border.[6] But this is not to say that the proposed Gorleben facility aroused no opposition.

The German Anti-nuclear Movement and Other Protest

The German anti-nuclear movement helped to put Gorleben on the map. Perhaps the best organized and most politically potent anti-nuclear

---

[3]Interview by the author with Karlheinz Hubenthal, Helmut H. Geipel, and R. Ollig, all of the Federal Ministry of Research and Technology, Bonn, December 15, 1980.

[4]"Nuclear Power in Central Europe," *The Ecologist*, July 1977; reprinted by the Political Ecology Research Group, Oxford, England, as Oxford Report no. 1A, January 1978.

[5]Letter of January 15, 1985, to the author from Horst W. zur Horst, who at the time of these events was an official in Lower Saxony's Ministry of Social Affairs, which was concerned with the Gorleben project.

[6]Interview with Hubenthal, Geipel, and Ollig, December 15, 1980.

movement in Europe, it began to assume its present form in the early to mid 1970s in the upper Rhine Valley, at Breisach and Wyhl in the state of Baden-Wurttemberg. There the wine growers and other farmers and villagers, determined to stop construction of a proposed reactor, organized initiatives and mounted spectacular demonstrations. On one occasion a march on Wyhl stretched along the road for six kilometers; on another a demonstration by some 20,000 people ended in a massive site occupation.[7]

From Wyhl and the upper Rhine Valley the citizens' initiative movement quickly spread to other places in Germany where nuclear projects were afoot, as at Brokdorf in Schleswig-Holstein. An astute observer of the events at Gorleben explained to me how the various local initiatives have taken shape and coalesced to mount large, and often effective, protests:

First, a local initiative by local people—farmers at Gorleben, wine growers at Wyhl, for example—is formed, then a configuration of outside groups forms around the local nucleus, bringing together students, intellectuals, workers, housewives, and so on, people of different ages, sexes, and occupations who normally have little to do with one another. Different initiatives then become interlinked....These people can be organized by telephone on short notice and will drive from all over Germany to be there, conspicuously showing solidarity. The saying is: "Gorleben is evil. Gorleben is everywhere."[8]

The local citizens' initiative at Gorleben was part of this national network, which time and time again has shown an impressive ability to mount large demonstrations in various parts of Germany, typically at reactor sites. From March on into early June of 1980 some 2,500 protesters occupied borehole site no. 1004, thus denying access by drilling crews. The protesters erected more than fifty wooden huts and tents, calling their village the Free Republic of Wendland, after the Wends, an ancient tribe of the region. Flags were hoisted and Wendland "passports" were issued as evidence of sovereignty over this ragtag—but proud and defiant—place. Organizers of the protest were the Citizens' Initiative for Environmental Protection of Luchow-Dannenberg (the county of which Gorleben is a part), the Farmers' Emergency Association, and the Greens, the environmentally oriented political party that has now become a significant new force in German politics.[9]

[7]Dorothy Nelkin and Michael Pollak, *The Atom Besieged* (Cambridge, Mass., MIT Press, 1981) pp. 60–64.

[8]Interview by the author with Hermann Graf Hatzfeldt, November 15, 1980, at Schloss Crottorf. See also Nelkin and Pollak, *The Atom Besieged*, for an account of the origins and methods of the anti-nuclear movement in Germany.

[9]*Republik Freies Wendland* (Frankfurt on Main, Zweitausendeins, August 1980).

Among the specific complaints voiced by the Gorleben protesters were that no criteria had been established for the disposal of waste in a salt formation (a charge denied by the German authorities); that criticisms of the site by independent experts had been ignored; and that with police and border guards proliferating in the Gorleben area, "the atomic state casts its shadow." The demonstrators promised to leave the site only if their conditions, which included abandonment of the Gorleben project, were met. On June 4, 1980, three months after the occupation of the site began, Lower Saxony's minister of interior responded with a force of 3,000 Lower Saxony and federal border police, who ousted the protesters and bulldozed their village. The encounter was for the most part nonviolent, with the protesters neither cooperating nor resisting; the last of the protesters had to be removed from tall platforms on which they were perched.[10]

Once the site was cleared, a veritable fortress was erected by the authorities. Portable walls were quickly put in place to form a compound within which drilling crews could work; water cannons were mounted on the walls to discourage any protesters. Altogether, 27 million deutsche marks (about 13.5 million dollars)—more than a quarter of the budget for the first phase of the Gorleben investigation—was earmarked for security against demonstrators.[11] The authorities in Hannover and Bonn, aware that the investigation would have to continue a number of years before enough was known to allow the repository to be built, were determined that the project should go forward.

When a citizens' initiative springs up in a community, the local population may find itself sharply divided over the project in controversy. In Gorleben the repository issue has cut across all elements of the local population, often bringing disagreement among members of the same family.[12] A majority of the county council supported the project, believing that by creating new jobs it would help lift the local economy from its depressed condition. But at the time the protesters took over the borehole site the council clearly was not speaking for all of the Luchow-Dannenberg farmers, a number of whom were bringing the protesters food. Nor did the council speak for Count Andreas Bernstorff, a cultivated aristocrat and the largest landowner in the district (as well as one of the largest in Germany). Although not a participant in the citizens'

[10]Ibid. See also Bradley Graham, "Germans Destroy Makeshift Village Protesting A-Site," *Washington Post*, June 5, 1980.

[11]Interview by the author with Helmut Röthemeyer and Werner Heintz of the Physikalisch-Technische Bundesanstalt, November 14, 1980.

[12]Interview by the author, April 22, 1982, with Stephen von Welck, nuclear attaché at the Embassy of the Federal Republic of Germany, Washington, D.C. Von Welck was originally from Luchow-Dannenberg.

initiative, Count Bernstorff was in sympathy with it and was a strong force behind the events which, as early as 1977, were to make Gorleben a name known around the world among followers of the nuclear debate.

The Gorleben International Review

As the conservative Christian Democratic Union's preeminent figure in Lower Saxony, Albrecht had reason to proceed with care in supporting a potentially controversial facility favored by federal chancellor Helmut Schmidt's Social Democratic government. Moreover, Albrecht did not want to bruise the sensibilities of his fellow political conservative and social acquaintance, Count Bernstorff. As a courtesy to Bernstorff, Albrecht had telephoned the count on February 21, 1977, the day before the public announcement of Gorleben's selection as a candidate site, to give him advance word that some of his land was within the twelve square kilometers needed for the integrated fuel cycle center. Bernstorff, anything but pleased at this prospect, in turn called his friend Count Hermann Hatzfeldt for advice.[13] At that time Hatzfeldt had little help to offer Bernstorff, for he had not been involved in any citizens' initiatives or other anti-nuclear activities, and was not connected with any environmental groups. But in the fall of 1977 Hatzfeldt, attending a conference in the United States, came in contact with Amory Lovins of Friends of the Earth, prominent proponent of solar energy and other soft technologies as alternatives to nuclear power. From Hatzfeldt's discussions with Lovins there emerged the idea for a Gorleben international review at which plans for the integrated fuel cycle center would be subjected to analysis and criticism by foreign and German experts knowledgeable about reprocessing and waste management.

Count Bernstorff, receptive to the idea for an international review, went to Albrecht with the proposal after Hatzfeldt and Lovins had found that a group of critics of sufficient size and competence could be assembled. Albrecht agreed to it and asked that the proposal be formally submitted, along with a suggested list of critics to be invited and topics to be covered. By agreement between Hatzfeldt and Albrecht's ministry of social affairs (*Sozialministerium*), a full-time coordinator for the critics' side of the review was appointed. He was Helmut Hirsch, who as a physicist with the Austrian ministry of industry had recently been scientific coordinator for the technical review that preceded a national

---

[13]The account that follows of Bernstorff's and Hatzfeldt's involvement in events related to the Gorleben project is drawn from an interview by the author with Count Hatzfeldt, November 15, 1980, unless otherwise noted.

referendum on Austria's Zwentendorf reactor (see chapter 12). The critics numbered twenty-five altogether;[14] five were German, the rest American, British, French, and Scandinavian. Besides Lovins, the Americans included Karl Z. Morgan, former head of health physics at the Oak Ridge National Laboratory; Gene I. Rochlin, a physicist at the Institute of Government Studies, University of California at Berkeley; Dean Abrahamson, a physicist at the University of Minnesota and at the time an adviser on nuclear policy issues to the Swedish government; and physicist Thomas Cochran of the Natural Resources Defense Council. The British critics included several, such as physicists Walter C. Patterson of Friends of the Earth and Gordon Thompson of the Political Ecology Research Group, who also took part in the Windscale inquiry. The five German critics included three faculty members from the University of Bremen, an institution which many establishment-oriented Germans look at askance.

From the outset there was a basic misunderstanding between the critics' group, on the one hand, and Prime Minister Albrecht and officials in Lower Saxony's ministry of social affairs, on the other. The critics wanted to review the project as proposed by DWK. But ministry officials saw no need for a review of the Gorleben proposal, which would go forward only if demanding licensing requirements could be met. They felt that the critics should address the relevant health and safety issues in a more general way, bringing to bear the knowledge and experience gained in their own countries and internationally.[15] According to Hirsch, his group got very limited and grudging cooperation from DWK and the relevant federal and state ministeries; requested information and documents often were provided late or not at all, and the contract under which the review group could be paid was long delayed.[16] Some critics felt they were playing against a stacked deck.

Several weeks before the review was to take place, the state announced the list of panelists who supported the Gorleben proposal. There were thirty-seven of them, twenty from outside Germany, and most were nominated by DWK. The foreign panelists came from a half-dozen countries and included nuclear physicists, fuel cycle specialists, and experts on health physics. Among them were such leading figures as the International Atomic Energy Agency's former inspector general, Rudolph Rometsch of Switzerland, and the director of Denmark's Riso

[14]H. Hatzfeldt and coeditors, *Der Gorleben-Report* (Frankfurt on Main, Fischer, 1979). All participants in the Gorleben International Review are identified on pp. 199–205 of the report.

[15]Letter to the author from Horst W. zur Horst, January 15, 1985.

[16]Interview by the author with Helmut Hirsch, November 1980, at Hannover.

National Laboratory, Niels W. Holm. The distinguished physicist and philosopher Carl Friedrich von Weizsacker, then director of the Max Planck Society's Institute for Research on Life in the Scientific Technical World, was asked by Albrecht to preside over the hearings. Von Weizsacker eventually accepted at the urging of Chancellor Schmidt, who told him the review was in the national interest.[17] Although known as a proponent of nuclear energy, von Weizsacker had reservations about it, as evidenced by his later proposal that all reactors be built underground as a precaution against military attack.

Before the review was to open in Hannover, Hatzfeldt invited Lovins and a number of other Gorleben critics to meet von Weizsacker during a weekend retreat at Schloss Crottorf, Hatzfeldt's ancestral estate some fifty miles east of Cologne. Hatzfeldt respected von Wiezsacker even though not sharing his acceptance of nuclear technology. The retreat seems to have been an astute move, for two reasons: first, the anti-nuclear movement in Germany had been marked by so much turbulence that it was useful to reassure von Weizsacker that the Gorleben critics were civilized and responsible; second, there was so much distrust of the coming review proceedings on the part of the critics that some of them were on the point of dropping out. Hatzfeldt believed that if the critics came to know von Weizsacker at least part of the suspicion would be relieved.

Over the weekend the critics gave von Weizsacker, in broad outline, a preview of their critique of the Gorleben project. Lovins, leading the discussion on the crucial question of whether Germany needed repro-cessing and an expanded program of nuclear energy, argued as follows: The German economy did not necessarily require more energy; if more were needed, it probably could be obtained by further gains in energy efficiency; if a convincing need for new energy supplies should arise, it would probably not be for electrical energy, and if it were the best way of meeting it would be not from nuclear reactors but from more benign sources.[18] Lovins and von Weizsacker are said to have established a rapport. Von Weizsacker seemed much engaged, for example, when in an informal evening recital that Hatzfeldt arranged, Lovins played Chopin nocturnes and Beethoven sonatas on one of Crottorf's Bechstein grand pianos. Jonathan F. Callender, a Gorleben critic and a geologist from the University of New Mexico, told me that through this recital Lovins established a new level of communication with von Weizsacker, who "was very taken by this other dimension of Lovins."[19]

[17]Letter to the author from Hermann Graf Hatzfeldt, December 6, 1982.

[18]Interview by the author with Lovins, November 13, 1981.

[19]Callender was interviewed by the author on June 26, 1982. Following the weekend at Schloss Crottorf, von Weizsacker twice invited Lovins to his home at Starnberg, near

An aristocrat such as Count Hatzfeldt, with 7,500 hectares of land and a two-moat castle, has standing even in an egalitarian society, or especially in an egalitarian society. So it was that the high point of the weekend was the arrival of federal chancellor Helmut Schmidt from Bonn, by helicopter. Schmidt remained about four hours to discuss German energy policy and the Gorleben project. According to Callender and others present, Schmidt argued that nuclear technology was important to Germany for three major reasons: first, nuclear energy figured vitally in the national strategy for energy independence and freedom from an excessive reliance on Persian Gulf oil; second, nuclear power was cleaner and safer than some of the major alternative energy sources; and third, nuclear development was important to the nation's technology export program in that only Germany and a few other countries would be able to offer foreign customers a full range of nuclear facilities and services. In making this last point, Schmidt is said to have referred explicitly to the possibility, over the longer term, of not only reprocessing foreign spent fuel but even of disposing of the radioactive waste arising from such reprocessing. Schmidt listened to what the Gorleben critics had to say, but Jonathan Callender sums up the encounter this way: "I didn't sense that we had moved him in any way. He didn't come to learn. He came to see if we were a bunch of crazies. He wanted to know who this group was and whether we were going to cause his government trouble."[20]

The Gorleben International Review opened in Hannover on March 28, 1979—the same day that the Three Mile Island accident began grabbing headlines around the world. To say the least, this coincidence affected the atmosphere of the hearings, and no doubt had a bearing on its outcome. As Lovins has put it, "simultaneously you had on page one stories about the Gorleben hearings and stories about whether the Three Mile Island reactor was going to do itself in. That certainly helped create some atmosphere of skepticism about [the project proponents'] safety claims."[21] But the impact of the events at Hannover and Harrisburg went beyond news coverage. The Gorleben Review had opened in the middle of the week; by the week's end up to 140,000 anti-nuclear protesters from all over Germany had converged on Hannover (peacefully, as it turned out), according to police estimates.[22]

Munich, for informal energy policy debates with Wolf Haefele, an articulate proponent of nuclear energy and at the time director of the energy program at the International Institute for Applied Systems Analysis near Vienna.

[20]Interview with Calender, June 26, 1982.

[21]Interview with Lovins, November 13, 1981.

[22]Walter C. Patterson, "Harrisburg ist überall," *Bulletin of the Atomic Scientists*, June 1979, pp. 9–11.

The Gorleben International Review was conducted in a way that displeased the critics. They felt that the review was focused not on the soundness of the Gorleben project, but on the competence and validity of their own report—a massive, 2,200-page document that addressed a variety of potential problems ranging from the unfavorable economics of reprocessing to the possibility that the security measures taken to prevent theft of plutonium and terrorist incidents could compromise civil liberties.[23] Furthermore, neither the fuel cycle company, DWK, nor the federal nuclear safety and radiation protection commissions were included as parties to the review proceedings. The proceedings took the form of a debate in which the critics' statements were always followed by rebuttals by the pro-Gorleben panel of experts. Inasmuch as Prime Minister Albrecht did not want the debate focused on the details of the Gorleben project, the critics found themselves frustrated in presenting their case, especially as it related to the merits of the Gorleben salt dome as a repository site.

Yet Albrecht was to find reason not to allow licensing to proceed for the integrated fuel cycle center DWK had proposed. The tens of thousands of protesters who had gathered in the streets of Hannover could hardly be overlooked, and the prime minister had to avoid putting himself and his Christian Democratic government at a disadvantage vis-à-vis the Social Democrats in the Lower Saxony parliament. In Bonn, the Social Democrats, under federal chancellor Schmidt's leadership, supported nuclear power and the Gorleben project; but the Social Democrats in Hannover were uneasy about nuclear power and were against the proposed fuel cycle center.

Sometime after the Gorleben International Review ended in early April, von Weizsacker, as presiding officer, submitted to Albrecht a report that was never made public. To judge from Albrecht's subsequent decision about Gorleben, it may well have reflected at least some of the concerns expressed by the critics.

The Nuclear Policy of 1979

About six weeks after the review Albrecht announced that while the fuel cycle center would be technically feasible with certain safety modifications, it would not be needed for a number of years and was not

---

[23]See *Bericht des Internationalen Gutachten Gorlebens* (Hannover, Niedersächsiches Socialministerium, April 1979). Adversarial hearings on this report are given in Deutches Atomforum e.v., *Rede-Gegenrede*, 7 vols. (1979) (remarks made originally in English or French are printed in this work in an unauthorized and not always reliable translation). The report and hearings are summarized in Hatzfeldt and coeditors, *Der Gorleben-Report*.

politically feasible at that time. Albrecht said, "It is not possible to expect the population to gain confidence in the nuclear Entsorgung-Zentrum if the politically responsible hold different opinions in this matter. Exactly that, however, is the case today."[24]

But while Albrecht declared that the application for the reprocessing plant could not be entertained, he proposed that the geologic investigation of the Gorleben dome for a radioactive waste repository go forward and that DWK submit applications to build interim spent fuel storage facilities designed to be "inherently safe." By this he meant facilities in which the fuel would be cooled by natural convection rather than by a forced-cooling system dependent on "the functioning of technical equipment or on human reliability."[25] In Albrecht's view, the direct disposal of spent fuel as nuclear waste was feasible in principle, as the Gorleben critics had argued. The case for reprocessing, as he saw it, turned largely on whether Germany was going to go to breeder reactors; this decision could not be made for a number of years, pending successful operation of an experimental breeder being built at Kalkar. Albrecht was saying, in sum, that there was no urgent need for reprocessing; that there was the alternative of storing and disposing of spent fuel; and that it was too much to expect him and his Christian Democratic Union party to take the political heat for approving a project about which there was so much public apprehension and over which the major opposition party was itself seriously divided.

Since May 1979, when Prime Minister Albrecht announced his decision on the Gorleben project, Germany's nuclear energy policy and the attendant political environment have undergone significant changes. Decisions made at a September 28, 1979, meeting between Chancellor Schmidt and the heads of the state (*lander*) governments signified a new and much more cautious approach to the entire nuclear center facility. Reprocessing would go forward, but on a smaller scale and with an effort to optimize the technology from a standpoint of safety. At the same time, there was to be a major study of an alternative—the direct geologic disposal of spent fuel. It was expected that by the mid 1980s a decision could be made as to which method of closing the nuclear fuel cycle offered a decisive advantage.

Exploration of the Gorleben salt dome for a repository would be accompanied by continued efforts to increase the interim surface storage of spent fuel. The latter efforts would involve more concentrated pool storage at the reactor site; away-from-reactor storage facilities not only

[24]Ernst Albrecht, "Declaration of the state government of Lower Saxony concerning the proposed nuclear fuel centre at Gorleben," May 16, 1979 (official translation).
[25]Ibid.

at Gorleben but also at Ahaus in North Rhine-Westphalia; and substantial buffer storage at any new reprocessing plant to be built. Early commitments to either large-scale commercial reprocessing or to a particular geologic site for disposal of reprocessing wastes or spent fuel thus would be unnecessary. Altogether, there now promises to be enough interim storage capacity to accommodate more than 85 percent of all spent fuel arising by the year 2000, even if there should be no further reprocessing of German fuel at home or abroad.

The more cautious and tentative fuel cycle policy arrived at in September 1979 did not mean that the scientists, engineers, and government officials of the German nuclear establishment were any less determined than before to push on with reprocessing and breeder development. But it did mean that the political leaders were now insisting on greater restraint. The following year Ulrich Steger, then a prominent Social Democrat in the Bundstag (now the minister of economics in the Social Democratic government of Hesse), told me that nuclear power had suffered from being oversold and pushed too rapidly. "I personally believe that nuclear power would not have come to a crisis of public acceptance if it had not been overstated as an answer to the first oil crisis," he said. "It is much better to start slowly and proceed steadily and allow the public to get used to the technology." The French nuclear program, he felt, was developing "much too quickly," and certainly the German program should not be following suit.[26]

The Changing Political Environment since 1982

From the fall of 1982 through the winter of 1983 the political environment in Germany underwent important changes that may have affected support for nuclear energy in somewhat contradictory ways. The net effect will not be known for some years. The new coalition government of Christian Union parties and Free Democrats under Helmut Kohl, which succeeded to power in October 1982, was more staunchly pronuclear than the Social Democrat-Free Democrat coalition government under Schmidt that had just fallen apart. Although Schmidt himself was dedicated to nuclear development, his party was deeply divided over nuclear power as well as nuclear weapons issues. With respect to nuclear weapons, the Social Democrats soon abandoned Schmidt's position supporting deployment of intermediate-range ballistic and cruise missiles, and it was with this radically altered posture on a key NATO issue that the party went into the national election in March 1983.

[26]Interview by the author with Ulrich Steger, November 1980, at Bonn.

The election produced decidedly mixed results from the standpoint of gains and losses by the supporters and opponents of nuclear energy. The Social Democrats did poorly, the Christian Union-Free Democrat coalition emerging with a better than 55 percent majority. On the other hand, both the Social Democrats and the Free Democrats lost voters to the new Green party, which by winning more than the required 5 percent of the total vote (a formidable barrier for any new party) entered the Bundestag.[27] This situation invited speculation about an eventual coalition between the Social Democratic party (SPD), now turned cool toward nuclear power, and the Green party, to whom nuclear power was anathema.

The emergence of the Greens had come about through what might be described as a gradual organic process. The Green party, which was not formally established at the national level until January 1980, represented a coalescence of local-initiative groups and citizen campaign committees which had sprung up since the early 1970s to oppose a variety of nuclear and non-nuclear development projects, from the fuel cycle center at Gorleben to a controversial runway extension at the Frankfurt airport. The introduction to the party platform expresses the Greens' gloomy view of existing trends:

The people of the Federal Republic of Germany are today threatened by the ecological and economic crisis of industrial society....The foundations of life are endangered by atomic power stations, pollution of air, water and earth, storage of dangerous waste products and squandering of raw materials.... Production is determined not by the needs of people, but by the interests of big capital....The ecological balance is being sacrificed to economic growth, competition and profit, thereby threatening total contamination and destruction of the basis of human life, rising unemployment and growing social and emotional misery.[28]

The Greens were—and are—predominantly young, educated, urban, and middle class, but they also include some older rural conservatives. The initiative movement began, it will be recalled, with the support of the wine growers of Baden-Wurttemberg, and it was in this same conservative region in the upper Rhine Valley that in March 1980 the Greens for the first time won enough votes to enter a state parliament. Over the next three years the Greens gained representation in the parliaments

[27]*Facts About Germany* (Gütersloh, Federal Republic of Germany, Bertelsmann Lexikothek, 1984) p. 120.

[28]Richard Davy, "West Germany's Green Party," undated excerpt from the *Journal of the Institute for Socioeconomic Studies*, received by the author in January 1985 from the press office, Embassy of the Federal Republic of Germany, Washington, D.C.

of five other states, including Lower Saxony. The emergence of the Greens as a serious presence at both the federal and state levels was significant because of the important role a party with even a small number of seats can play in parliamentary coalition politics.

The Social Democrats also represent something new in the German experience in that today, as the principal opposition party, they are highly critical of the nuclear program and have now moved from wanting to block some of its important features to wanting to put an end to nuclear power. At the Social Democrats' national convention in May 1984 at Essen they adopted a resolution opposing any reprocessing other than that done experimentally at a small 35-ton facility at Karlsruhe. At their convention in August 1986 at Nuremberg they called for phasing out all nuclear power plants within ten years. Coming only four months after the nuclear accident at Chernobyl, the 1986 convention was no doubt influenced by public opinion polls showing a majority of Germans no longer in support of nuclear power.[29] Leadership of the party has been taken over by Johannes Rau, a strong personality who is prime minister of North Rhine-Westphalia, and who governs with an absolute SPD majority. Opposed to plutonium recycling, Rau has been delaying a licensing decision on the Kalkar breeder reactor until after the January 1987 national elections, in which he will be leading the Social Democrats' effort to oust Chancellor Kohl's Christian Union-Free Democrat coalition from power. If Rau and the SPD should unexpectedly prevail, both the Kalkar breeder project and a DWK project to build a demonstration reprocessing plant at Wackersdorf in Bavaria may well be at an end.

The controversy surrounding the Wackersdorf project, which is a direct spin-off from the aborted fuel cycle center project at Gorleben, is worth examination.

The Wackersdorf Demonstration Reprocessing Plant

From the end of 1979 through the first half of the 1980s the utilities' fuel cycle company, DWK, worked doggedly at obtaining a site for its demonstration reprocessing project and at perfecting plans to make this plant—called WA 350—a state-of-the-art facility. Early on a site in Hesse, where there were several fuel cycle and nuclear materials transport companies at Hanau, was considered a possibility, but various problems

---

[29]Mark Hibbs, "Germany's SPD Will Ban Reprocessing, Phase Out Nuclear if It Wins in 1987," *Nuclear Fuel*, July 28, 1986; "Pledge to Get Rid of Nuclear Energy," *The German Tribune*, September 7, 1986; interview by the author with Bernd Jahn, energy attaché, Embassy of the Federal Republic of Germany, Washington, D.C., October 27, 1986.

were encountered, not the least of them political. Although the Greens did not enter the Hessian parliament until 1983, they had already made their influence felt locally; in the communal elections of March 1981, for example, the Greens got more than 40 percent of the vote at Volkmarsen, a small community high on DWK's list of potential sites.[30]

Several states were at different times considered possible hosts for the WA 350 plant, but in the end the candidate sites came down to two: one was in Lower Saxony, the other in Bavaria. The Lower Saxony site was Dragahn, thirty miles from Gorleben but in the same Luchow-Dannenberg district. Prime Minister Albrecht had fairly early come around to favoring the project, evidently for economic reasons, saying that his rejection of the original plan for a fuel cycle center did not go beyond the specific project at the Gorleben site.

In Luchow-Dannenberg, a district hungry for new jobs, the demonstration reprocessing plant project commanded majority support even though many farmers, small businessmen, and craftsmen have feared that the changes brought on by any large nuclear facility might not be for the better.[31] It was not altogether clear how Albrecht had decided that a reprocessing venture was politically feasible now in the 1980s when only a few years earlier it had not been. The tumult that accompanied Three Mile Island had long subsided, but the Greens were a new presence in the state legislature and the Social Democrats there were now flatly opposed to a reprocessing project in Lower Saxony, as they announced in June 1984.

DWK's candidate site at Wackersdorf in Bavaria, like the Dragahn site in Lower Saxony, was close enough to the East German border to be eligible for a development subsidy from Bonn. Whatever else DWK may have found appealing about this location, its political advantages were manifest. The strongly pro-nuclear Christian Social Union party led by Franz Josef Strauss held sway in the state parliament, where it had a comfortable absolute majority. Strauss not only wanted the WA 350 project, but was eagerly courting DWK to get it.

The decision on where the plant was to go was set for early February 1985, and Albrecht, evidently sensing that DWK was leaning to Wackersdorf, in September 1984 warned publicly that Lower Saxony could

[30]Silke McQueen, "License Procedures for the 350-Tonnes/Yr. Reprocessing Plant," *Nucleonics Week*, May 7, 1981. Following the 1983 Hessian elections, the Social Democrats lacked a parliamentary majority and could rule only with the toleration of the Greens. In early 1986 the Social Democrats and the Greens in Hesse formally entered into a coalition, and a leader of the Greens was brought into the cabinet as minister of the environment. This has further complicated and worsened the licensing problems faced by such nuclear fuel manufacturers in Hesse as Alkem, a producer of plutonium fuel. Mark Hibbs, "Nukem Group Emerging—Slowly—from Thicket of Licensing Woes," *Nuclear Fuel*, February 10, 1986.

[31]Interview with Stephen von Welck, April 22, 1982.

not be expected to accept nuclear waste if it would not also get the economic benefits of the reprocessing facility.[32] This implied threat to the Wackersdorf project seems not to have disturbed DWK overmuch, however, because as Albrecht had feared the Bavarian site was selected. Albrecht called the decision "economically wrong and politically short-sighted." He pointed to the security and technical advantages that the Dragahn site would have because of its proximity to the spent fuel storage facility at Gorleben and the high-level waste repository that might be built there.[33]

Whatever the merits of the siting choice for the demonstration plant, DWK has sought to advance the state of the art of reprocessing. The reprocessing plant proposed as part of the original *Entsorgung-Zentrum* plan was to have essentially the same mix of remote and contact main-tenance that is found at Barnwell and La Hague. But the WA 350 plant will be entirely remotely maintained.[34] DWK is looking to innovative designs, developed by engineers at Oak Ridge National Laboratory in the United States, for remote maintenance by robots; backup would be provided by the long-proven method, developed at Hanford and Savan-nah River, of removing entire modular units by remotely operated over-head cranes, should any component fail. If the plant runs as it is supposed to, both exposure of workers to radiation and plant downtime to cope with leaks, plugged pipes, and other problems should be less than at other commercial plants. (Oddly enough, even as DWK was seeking major advances in the state of the art through its Oak Ridge consultants in 1981 and 1982, the company was trying to buy into the Barnwell reprocessing plant, about which these same consultants had grave doubts. See chapter 3.)

The design of WA 350 is intended to keep releases of radioactivity to lower levels than at any other existing or proposed commercial repro-cessing plant. The maximum releases expected to be authorized are (with certain exceptions) well below 1 curie per year for both alpha and beta emissions (taken together),[35] or substantially lower than even the much-reduced limits now planned for Sellafield. The German limit on beta emissions, however, does not apply to tritium, carbon-14, or

[32]Silke McQueen, "Mont Louis Incident Could Affect Critical German Nuclear Deci-sions," *Nucleonics Week*, September 20, 1984.

[33]Mark Hibbs and Ann MacLachlan, "DWK Chooses Bavarian Site for First Big German Reprocessing Plant," *Nucleonics Week*, February 7, 1985.

[34]Carsten Salander, "Closing the Fuel Cycle in the FRG," *Nuclear Europe*, April 1983, pp. 36–37; Joachim Mischke and Klaus Hendrick, "Remote Maintenance Concept of the DWK Spent Fuel Reprocessing Plant," in *Proceedings of the 1984 National Topical Meeting on Robotics and Remote Handling in Hostile Environments* (n.p., n.d.).

[35]Letter to the author from Carsten Salander of DWK, January 30, 1985.

the noble gas krypton-85. Germany's Radiation Protection Commission has concluded that retention of krypton and carbon-14 is not necessary for operations at the level presently planned, although the commission has called for construction of a demonstration krypton-retention facility in case there is a major future expansion of reprocessing.[36] The greater part of the tritium (H-3), the release of which to fresh and potentially potable water is controversial, is expected to be retained; the annual release limit is set at 40,000 curies (a small fraction of the amount to be released to salt water from the THORP plant under construction at Sellafield).[37] But tritium removal is a difficult and demanding process that is not cheap, and according to at least one reprocessing expert, is not feasible; its cost has been estimated by an American specialist at 10 percent of the total cost of reprocessing.[38]

It should be noted that all of the low- and intermediate-level wastes captured at the WA 350 plant, like those captured at German reactors and other fuel cycle facilities, are destined for geologic disposal. The Germans do not consider near-surface disposal, the practice favored by the United States and France, to be sufficiently robust, and it is not done. Low- and intermediate-level wastes were disposed of in a salt mine at Asse until that facility was closed in 1978; in the future these wastes are expected to go to a former iron mine near the Lower Saxony town of Salzgitter.[39]

But while the present plans for reprocessing and waste management have their strong points, the WA 350 plant will contribute significantly to an inventory of separated plutonium in Europe that will be far greater than needed for the ongoing experimental and demonstration work with fast breeders. If the WA 350 facility meets its nominal reprocessing capacity of 350 tons of fuel a year (its actual capacity is expected to be greater), that plant alone will produce up to 3 tons of plutonium a

[36]Jeffrey Schevitz, "The Backend of the Nuclear Fuel Cycle in the Federal Republic of Germany," draft, December 1984. Schevitz is with the Karlsruhe Nuclear Research Center.

[37]DWK, "Kurzbeshreibung für die Wiederaufarbeitungsanlage Wackersdorf," brochure describing the WA 350 plant (August 1983), table 5, p. 57.

[38]Interview by the author on June 24, 1985, with R. W. McKee, technical leader in waste management at the Battelle Pacific Northwest Laboratories, Richland, Wash. The tritium removal process is called voloxidation. It would involve heating the chopped-up fuel at the start of reprocessing, then capturing the tritium and other volatile elements as they come off as gases. The gaseous tritium would probably be condensed and captured as tritiated water. The process requirements are exceedingly demanding: high temperatures (up to 500° C) and rotating, remotely maintained equipment that must be kept air-tight so as not to allow any dilution of the off-gases from the heated fuel. Walton Rodger, at one time manager of the West Valley reprocessing plant, doubts that voloxidation is feasible; some other fuel cycle experts think the process would be feasible but expensive.

[39]Salander, "Closing the Fuel Cycle in the FRG."

year—several times the amount needed for the demonstration breeder awaiting licensing at Kalkar. In addition, if the production goals of the French and British reprocessors are met, by the late 1990s there will be some 3,600 kilograms of plutonium returned to German utilities each year from abroad.[40]

The startup of the WA 350 plant, if it occurs, will be hard to justify unless the plutonium recovered is used in existing reactors. The high reprocessing costs that will have been incurred and the substantial further costs associated with safely storing plutonium indefinitely will push reprocessors and their utility customers in all countries to go to such recycling (see chapter 12). In fact, the twelve utilities that own DWK already have agreed to pool the plutonium returned from France and Britain for recycling in their light water reactors.[41] Yet because plutonium is an explosive as well as a fuel, the prospect of recycling allows the Social Democrats, the Greens, and others to point to the link between nuclear power and nuclear bombs.

This exacerbates the German nuclear industry's political problem at a time when, following Chernobyl, it is very much on the defensive, as evidenced by the shift of the Social Democrats from being critics of nuclear power to being its emphatic opponents. In the elections in Lower Saxony and Bavaria in June and October of 1986 the Christian Union parties retained their control of the *lander* parliaments; but in Lower Saxony it was a near thing, with Ernst Albrecht losing his absolute majority and seeing the Social Democrats and Greens win 49 percent of the seats. In Bavaria, where the Christian Social Union party is led by the strongly pro-nuclear Franz Josef Strauss, the Social Democrats lost ground—but to the Greens, who entered the Bavarian parliament for the first time.

In the spring of 1986 as many as 50,000 protesters gathered at the Wackersdorf site.[42] Such turbulence may offend more voters than it persuades, especially in Bavaria, but it has nonetheless contributed to the palpable evidence that nuclear power in Germany sorely lacks the trust of a wide public.

[40]David Albright and Harold Feiveson, "Plutonium Recycle and the Problem of Nuclear Proliferation," PU/CEES Report no. 206, Center for Energy and Environmental Studies, Princeton University (April 1986). See also David Albright, "World Inventories of Plutonium," Research Report 195, Center for Energy and Environmental Studies, Princeton University (March 1986).

[41]Ann MacLachlan, "German Utilities Pooling Their Plutonium to Promote Economics for Recycle in LWRs," *Nuclear Fuel*, November 22, 1982.

[42]Mark Hibbs, "Germany's DWK, Bavarian Officials Agree to Work for Public Support of Reprocessing," *Nuclear Fuel*, April 7, 1986.

Door Ajar: The Geologic Disposal Alternative for Spent Fuel

On January 25, 1985, a month before the scheduled announcement by DWK of its siting decision for the demonstration reprocessing plant, Chancellor Kohl and his cabinet took an action carefully timed to ensure a favorable policy setting for that announcement: refinement of the government's nuclear fuel cycle policy in light of the findings of the four-year, $20-million study of alternative fuel cycle options which had been mandated by the heads of the federal and state governments in September 1979. In its 1985 policy statement the cabinet's rhetorical emphasis was on reprocessing as the preferred and still legally required method of closing the cycle. This emphasis was entirely predictable; in the fall of 1983, long before the study of alternatives was completed, Interior Minister Friedrich Zimmermann had declared reprocessing to be essential and said that Germany should not be forever dependent on foreign reprocessors.[43] Now the Kohl cabinet was stating as its present view that under German atomic energy law direct geologic disposal of spent fuel was permissible only for that fuel for which reprocessing could not be economically justified—fuel, for instance, known to contain relatively little plutonium. But at the same time the cabinet left the door ajar for the direct disposal alternative. This alternative appeared realizable in principle, it said, and work would continue toward bringing it to technical maturity.[44] Moreover, the cabinet's judgment that direct disposal could not be the preferred choice is clearly subject to future reevaluation in view of German and foreign experience.

Despite official preference for reprocessing, the study of alternatives, carried out by the German nuclear industry under the leadership of a special group at the Karlsruhe Nuclear Research Center, placed direct disposal in a favorable light. This group found no decisive differences with respect to radiological safety between the two methods for closing the fuel cycle. For direct disposal, the occupational radiation dose was estimated at 15 percent less than the dose from the fuel cycle that includes recycling; the collective dose commitment to the public was put at 50 percent less. For both methods, the doses were deemed negligible compared to exposure to radiation from natural sources.[45]

[43]Silke McQueen, "Germans Say Reprocessing Plant Is a Must Despite Spent-Fuel Coverage Past 2000," *Nuclear Fuel*, November 7, 1983.

[44]Letter of Carsten Salander to the author, January 30, 1985.

[45]K. D. Closs and H. Geipel, "Some Preliminary Results for the FRG Alternative Fuel Cycle Evaluation," paper presented at the International Meeting on Fuel Reprocessing and

The cost comparison favored direct disposal by about 30 percent,[46] but that cost advantage was probably understated. The study assumed that the real cost of reprocessing will decline substantially over the next fifteen years as a result of advances in the state of the art. The cost for direct disposal for the year 2000 was put at DM 1,112 per kilogram of fuel, compared to DM 1,592 for the fuel cycle that involves reprocessing and recycling, or $556 compared to $796 (at 2 marks to the dollar). The economic edge of direct geologic disposal over reprocessing and thermal recycling was expected to continue even if the price of uranium (after inflation) increased by 3 percent a year between now and the turn of the century. Reprocessing would allow some saving of uranium, but according to the study no major reduction in uranium imports would occur without fast breeders.

The system for geologic disposal of spent fuel described in the Karlsruhe study was novel, but to all appearances straightforward and robust. The emplacement method proposed was quite different from that contemplated in the United States, where spent fuel assemblies (or the disassembled rods from several assemblies) would be inserted in canisters to be placed by a heavily shielded remote-handling device into vertical holes drilled in the tunnel floor, one canister to a hole. In the Karlsruhe concept, three intact fuel assemblies would be put at the center of a massive, 55-ton waste container, 6.0 meters long and 1.4 meters in diameter. The container would consist of modular cast-iron shielding surrounding a high-strength, corrosion-resistant steel canister in which a steel "drybin" containing the assemblies is inserted. Because of the heavy shielding from the iron, the waste containers could be safely handled by workers, with very little radiation exposure, and might eventually be licensable as transport casks. They would be placed end to end on the floor of tunnels in the salt dome repository, then covered over with crushed salt as each tunnel was filled. Retrieval of the containers was not contemplated in the study. The temperature at the surface of a canister would reach a maximum of 165° centigrade after thirty years, but would still exceed 100° C after a few hundred years. In the view of the authors of the study, gaining unauthorized access to the packages would pose "enormous technical problems."

While these technical studies were being made and while political and policy decisionmaking was going on at the federal and state levels,

---

Waste Management, Jackson, Wyo., August 26–29, 1984, figs. 5, 6, and 7. Closs is with the Nuclear Research Center at Karlsruhe, Geipel with the federal Ministry for Research and Technology.

[46]Project Group Alternative Entsorgung, "Systems Study Alternative Entsorgung," executive summary, final report KWA 2190/1 (Karlsruhe Nuclear Research Center, 1984).

the investigation of the salt dome itself was going on at Gorleben—and the findings were being criticized.

The Status of the Gorleben Salt Dome Investigation

The investigation of the Gorleben salt dome that began in the spring of 1979 has become a large and expensive undertaking, one which will stretch over more than twelve years and cost more than a billion dollars. The initial phase of the investigation, involving the drilling of boreholes from the surface and use of seismic surveying techniques, was completed by mid 1983.[47] By that time a total of more than 430 boreholes had been sunk in an area of some 300 square kilometers; most had been drilled to relatively shallow depths in order to study the regional hydrology and to map the top of the dome. Four holes were sunk to a depth of 2,000 meters in the dome's flanks, to take core samples; then a crucial step was taken when two pilot holes were sunk to almost 1,000 meters at the center of the dome.

On the strength of the first part of the investigation, Chancellor Kohl and his cabinet agreed in July 1983 that the undertaking should move on to the next phase, in which the dome is to be examined from two exploratory shafts and connecting tunnels.[48] But it would be the turn of the century before a repository could begin operating. The prolific aquifer overlying the dome has presented, at the start, a complication requiring several years to overcome. Shaft construction could proceed only after the freezing of the aquifer from top to bottom at the points where the shafts were to be dug. This task was still in progress in 1986, with the actual sinking of the exploratory shafts not expected to begin until the fall of that year. Altogether, excavating the shafts and tunnels and carrying out the investigation at repository depth will take six to eight years and will not be completed before 1992 or 1994. The cost of the site exploration and testing is put at DM 2.6 billion, or $1.3 billion.[49] If all has gone well, the project will then move on to final plan verification, licensing, and construction of the repository.

[47]Franz-Peter Oesterle, "Status of Site Investigations at Gorleben," paper presented at the American Nuclear Society meeting, Detroit, June 12–17, 1983. Oesterle is with the Physikalisch-Technische Bundesanstalt, Braunschweig.

[48]Schevitz, "The Backend of the Nuclear Fuel Cycle," para. 1.8.4.1.

[49]Hans-Anton Papendieck, "Salt-Vault Nuclear-Waste Storage Dispute Persists," *Stuttgarter Zeitung*, November 1985, translated and reprinted by *The German Tribune*, December 29, 1985. At each of the two spots where a shaft was to be excavated some forty holes were drilled to a depth of 300 meters and filled with a coolant (calcium chloride) kept at a temperature of −40° C by "gigantic refrigeration units with the capacity of 30,000 domestic freezers." Ibid.

A number of significant findings were made during the initial, or surface, phase of the investigation. It was discovered early on that the dome was narrower than first believed.[50] This would have posed major constraints on waste storage capacity except for the novel emplacement concept that was envisioned: holes 300 meters deep were to be drilled down from the floor of the tunnels, and long strings of canisters would go into each of them.[51] With this vertical emplacement configuration, there would be room for at least a fifty-year accumulation of high-level waste, with buffers of salt 200 meters thick between the repository and the dome's flanks and from 400 to 500 meters thick from the repository to the top of the dome. But should the waste ever consist predominantly of spent fuel, storage constraints might indeed arise, for as noted above the casks envisioned for spent fuel in the Karlsruhe study are massive containers that would be left on the floor of the tunnels, at the same elevation or horizon within the dome.

Although the Gorleben dome is no exception to the general rule that salt domes are complex, the limited deep borehole drilling that had been done convinced the technical agency in charge of the project that its structure is simpler than that of most of the domes where salt has been mined.[52] But one project consultant found it disturbing that in 1969 gas explosions had blown out an exploratory well about 1.5 kilometers away on East Germany's Rambow Dome, which is connected to the Gorleben dome.[53] Also, the fact that two of the deep exploratory boreholes at the dome's flanks had encountered brine influxes encouraged speculation that pathways for the flow of brine from the inside to the outside of the dome might open up after the introduction of heat-generating waste.[54]

In June 1984 the Bundestag's Interior Committee held a hearing on the technical merits of the Gorleben project. The criticims made by the Gruppe Okologie Hannover, of whom one of the leaders was Helmut Hirsch, were extensive and detailed, but fairly or not could be dismissed by project proponents as predictable. Not so easily dismissed, however, was the critical testimony by several independent academic experts. One of them was Klaus Duphorn, professor at the University of Kiel's

[50]Interview with Stephen von Welck.

[51]H. Röthemeyer, "Site Investigations and Conceptual Design for the Repository in the Nuclear 'Entsorgungszentrum' of the Federal Republic of Germany," reprinted from International Atomic Energy Agency, *Underground Disposal of Radioactive Wastes* vol. 1 (Vienna, 1980).

[52]Oesterle, "Status of Site Investigations at Gorleben."

[53]Schevitz, "The Backend of the Nuclear Fuel Cycle," summarizing the views of Klaus Duphorn of the Institute and Museum of Geology and Paleontology at the University of Kiel.

[54]Ibid.

Institute and Museum of Geology and Paleontology. As a consultant on the Gorleben project, Duphorn had in May 1982 submitted to the agency in charge a report recommending that the Gorleben dome be abandoned.[55] Now he was highlighting his reasons for believing the dome to be unsuitable: geologically recent processes—processes which might recur during the period of radioactive hazard—had affected the dome drastically, causing the loss of at least 4 cubic kilometers of salt to erosion and dissolution within the last million years. It might take as much as another half-million years for the salt above the repository to dissolve, Duphorn conceded, but in his view tension in the salt induced by heat-generating waste, together with certain selective processes of salt dissolution suspected at Gorleben, might bring about a hydraulic short circuit and an escape of radionuclides in a much shorter time.[56]

Another critic, A. G. Herrmann of the University of Gottingen's Georg-August Institute of Geochemistry, testified that borehole core samples had shown that the Gorleben dome and the overlying strata could not satisfy geologic isolation requirements.[57] According to his interpretation, the data not only confirmed that brine could penetrate the salt, but showed that some brine must have moved several hundred meters from its place of origin. From the work of one of his students Herrmann had concluded that heat from the waste would stress and deform the salt and allow brine to migrate on a "front" through the salt. Claims that salt always changes its form without fracturing, and is thus impermeable to solutions and gases, should not be accepted, he warned.

On two occasions officials in Bonn, first under Chancellor Schmidt and then under Chancellor Kohl, had considered but rejected expansion of the German investigation to include additional salt domes.[58] Both the technical agency in charge at Gorleben and the Reactor Safety Commission had concluded that the critics were unduly pessimistic and that there was nothing yet to indicate that the site was likely to be unsuitable for repository development. In July 1983 the cabinet therefore found no need to widen the investigation to one or more additional sites. On this point Helmut Röthemeyer, a leader of the investigation, has observed that if a second site were explored and found to be marginally better than the Gorleben dome, disagreements would inevitably arise over assessment criteria, possibly making investigation of a third site necessary.[59]

[55]"Atom Waste," Der Spiegel no. 28 (July 28, 1982) p. 146.
[56]Schevitz, "The Backend of the Nuclear Fuel Cycle."
[57]Ibid.
[58]Interview with Stephen von Welck, April 22, 1982; Schevitz, "The Backend of the Nuclear Fuel Cycle."
[59]Papendieck, "Salt-Vault Nuclear-Waste Storage Dispute Persists."

Conclusion

The Gorleben project has from the beginning symbolized a major gamble. The least of this gamble may turn out to be the heavy and exclusive bet that the German nuclear industry and the present government have placed on the Gorleben salt dome, although clearly it would be technically more prudent to have one or more back-up sites. The larger gamble is that, despite Chernobyl and the present defensive posture of the nuclear industry in Germany and around the world, the industry and government are betting heavily that the time has come to proceed with commercial reprocessing—and, inevitably, if the continued accumulation of burdensome stocks of separated plutonium is to be avoided, to use of plutonium fuel in light water reactors.

This gamble has affected nuclear waste management directly because disposal of high-level waste from reprocessing is the driving purpose behind the investigation and repository design work at Gorleben. Yet a likely outcome of this is a costly and embarrassing mismatch between repository and the waste requiring disposal. For many years, and possibly many decades, the greater part of the high-level waste on hand will be in the form of unreprocessed spent fuel. But the $1-billion-plus commitment to the exploration of the Gorleben salt dome has been made despite the fact that this site appears unsuited for spent fuel disposal as the Germans themselves have defined it, and is clearly unsuited for disposal in a retrievable mode.

The industry's present spent fuel management policy of expanding surface storage to fit the need, if continued long enough, will invite renewed charges that future generations will inherit the burden of coping with this generation's nuclear waste.

9
Sweden: Robust Solutions

Sweden has in recent years been in a period of relative calm with respect to nuclear power and nuclear waste. But for nearly a decade Sweden was gripped by nuclear controversy, with the waste issue at the heart of much of it. At one point a coalition government actually fell following an internal disagreement over these matters.

Despite past controversy, nuclear power has become an increasingly important Swedish energy source. With two more reactors connected to the grid in March 1985, there are now a total of twelve, located in four coastal stations. They generate 45 percent of Sweden's electric energy, most of the rest coming from hydropower.[1] During the 1970s the nuclear controversy seemed to increase in step with the growth of nuclear construction. The controversy abated following the national nuclear referendum of March 1980, by which Swedish voters decided that fission energy would be phased out over the life of the twelve reactors. This decision was subsequently interpreted by the Parliament to mean that all nuclear generation will stop not later than the year 2010, now less than twenty-five years away.

But, in truth, what Sweden's national policy for nuclear power ultimately will be remains uncertain. The Social Democratic government has been sincerely concerned to bring about the nuclear phase-out, particularly since the further loss of popular support that nuclear power has suffered following the Chernobyl accident. The phase-out is supposed to be made possible through increased efforts to improve energy efficiency and development of alternative sources. Yet the outcome of the present policy will almost surely be decided by such factors as the

---

[1]In 1986 the mix of sources for Swedish electricity generation was: hydropower, 50 percent; nuclear power, 45 percent; oil and coal, 5 percent. Interview by the author, October 20, 1986, with Inge Pierre, nuclear attaché, Embassy of Sweden, Washington, D.C.

future of the Swedish economy and its energy demand; the availability and acceptability of the alternatives to fission energy; public confidence (or the lack thereof) in reactor safety and waste management; and the fate of nuclear power worldwide. In fact, one interpretation of present policy, and quite possibly the most valid interpretation, is that it represents not so much an assured phase-out as a moratorium that could lead either to abandonment of nuclear energy or to further commitment to it. "The net outcome [of the referendum] is, I would say, a 10 to 15 year moratorium on the nuclear issue," I was told by Mäns Lönnroth, a prominent energy analyst at Stockholm's Secretariat for Future Studies.[2] Many in Sweden's nuclear industry who believe that conservation and alternative sources (apart from imported coal) cannot suffice to replace nuclear power also interpret present policy as a moratorium of sorts.

If in the end nuclear power should remain a major part of Swedish energy supply, the reason will lie partly in current efforts to establish spent fuel management and waste disposal practices that inspire public confidence. The schedule for ultimate disposal is protracted, with no repository operations contemplated before the year 2020;[3] but the chances for success in carrying out an independent national disposal program seem better for Sweden than for most countries. The Precambrian shield granite and other crystalline rock of the Scandinavian peninsula may offer the best medium for a repository to be found in Europe.

The Swedish approach to waste and spent fuel storage, transport, and disposal problems—both technical and political—hardly could be more cautious and conservative. Furthermore, Sweden is spared some vexing waste management and other fuel cycle problems because of the limited scope of its nuclear program. The Swedes take pride in being the only nation to develop light water reactors without relying on United States or Soviet patented technology. But the Swedes have done no uranium mining or enrichment, no reprocessing of spent fuel, and no breeder development. Consequently, there are no uranium mill tailings waiting to be buried, no Hanfords, West Valleys, and Sellafields waiting to be cleaned up, and no separated plutonium awaiting a foolproof system of protection against theft, diversion, or terrorist incidents. Most and perhaps all high-level and long-lived waste will be committed to a deep geologic repository.

[2]Personal communication from Mäns Lönnroth to the author, January 1983.
[3]Swedish Nuclear Fuel and Waste Management Co. (SKB), "SKB Activities" (Stockholm, November 1985), brochure.

Early Controversy: Nuclear Development and Its Opposition

The long controversy that preceded the 1980 referendum began in the early to mid 1970s. Nuclear power had been perceived as a major energy option as early as the 1950s, when proposals to expand hydropower in the north of Sweden encountered increasing opposition from environmentalists. Sweden had no fossil fuels except peat; but it had abundant, if mostly low-grade, uranium deposits, and the stable Precambrian shield rock was but one of several factors favorable to the siting of waste disposal facilities within Sweden. Siting was also aided by the fact that by European standards Sweden is large and sparsely populated; it is almost twice the size of the United Kingdom, but has only about a seventh as many inhabitants. By the 1970s the Swedish utilities, consistent with world nuclear optimism of that time, were forecasting a need for twenty-four light water reactors in 1990.[4] Breeder reactors and facilities for a complete breeder cycle had been envisioned, but commercial nuclear power was still at such an early stage that breeders were nowhere near realization. Moreover, because Sweden possessed large uranium resources, the breeding and recycling of plutonium were not considered pressing needs. The demonstration by students in 1970 at a west coast site acquired some years before for an experimental breeder was addressing more of a phantom than a reality, for there were still no firm plans to undertake this or any other breeder development project.[5] From 1965 on the focus of nuclear development was almost entirely on light water reactors.

By 1973, at the time of the first oil shock, the Social Democratic government had licensed eleven reactors for construction. But in that year the first important and sustained opposition to nuclear power began in Parliament, particularly in the Center party, largest of the opposition parties. Thorbjorn Falldin, the Center party leader, has been described as holding almost a religious conviction that dependence on nuclear power to sustain living standards is inherently evil. A sheep farmer from northern Sweden, Falldin saw nuclear waste as a potential environmental

[4]Mäns Lönnroth, "Nuclear Power in the Swedish Energy Futures—Commitments and Alternatives," paper presented at Nuclear Energy—Implications for Society, conference sponsored by the Groupe de Bellerive in Geneva, Switzerland, February 15–17, 1979.

[5]Letter to the author from Erik Svenke, former president of the Swedish Nuclear Fuel Supply Co., August 27, 1985.

poison and plutonium as a potential nuclear explosive subject to abuse.[6] In this view he is said to have been strongly influenced by the Swedish Nobel laureate Hannes Alfven, a physicist who was once a supporter of fission energy but since the late 1960s has been a severe critic.[7] The only other anti-nuclear party was the Communist party, which in Sweden is Marxist but not oriented toward Moscow; it has a modest following among workers and professionals.

An energy policy bill passed by Parliament in 1975 increased the nuclear commitment from eleven power reactors to thirteen, but in that bill conservation was emphasized to reduce the growth of energy demand.[8] And while reprocessing was still seen as essential in the long run, decisions were deferred both on this matter and on the sensitive question of mining Swedish uranium instead of continuing to import uranium from abroad.[9]

The next major development came during the 1976 election campaign. The Social Democrats, a party of low- and middle-level professionals and the trade unionists, had held office for forty-four years. But in this election year Falldin's Center party and the other two nonsocialist parties, the Conservatives and the Liberals, joined in a coalition in hope of finally ousting the Social Democrats from power. The coalition covered a remarkably diverse lot of constituencies. The Center party represented farmers, some workers, and some low-level professionals; the Liberal party represented mainly mid-level professionals; and the Conservative party represented industry and the higher professional and civil service cadres.

The idea that this odd and loosely coupled assortment of interests might soon be challenged to maintain a semblance of unity on the emotionally divisive nuclear power issue seems not to have occurred to anyone. Indeed, up until several weeks before the election the principal issue in the campaign was not nuclear power, but a proposal by the Swedish Labor Organization, a powerful element within the Social Democratic party, to have 20 percent of all corporate profits go to a fund run by the trade unions. But then Falldin, to the surprise of the strongly pro-nuclear Conservatives and the mildly pro-nuclear Liberals, stirred things up by promising that if he became prime minister an impartial

[6]Marquis Childs, *Sweden: The Middle Way on Trial* (New Haven, Yale University Press, 1980) pp. 66–67.
[7]Torsten Kalvemark, "Swedish Public Discussion on Nuclear Power," *Current Sweden* no. 245 (The Swedish Institute, February 1980); Kerstin Niblaeus, "National Referendum on Nuclear Power in Sweden," *Current Sweden* no. 246 (February 1980).
[8]Lönnroth, "Nuclear Power in the Swedish Energy Futures."
[9]Sweden has been importing uranium from Australia, Canada, the United States, and France (the uranium from France originates in Niger and Gabon). SKB, "SKB Activities."

study of the nuclear hazards—especially the waste hazard—would be commissioned, and that a popular referendum on the future of nuclear power might follow. Falldin favored a nuclear phase-out by 1985, and suggested that absent a safe solution to the problem of waste disposal, the five reactors already operating should be shut down. Some observers think that Falldin's strong position against nuclear power tipped the election in his favor. In any case, the coalition won a parliamentary majority and took over the government.[10]

The Stipulation Act: Political Compromise—with Political Fallout

Within the victorious coalition, sharp differences over the nuclear issue soon became evident. If the cabinet would not permit Falldin to honor his anti-nuclear campaign commitments, his choice would be either shamefacedly to wriggle out of them or to resign. But the Conservative and Liberal leaders were under pressure to help Falldin out of his political predicament: if he resigned, the government would collapse and the new one that would then have to be formed might exclude them. The upshot was a compromise that led to a major study of energy options, and to passage of the Stipulation Act by Parliament in April 1977. In the main, the act required that before any new reactor could be loaded with fuel, the utility would have to either present an acceptable contract for the reprocessing of the spent fuel and show how and where the terminal storage of the resulting high-level waste could be achieved with "absolute safety," or show how and where the spent fuel itself was to be finally stored with absolute safety.[11]

These conditions applied to the last six reactors in Sweden's twelve-reactor program, and with most immediacy to Ringhals-3 and Forsmark-1, which were expected to be ready for loading in 1978, then only a year away. The first five power reactors, already operating, did not come under the act. But the sixth reactor, Barseback-2, expected to start up in 1977, was covered by a special provision holding, in effect, that this reactor would not be permitted to operate after the end of the year unless an acceptable reprocessing contract could be obtained to take care of its spent fuel.

---

[10]Childs, *Sweden*, pp. 56–70; Mäns Lönnroth, "The Politics of the Back End of the Nuclear Fuel Cycle in Sweden," paper presented at the workshop on International Implications of Radioactive Waste Management, Keystone, Colo., September 15–17, 1979 (hereafter cited as Lönnroth, Keystone paper).

[11]Lönnroth, Keystone paper.

The Stipulation Act had the utilities and their Swedish Nuclear Fuel Supply Company (SKBF) in a bind. Anticipating the shortage of commercial reprocessing capacity that had developed in Europe, SKBF and the utilities had adopted a wait-and-see policy and were already planning for a central spent fuel storage facility.[12] This policy had been expected to make it unnecessary to rush into reprocessing contracts. The fact that Sweden had enormous reserves of low-grade uranium also offered reassurance. But with the new legislation the pressure was on, and SKBF was forced to seek reprocessing contracts. The Stipulation Act's Barseback-2 provision permitted no other option. Also, given the time constraints imposed by the act, the Swedish utilities had to consider how the requirements for showing plans for safe terminal isolation could be most quickly met. Planning concepts and some technology development were considered more nearly available for isolation of reprocessing wastes than for isolation of spent fuel.

In any case, SKBF officials persuaded COGEMA, the French firm, to give them a contract for the reprocessing of Barseback-2 fuel at the existing plant at La Hague. A contract was then negotiated to have fuel from the Ringhals and Forsmark nuclear power stations reprocessed at the new COGEMA plant to be built in the 1980s. The Swedish utilities also hurried to develop a plan for the management and terminal geologic disposal of the reprocessing waste that would be returned from La Hague.[13]

Falldin and the Center party believed the Stipulation Act imposed a moratorium on further nuclear development. Falldin was convinced that the terms of the act could not be met, at least not before the next regular election in September 1979. But the Conservative and Liberal party leaders viewed the nuclear waste problem as readily solvable. As one astute observer has characterized their attitude, they saw the Stipulation Act as a "spur for the utilities to speed up a program that would settle the issue once and for all."[14] The Social Democrats, for their part, had opposed the act as unnecessary and likely to engender hurried and possibly unsafe responses by the utilities. Their proposal, which Parliament had rejected, was for an accelerated program of away-from-reactor storage and for a long deferral of decisions on reprocessing.[15]

The Stipulation Act set forth no criteria for judging the adequacy of either the reprocessing contracts or the plan for geologic isolation that SKBF and the utilities were developing. The Center party argued that

[12]Erik Svenke, letter to the author, January 25, 1985.
[13]Svenke, letter to the author, August 27, 1985.
[14]Lönnroth, Keystone paper.
[15]Ibid.

the contract obtained from COGEMA offered no assurance that reprocessing would take place, and that it therefore did not fulfill the act's
requirements. But the cabinet decided that it was enough to have a
plausible contract with a reprocessor who could be expected to deliver
on its commitment.[16]

To meet the requirement for showing that a method and a site for
geologic disposal of waste were available, a new organization known as
KBS (Kärn-Bränsle Säkerhet, or Nuclear Fuel Safety) was created as an
appendage to SKBF. It was forced to attempt to accomplish within a
year what still eluded the nuclear enterprise in the United States and
other countries after many years. The first KBS report (KBS I) was
published in late 1977. It contemplated a comprehensive series of steps
for spent fuel management and waste disposal: removal of spent fuel
from the reactors, and temporary storage at the reactor site; transportation of the spent fuel to a near-surface away-from-reactor facility, to
be built in caverns excavated from hard rock, where the fuel would be
further stored for up to ten years; eventual transportation of the spent
fuel to France for reprocessing by COGEMA at the planned UP-3A plant
at La Hague; vitrification of the liquid high-level waste at La Hague and
return of this waste to Sweden in stainless steel canisters, which would
be stored for at least thirty years in a near-surface facility similar to, and
co-located with, the AFR facility for spent fuel; encapsulation of the high-
level waste cylinders in canisters of titanium, lead, and stainless steel
expected to last at least a thousand years; and, finally, commitment of
the canisters to a repository built 500 meters deep in granite or other
hard rock, beginning in the year 2020.[17]

Selection of a site for a repository was not expected until about the
year 2000, but to satisfy the Stipulation Act evidence of the feasibility
of geologic isolation was needed right away. To that end, several boreholes were drilled to a depth of 500 meters at each of three different
places in southeastern Sweden; the crystalline rocks present at these
locations—granite, gneiss, and gneiss-granite—were deemed representative of that part of the country. All three places were declared at the
time to be satisfactory for final storage of the waste, although it was
emphasized that the actual designation of a repository site was still years
away.[18]

In keeping with Swedish custom, the KBS report was sent out to
interested groups, universities, and research laboratories for review. But

[16]Lönnroth, Keystone paper; Svenke, letter to the author, August 27, 1985.
[17]Kärn-Bränsle Säkerhet (KBS), *Safe Handling and Storage of High-Level Radioactive
Waste* (Stockholm, Swedish Nuclear Fuel Supply Co., 1979).
[18]Ibid.

in an unprecedented extension of the usual procedure, reviews were also solicited from twenty-five selected foreign experts and laboratories. The reviews were in hand by mid 1978. A number of them, including some of the more critical, praised the report as a substantial contribution to the formulation and analysis of nuclear waste management and disposal concepts. For instance, the Jet Propulsion Laboratory at the California Institute of Technology called it a "major step forward" in the analysis of high-level waste disposal systems.[19] But that review, among other important reviews, made clear that for the KBS plan to meet Sweden's "absolutely safe" standard much more research and verification would be required.

Hard as it was for the Swedish utilities to show that the proposed high-level waste disposal system would be absolutely safe, it was scarcely less difficult for the Center party and scientific critics of the plan to show that the system would be unsafe. While in principle the burden of proof was on the utilities, the political reality was that no reactors were going to be abandoned except for compelling reasons.

An odd situation came about within the governing coalition in the late summer of 1978. Falldin and other Center party leaders, such as Energy Minister Olof Johansson, passionately believed that the test of the Stipulation Act had not been met. But the Conservative and Liberal leaders disagreed. If Falldin could not win them over to his point of view, his party's position on nuclear power would be not merely undercut, but dramatically reversed. The fueling of new reactors could leave the way clear for an open-ended expansion of nuclear power. Furthermore, from press accounts the public could only think that Falldin was surrendering on a point of principle. A "rank-and-file upheaval within the Center Party" is said to have ensued.[20]

In an effort to hold the coalition government together, the cabinet agreed that the terms of the Stipulation Act could not be met without further investigation of one or more of the three sites where test drilling had been done. But the expectation was that after the drilling of a few more boreholes—former prime minister Palme called them "political drilling holes"—the utilities would be back asserting that a demonstrably suitable site had indeed been found. Meanwhile, the report of a commission which had been appointed after the 1976 election to review the Swedish energy situation turned out to be of no help to Falldin either, for the commission reaffirmed the advantages of nuclear energy. Falldin's position was now untenable. In October 1978 he resigned, and

[19]Ministry of Industry, "Report on Review Through Foreign Expertise of the Report *Handling of Spent Nuclear Fuel and Final Storage of Vitrified High-level Reprocessing Waste* [KBS I report]," Ds I 1978:28 (Stockholm, 1978) p. 1-1.

[20]Lönnroth, Keystone paper.

the Liberal party formed what was to be a short-lived minority government.[21]

In March of 1979, as expected, the KBS project sponsors, with new data in hand from the additional test drilling, were again asking that the fueling of Ringhals-3 and Forsmark-1 be approved as being in compliance with the Stipulation Act. The KBS plan now identified what came to be known as the Sterno site. On utility-owned land in a well-populated area near Karlshamm in southern coastal Sweden, this site was represented as suitable for accommodating high-level waste. The Swedish Nuclear Power Inspectorate (SKI) was charged by the Liberal government with determining whether the site could be accepted, so that the two reactors could be started up. In turn, the inspectorate appointed a panel of consultants, most of them geologists, to review the Sterno site data. But an awkward pass was reached when seven of the eight geologists concluded from test borehole samples that if the site could be considered usable at all, it could not accommodate a repository large enough to receive the high-level waste from the planned nuclear program—which, as recently modified by the government, was to consist of twelve reactors. Besides finding the site too small, these advisers questioned the KBS method of assessing the rock's permeability, or characteristics with respect to groundwater movement.[22]

SKI and the Liberal government were now in a dilemma. To accept the site would be to go against the advice of the inspectorate's own panel of experts; but to reject the site and the assessment methodology would be to delay indefinitely the use of two expensive reactors ready for fueling, and perhaps that of several more in various stages of construction. In the end, the government declared that the KBS methodology was acceptable; then SKI's board, by a vote of 6 to 2, decided that the Sterno site was adequate for a repository for the high-level waste from the two reactors in question. Remarkably, the corner of the Sterno site deemed acceptable by the board was a portion in which no test holes had been drilled.[23]

The inspectorate's decision came after midnight at the end of a tense thirteen-hour meeting, with scores of reporters waiting outside the door. Although this affair had aspects of a charade, the way now seemed clear for the government to allow Ringhals-3 and Forsmark-1 to go on line.

[21]The events leading to the collapse of the Falldin governments are described by Childs, *Sweden*, pp. 101–109; Lönnroth, Keystone paper; Kalvemark, "Swedish Public Discussion on Nuclear Power."

[22]Personal communication to the author, November 22, 1982, from Thomas B. Johansson, one of two dissenting members of the SKI board and formerly a scientific adviser to the Ministry of Industry on the KBS project. Johansson is a professor of environmental studies at the University of Lund.

[23]Ibid.

The Nuclear Referendum Campaign

Three Mile Island intervened the day after the SKI decision, and there are Swedes who say, only half in jest, that this accident's political impact was greater in Stockholm than in Harrisburg. The nuclear waste issue was immediately eclipsed by the problem of reactor safety. Public concern was widespread and none of the political parties could ignore it. Falldin, frustrated in the past by his inability to have the government call a referendum even when he was prime minister, now suddenly found himself joined by Palme and the Social Democrats in urging that the voters be allowed to decide the future of nuclear power. The Liberal government and the Conservative party also now favored a referendum, and one was subsequently called for March 1980. It was decided, too, that pending the outcome of the referendum, the start-up of Ringhals-3 and Forsmark-1 would not be approved. With the future of nuclear power to be left to a popular vote, the nuclear issue was effectively removed from the upcoming September 1979 election. As it turned out, Falldin was narrowly returned to power as the head of a new coalition government.[24]

To shorten a long and complicated story, the voters were not given a straightforward choice between the early phase-out of nuclear power favored by the anti-nuclear Center party and the open-ended continuation of nuclear power favored by the pro-nuclear Conservative party. The Center party, followed by the Communists, adopted a ballot proposition that staked out its anti-nuclear position unambiguously. But the Conservatives saw the Social Democrats and the Liberals trying to claim a middle-ground position, and they chose to follow suit. After the Social Democrats (followed by the Liberals) adopted a proposition calling for the twelve-reactor program to be completed and then phased out over the life of the reactors, the Conservatives adopted a similar one.[25]

Thus it became more important for voters to know the motivation and intention of the sponsoring parties than to know the wording of the three propositions on the ballot. The referendum campaign, with public financing by the government for all the participants, continued for a year at a remarkably high level of intensity. Although strongly presented, the Center party campaign, aided by environmentalists and many artists, musicians, and other cultural and intellectual figures, won just under 40 percent of the vote. As expected, those holding the middle ground

---

[24]Lönnroth, letter to the author, January 1983; Per Ragnarson, "Before and After: The Swedish Referendum on Nuclear Power," *Current Sweden* no. 255 (July 1980); Childs, *Sweden*, pp. 166, 167–168.

[25]Lönnroth, letter to the author, January 1983.

prevailed, if somewhat ambiguously. The Social Democratic and Conservative propositions together claimed nearly 60 percent of the vote. But the votes for the Social Democratic and Center party propositions could be totaled together just as legitimately and found to represent an 80-percent majority for a phase-out of nuclear power, with about half favoring an early phase-out and half favoring a later one. Four of the five political parties were behind this consensus position.[26]

The referendum seemed to serve as an emotional and intellectual catharsis, and the public welcomed a relief from the nuclear issue, which for nearly five years had dominated Sweden's political scene. "I don't think they [the public] could take any more," a Swedish embassy official told me.[27] "They had heard all the questions and all the answers. They are saturated."

In the spring of 1981, a year after the referendum, the Parliament moved to develop a plan for phasing out nuclear power by the year 2010. It also moved to meet the other objectives of Swedish energy policy by reducing Sweden's dependence on foreign oil through improving the end-use efficiency of energy and increasing the use of renewable sources. But a parliamentary commission has since indicated, in a 1984 report, that to exclude nuclear power from the energy mix, while feasible, could present painful dilemmas. The commission identified two principal substitute energy sources—additional hydropower, and fossil-fuel generating stations burning imported coal.[28] Yet it was in seeking to avoid more hydropower that Sweden went to nuclear power in the first place; and even the most advanced combustion technology and emissions controls will not make coal pollution-free. Other alternatives may help—the Center party is demanding a greater conservation effort—but to what degree these dilemmas can be avoided is by no means clear. For these reasons, Swedish policy appears really to represent a nuclear moratorium during which the acceptability of fission energy in relation to future needs and available alternatives will undergo an extended period of testing.

Spent Fuel and Waste Management: Facilities and Technology

Leaders of the nuclear enterprise in Sweden have seemed determined not to allow the waste issue to diminish prospects for a continuation

[26]Ibid.; Ragnarson, "Before and After: The Swedish Referendum."

[27]Interview by the author with Lief Ericson, nuclear attaché, Embassy of Sweden, Washington, D.C., June 29, 1982.

[28]Robert Skole, "Swedish Commission Says Nuclear Phase-out Will Not Cause Major Impact," *Nucleonics Week*, August 30, 1984.

and perhaps eventual expansion of nuclear power. The Swedish Nuclear Fuel and Waste Management Company, or SKB as it is now called, has been given the responsibility by the government and Parliament to carry out a comprehensive policy for spent fuel and waste management. This policy, which is essentially of the industry's own devising, is backed by a fee on nuclear generation similar to the one imposed by the Nuclear Waste Policy Act in the United States, except that Sweden's, at almost 3 mills per kilowatt hour, is nearly three times higher. The fee is meant to cover all costs related to waste processing and disposal, including the cost of the final decommissioning of nuclear facilities (an item not now covered by the U.S. fee).

CLAB: Swedish Monitored Retrievable Storage

In July 1985, SKB brought into operation at Oskarshamm, on the Baltic coast 150 miles south of Stockholm, its central storage facility for spent fuel, known by its Swedish acronym, CLAB. Built in caverns excavated in hard rock some 25 meters below the surface, CLAB consists of four pools with an overall capacity of 3,000 tons of spent fuel; it can later be expanded, by creating more caverns and pools, to a capacity of 9,000 tons—an amount substantially greater than the 7,500 tons expected to be discharged from the twelve reactors by the year 2010. According to present policy, the great bulk of Sweden's spent fuel will be shipped to CLAB after about one year of cooling in pools at the reactor sites, and will remain there for thirty to forty years. Transport to a deep geologic repository is not to begin until about 2020.[29]

Why this long period of interim storage? There appear to be a variety of reasons. Both the spent fuel and the political passions that have been aroused by the nuclear controversy will have more time to cool; there will be more time to investigate geologic disposal sites, and thus avoid the accusation that alternative sites have not been examined.[30] In addition, in the industry's perspective, Sweden, as a small country with a limited nuclear program, will have longer to learn from the experience abroad whether nuclear power has a future, and if so what that future is likely to be. In the event use of plutonium fuel were to become important, for instance, the question of whether more Swedish fuel should be reprocessed could be raised anew.

The CLAB project, unlike proposals for monitored retrievable storage facilities in the United States, appears to have given rise to little or no

---

[29]SKB, "SKB Activities" (Stockholm, January 1985), brochure.
[30]Svenke, letter to the author, August 27, 1985.

SWEDEN

301

controversy. One explanation is that with the nuclear phase-out or moratorium policy in place, anti-nuclear activists seem to find less incentive to take issue with spent fuel and waste management practices. Another is that SKB chose to build CLAB at an existing nuclear station rather than at a new site. Then, too, there is the fact that all spent fuel transport is by sea aboard a ship specially designed for the task, the *Sigyn*.[31] That there is no need to move fuel by truck or rail through or near cities and towns is an important advantage.

Offshore Caverns: Low- and Intermediate-level Waste

The unusual concern of SKB for secure permanent isolation of low- and intermediate-level waste makes Swedish disposal practices almost as demanding as those in Germany, where all radioactive waste is to go to deep geologic disposal (see chapter 8). Sea dumping is forbidden by Swedish law. Near-surface disposal is allowed at reactor sites for very low-level reactor plant wastes; but all other low- and intermediate-level wastes (including a small amount of waste from Swedish medical and industrial uses of radioactive materials) will go to a unique "final repository for reactor wastes," known by its acronym, SFR. The SFR is being built just off shore from the Forsmark reactor station on the Baltic coast north of Stockholm. To cost about $200 million, the facility will be 50 meters beneath the seabed, but connected by tunnels to a surface terminal on land where the waste will be received. The intermediate wastes will be committed to large concrete-lined silos, with concrete slurry poured around them to fill the empty spaces. The low-level waste will be deposited in tunnels or caverns; some of this material would also be covered with concrete before the storage areas are sealed.[32]

Building the SFR facility off shore beneath the seabed is believed by the Swedes to offer three advantages: first, the hydraulic gradient there is low, and groundwater flow through the rock is expected to be slight; second, any radionuclides that escape confinement would be diluted by

[31]The *Sigyn* suffered an embarrassment just as she was coming into service in late 1982. This vessel had been built with double bottoms and double rudders to make her one of the safest ships afloat. But on her maiden voyage she went aground at the entrance to the Barseback power station harbor; the grounding was caused primarily by an incorrectly placed channel buoy. Since this early embarrassment, however, the *Sigyn* has served without mishap. Swedish Maritime Investigation Commission, "Investigation Report Concerning the Grounding of the French-Registered Cargo Ship *SIGYN* ... on 25 November 1982" (May 1983).

[32]Lars B. Nilsson, "A Central Repository for Final Disposal of the Swedish Low and Intermediate Level Wastes" (n.d.). Nilsson is with the Swedish Nuclear Fuel Co.

an enormous volume of seawater (no transuranic or plutonium-contaminated waste would go to the SFR); and third, nobody is going to drill a water well into the facility, as might happen in the case of a repository beneath the land surface if knowledge of the facility's location should ever be lost or forgotten.[33] As in the case of spent fuel transport, all movement of low- and medium-level wastes would be by sea, aboard the *Sigyn*.

Geologic Disposal: High-level Waste and Spent Fuel

Under the SKB plan, all high-level waste and spent fuel is to go into a repository built about 500 meters deep in the crystalline bedrock. As matters have shaken out, only about 12 percent, or 729 tons, of the 7,650 tons of spent fuel to be generated over the life of the Swedish nuclear program is covered by COGEMA and British Nuclear Fuels contracts for reprocessing at La Hague or Sellafield. The SKB managers, preferring to exercise direct control over the fuel themselves, seek to make the percentage smaller still.[34] Thus the Swedes have worked out an arrangement for three German utilities to assume ownership of the 57 tons of Swedish fuel that has already been shipped to La Hague, while SKB takes an equivalent amount (in fissile content) of spent German mixed oxide fuel not suitable for conventional reprocessing.[35] A Japanese utility has taken over the Swedish contract obligation for another 178 tons.[36]

The SKB plans for beginning deep geologic disposal in the year 2020 represent a long deferral, but unlike the British, the Swedes have not stopped their field investigations.[37] Indeed, they have already carried out extensive borehole drilling at four sites, are now investigating a fifth, and plan to investigate another five before narrowing the choice in the 1990s and selecting one site for the repository at the turn of the century.

[33]Ibid.

[34]Interview by the author with Erik Svenke, November 17, 1982. Svenke believed it desirable, however, that Sweden have at least a small amount of fuel under contract so as to be directly represented in development of reprocessing technology.

[35]Ann MacLachlan, "Swedish and German Utilities Exploring Spent Fuel Swap," *Nucleonics Week*, February 14, 1985.

[36]Personal communication to the author from Erik Svenke, August 27, 1985.

[37]Fieldwork in Sweden related to geologic disposal has been of international importance since 1977, when the Stripa research project was begun. Located in an abandoned iron mine in crystalline rock some seventy miles west of Stockholm, this project has involved the participation of the United States and a number of other countries. Research has included extensive heater tests and studies in rock mechanics, hydrology, and geochemistry. See P. A. Witherspoon, N. G. W. Cook, and J. E. Gale, "Geologic Storage of Radioactive Waste: Field Studies in Sweden," *Science*, February 27, 1981, pp. 894–900.

But responsible officials have not forgotten the lesson taught in 1980 at a lonely place called Kynnefjall. Situated just inland from Sweden's west coast and only about fifteen miles from the Norwegian border, Kynnefjall is a broad, plateaulike mountaintop mostly covered by evergreen and alder, but with numerous rocky outcroppings. When I visited Kynnefjall in October 1980, seven months after the national referendum, I found it deserted except for the keepers of the few isolated farmsteads, some moose hunters, and curiously, two couples occupying trailers parked at crossroads on the mountaintop. The trailers were lookout posts that the local organization Radda Kynnefjall (Save Kynnefjall) had established the previous spring, and had continuously manned to watch for any borehole drilling crews. If a crew showed up, the plan was to get the word out immediately so that people from the neighboring towns and villages could rush to the scene to obstruct the drilling, if necessary by lying down in the path of equipment. The responsible officials in Stockholm had chosen early not to take up the challenge after one of its contractors was turned back by the Radda Kynnefjall activists.

Politically, the Kynnefjall incident was particularly significant at the time because the rural villagers and farmers who made up Radda Kynnefjall were mostly Center party members and could count on the personal support of Prime Minister Falldin. But the present Social Democratic government, which was returned to power in the September 1982 election (a contest in which nuclear issues played no part), is no more eager than its predecessor to stir up trouble in rural Sweden. According to Erik Svenke, SKB's president until a few years ago, the company's policy is to do test drilling only at sites where it has the landowner's permission—at Kynnefjall this was not a problem, for the Kynnefjall site was government-owned—and only where it does not encounter significant local opposition.[38] This policy has thus far proved workable. According to Swedish legal tradition the construction of a repository apparently will require local consent.[39]

To satisfy the national Stipulation Act, SKB has had to convince the Swedish Nuclear Power Inspectorate that spent fuel can be safely and permanently isolated, even though in no instance has the suitability of a specific site been confirmed by exploratory shafts and tunnels at the

[38]Interviews by the author with Svenke, November 16–17, 1982.

[39]Erik Svenke, "Policy and Planning for Sweden's Nuclear Fuel Cycle," paper presented at the Atomic Industrial Forum/Forafom Meeting, Geneva, Switzerland, June 1, 1983. "Site approval for most kinds of nuclear installations ... can only be granted provided that the local community gives its consent (a local community has the right to veto on siting within the community of various specified industrial undertakings)," Svenke observed. Svenke's comment was made in discussing the siting and approval of reactor projects, but it would appear also to apply to repository siting.

repository horizon. The first plan for direct disposal of spent fuel, KBS II, was issued in 1978, but then revised in 1983 and sent out for review by Swedish and foreign experts as KBS III (the KBS I plan approved in 1979 was concerned only with disposal of waste from reprocessing). The reviews were generally favorable,[40] and in June 1984 SKI granted permission for fueling the last two reactors in the Swedish program, Forsmark-3 and Oskarshamm-3.[41] Approval of KBS III eliminated the possibility that reactors put on line earlier might have to be shut down prematurely.

A strikingly unusual feature of KBS plans II and III is that they rely primarily on an engineered barrier—the waste canister—rather than on the geologic formation to prevent radionuclide release over the period of hazard.[42] This feature can be laid neither to concern that good sites will be lacking nor to the exigencies of the Stipulation Act. The act has been amended to eliminate the "absolutely safe" language in favor of a reactor licensing requirement for proof that "there is a method for the handling and final disposal of spent fuel [and other wastes from the reactor] acceptable with regard to safety and radiation protection."[43] And while none of the four sites investigated in recent years has been found to be ideal, in the opinion of American reviewers at the National Academy of Sciences any of them could be "made usable by proper choice of locations between fracture zones and by grouting of minor fractures."[44] The principal explanation for the emphasis on engineered barriers seems to be the Swedes' concern to hedge against technical uncertainty.

---

[40]Ministry of Industry, *Review of the KBS III Plan for Handling and Final Storage of Unreprocessed Spent Nuclear Fuel*, Ds 1984:17 (Stockholm, 1984). This document contains reviews by the National Academy of Sciences in the United States; agencies in the United Kingdom, France, and Canada; and the International Atomic Energy Agency and the Nuclear Energy Agency of the Organization for Economic Cooperation and Development.

[41]"Swedes Accept Spent Fuel Disposal Plan and Okay Fuel Load at Two Plants," *Nucleonics Week*, July 5, 1984.

[42]The U.S. National Academy of Sciences' review panel has observed that "the engineered barriers called for in the KBS plan are the important ones, whereas in most geologic disposal plans the engineered barriers are assumed to fail within short times (i.e., compared with radionuclide half-lives), and the natural barriers are then depended on for slowing long-term radionuclide release." National Academy of Sciences, *A Review of the Swedish KBS-3 Plan for Final Storage of Spent Nuclear Fuel* (Washington, D.C., 1984) pp. 2, 6-1, and 6-2.

[43]Ministry of Industry, *New Swedish Nuclear Legislation*, Ds I 1984:18 (Stockholm, 1984).

[44]National Academy of Sciences, *Review of the Swedish KBS-3 Plan*, pp. 5–6.

KBS III represents a more modest and practical approach than KBS II. In the KBS II design, the fuel assemblies were to be dismantled and their rods encapsulated in canisters that would have had 8-inch-thick walls of pure copper and would weigh 20 tons each. Voids within the canister were to be filled with molten lead, pumped in under a high vacuum. One American reviewer of this early plan said that the intricate, remotely operated system for dismantling the forty-year-old fuel assemblies and repackaging the rods seemed "to add up to a disaster waiting to happen."[45] Such criticism struck home, for in the KBS III design the thickness of the canisters was reduced by half, and the fuel assemblies were to be left intact and not dismantled. Voids might be filled by powdered copper instead of lead. The life of this canister has been put by Swedish corrosion experts at more than a million years. Elaborate as this plan is, the total estimated cost for disposal of all spent fuel and high-level waste for the life of the Swedish nuclear program until phase-out in 2010, including the cost of research and development, transportation, encapsulation, and final commitment to a repository, is only about $1.7 billion (1984 dollars).[46] This works out to about $250 per kilogram of uranium for the some 6,780 tons of spent fuel to be disposed of, a sum roughly consistent with present estimates by the U.S. Department of Energy for its disposal costs.[47]

Inasmuch as construction of a Swedish repository is not to start before the year 2010, the disposal plan actually followed may turn out to be different from the KBS III plan. In 1984 the Swedes began studying the new concept of dry retrievable storage, or final disposal of spent fuel inside a huge, up-ended, egg-shaped "cave" or cavern from which groundwater has been diverted and heat can be removed.[48] It should also be noted that given Sweden's planned phase-out of nuclear power and the long interim storage to be provided by CLAB, there is not now any place in the KBS repository plans for retrievability.

[45]William C. McClain, letter of November 28, 1979, to Larry D. Ramspott of the Lawrence Livermore Laboratory.

[46]Calculated by SKB, this cost estimate was given to the author in a letter of February 5, 1985, from Inge Pierre of the Swedish Embassy, Washington, D.C.

[47]U.S. Department of Energy, *Nuclear Waste Fund Fee Adequacy: An Assessment*, DOE/ RW-0020 (Washington, D.C., February 1985). Cost estimates falling in the $200- to $300-per-kilogram range can be derived from tables 1 and 2, p. 7, and the program assumptions shown on p. 3.

[48]Christer Svemar, "Swedes Probe New Cave Concept for Storage of Spent Fuel," *Nuclear Europe*, October 1984, p. 43. SKB announced plans for an underground laboratory to be built in granite, at a depth of at least 400 meters, near the Oskarshamm nuclear station south of Stockholm. Robert Skole, "Sweden's SKB to Build New Lab to Test Granite as Nuclear Waste Disposal Medium," *Nuclear Fuel*, October 6, 1986.

Conclusion: Meeting the Imperatives

According to Erik Svenke, who has been a major figure in the Swedish nuclear industry, the industry worldwide would benefit most from a patient, careful effort to build public trust by showing nuclear power to be the reliable, safe, and clean energy source he believes it can be. This, he indicates, must be done first, and done well, by improvements to the existing system of light water reactors. The next generation of reactors, in his view, should be made inherently safe, or immune to the possibility of containment failures resulting from core meltdowns. Several years ago when the funding and completion of the now dead Clinch River breeder project were still a prime objective of the nuclear industry in the United States, Svenke observed to me, "Wouldn't that money be better spent on the Three Mile Island cleanup?"

The Svenke philosophy appears to be strongly expressed in the Swedish nuclear program, and in the waste program in particular. No nation has done better in the past than the Swedes, or seems likely to do better in the future, to honor the political imperative to contain radioactivity in nuclear operations. Sweden's past success in this respect rests partly upon its decision long ago not to undertake reprocessing for the recovery of plutonium for use as either a nuclear explosive or a fuel. Its future success seems likely to continue to depend upon the exceptional measures the Swedes are willing to take to assure that radionuclides remain isolated and contained. The nuclear phase-out, which may yet prove to be no more than a long moratorium, has put the Swedish industry on its mettle, and to date it is responding well.

10

France: Commitment to Plutonium Fuel

The French nuclear program, second only to that of the United States in the number of reactors built or under construction, is now second to none in the size and scope of its development overall. By the 1990s France will have light water reactor stations in every quarter of the republic, with fifty-six large Westinghouse-type pressurized water reactors altogether; a huge uranium enrichment center in the Rhone Valley (Eurodif, a French-controlled, multinational venture already in place); a large center for the reprocessing of commercial fuel at La Hague; and the big Superphenix breeder reactor at Creys-Malville east of Lyon.

France's extraordinarily ambitious commitment *électronucléaire* appears to rest in part on strong, rocklike foundations. Its program for developing, building, maintaining, and operating reactors, while certainly no model of decentralized democratic procedure with opportunity for public intervention, is unsurpassed in its managerial and technical coherence and has produced results that are the envy of the nuclear industry worldwide.[1] The French program, moreover, appears to be backed up by a substantial, persevering program of research and development. In nuclear waste management, for example, the French are far

---

[1]But French reactor programs have had their own technical troubles For instance, two accidents involving fuel melting at the Saint Laurent A1 and A2 gas-cooled, graphite-moderated reactors caused prolonged shut-downs. In 1969 reactor A1 was shut down for a year after five fuel elements melted. Beginning in 1980, reactor A2 was down for two and a half years after two fuel elements melted where six to eight stack channels had become obstructed by a metal plate. Ministry of Industry and Research, Information Document no. 36 (June 1983) on nuclear safety, in an English translation. To a request by the author for more information about the 1980 incident and subsequent cleanup at Saint Laurent A2, André Gauvenet, Inspector General for Nuclear Safety and Protection, provided none, reporting only that the accident resulted in no release of radioactivity to the environment. Telex from Gauvenet to Daniel Chavardés, nuclear attaché at the Embassy of France in Washington, D.C., November 30, 1984.

ahead of other countries in bringing high-level waste solidification efforts to practical industrial application. Yet the French have in some respects been even more heedless than others in allowing their program of nuclear development to get ahead of itself.

Until recently France had done little toward the geologic isolation of its nuclear wastes, even lagging behind Switzerland in this field, to say nothing of Sweden, Germany, and the United States. But evidence that the French nuclear program has leaped ahead of itself is most striking in the commercial reprocessing venture at La Hague—a large program undertaken in the absence of an economic need for the plutonium that will be separated. If the need for plutonium fuel and breeder reactors ever becomes great enough to outweigh the risks of nuclear proliferation and terrorism associated with reprocessing and separated plutonium, it will not be until well into the next century.

Early Developments and Expansion

The nuclear power program in France, like the programs in the United States and Great Britain, was initiated with the start-up of facilities for producing plutonium for bombs. In 1957 a center for this purpose began operating at Marcoule; it consisted of gas-cooled, graphite-moderated reactors and the reprocessing plant Usine Plutonium-1, or UP-1, for recovery of the plutonium from the reactors' lightly irradiated natural uranium fuel. This achievement, done on the strength of France's own scientific and industrial resources and without technical assistance from the United States, marked the beginning of a strong sense of national pride in the French program of military and civil nuclear development— a pride that probably helps to explain why, a month after Chernobyl, better than half the French population was found to support continued development of nuclear power.[2] The early French reactors, like Britain's Calder Hall reactors at Windscale, served to produce both plutonium and electricity for the national grid. The reprocessing of the fuel, irra- diated only 3,000 to 5,500 megawatt days per ton, was carried out effectively, and during the first twenty years or so of its operation UP-1 reprocessed a total of about 10,000 tons from the three reactors at

[2]In a Louis Harris poll conducted May 26–28, 1986, 53 percent of the respondents were found to believe that continued nuclear development was necessary in France. But 61 percent would be "rather worried" or "very worried" if a nuclear station were built near them. The poll found a distrust of official sources of information. Embassy of France, Washington, D.C., to the author, 1986.

Marcoule.[3] In 1966, UP-2, a second reprocessing facility, went into operation at La Hague on the narrow Cotentin Peninsula that juts into the English Channel.

In 1969 France decided not to build more gas-graphite reactors. Heavily influenced by the American example,[4] the switch was made to the light water reactor, specifically to the Westinghouse-designed pressurized water reactor. There was only one domestic customer, Electricité de France (EDF), to buy, own, and operate the nuclear generating stations. To match this one customer, there was a single reactor manufacturer, Framatome (in which the government held a minority interest), offering but one kind of reactor. Compared to the situation in the United States, where there were sixty utilities, four reactor builders, numerous reactor types, and innumerable modifications of standard designs, the French nuclear program was a model of simplicity and good order.

But the switch to the light water reactor meant major problems for La Hague because the level of irradiation of the fuel to be reprocessed would be increased by more than sixfold, from a maximum of 5,500 megawatt days per ton for the plutonium-production reactors to about 33,000 for the light water reactors.[5] Yet if the Americans, leaders in reprocessing and virtually every other aspect of nuclear technology, were not worried that reprocessing of the higher burnup fuel might prove problematic, why should the French themselves be worried? In any case, the switch to pressurized water reactors was made, and following the oil shock of 1973–1974 the French nuclear commitment became ambitious indeed, with program goals beginning the steep ascent to their present levels. Nuclear development was seen as the principal thrust of a French policy of energy independence.

Nowhere was the commitment to nuclear power more evident than at La Hague, which was destined to become a major center for the reprocessing of oxide fuel from French and foreign reactors. The first step came with the start-up of the high-activity oxide (HAO) fuel receiving unit in 1976, which gave the UP-2 plant a nominal capacity to reprocess 400 tons of oxide fuel a year. Then, through contracts between the Compagnie Générale des Matières Nucléaire (COGEMA) and utilities in Japan and other countries, came the commitment to bring on line by

[3]François David and Jean-Paul Schapira, "Le retraitement des combustibles nucléaires," *La Recherche* vol. 11, no. 111 (May 1980) p. 526.

[4]Irvin C. Bupp and Jean-Claude Derian, *The Failed Promise of Nuclear Power* (New York, Basic Books, 1981), first published in 1978 as *Light Water: How the Nuclear Dream Dissolved.*

[5]David and Schapira, "Le retraitment des combustibles nucléaires," pp. 523–524.

the end of the 1980s the 800-tons-a-year UP-3A plant. At the same time, the capacity of UP-2 would be increased to 800 tons, enough to accommodate a bit more than half the fuel to be discharged annually in the 1980s by EDF's reactors.[6]

La Hague's total nominal capacity of 1,600 tons would be enough to serve some sixty French and foreign reactors altogether, and to produce on the order of 10 or 12 tons of plutonium a year. The still slightly enriched uranium recovered from the French fuel would be further enriched, then recycled in the pressurized water reactors (PWRs),[7] while the recovered plutonium—according to the original strategy— would be saved for the start-up of a succession of commercial breeders, of which Superphenix would be an early prototype. It was only later, after it became apparent to all that breeder commercialization was to be long delayed, that COGEMA and the Commissariat à l'Energie Atomique (CEA) found themselves facing the question of what to do about a growing inventory of plutonium in excess of breeder needs. Carried along by the momentum of their commitments at La Hague, these agencies were not going to consider cutting back on reprocessing to reduce plutonium production. Instead, their predictable recourse was to begin edging toward a commitment to use of plutonium fuel in conventional light water reactors.[8]

Sources of Opposition and Support

The rapid expansion of the nuclear program during the 1970s led to conflict and resistance. Thirty-three reactors were ordered for thirteen different sites from the beginning of 1973 to the end of 1979, with as many as six new projects getting under way each year. The reactor construction program stirred up opposition enough to make it appear

---

[6]An overview of French experience in reprocessing from the late 1950s through the 1970s is provided by David and Schapira, "Le retraitement des combustibles nucléaire." See also the interviews conducted by Mark Reynolds with Bertrand Barré, nuclear attaché at the French Embassy in Washington, D.C., in *EPRI Journal*, July–August 1984, pp. 32–37.

[7]According to fuel cycle specialist Clint Bastin of the U.S. Department of Energy, some 6 to 9 kilograms of plutonium can be recovered from every metric ton of light water reactor fuel reprocessed. About 80 percent of the plutonium recovered is made up of the fissile isotopes Pu-239 and Pu-241. Bastin was interviewed by the author on October 22, 1986.

[8]In June 1985, Electricité de France, the government-owned utility, decided that plutonium was to be recycled in some of its 900-megawatt PWRs. Daniel Chavardés, "Brief Update of the French Nuclear Program" (November 1985).

for a time that the nuclear program might become caught up in political controversy sufficiently intense to cripple it.[9] Dorothy Nelkin and Michael Pollak, summarizing for *Technology Review* findings arrived at in their book *The Atom Besieged*, have described the anti-nuclear movement of the 1970s this way:

The main constituency was young, educated, and middle class, including many students but also scientists and other professionals socialized during the period of dissent in the 1960s.... In addition, farmers concerned about the industrialization of rural areas, or directly affected by proposed construction, joined the local and regional organizations to oppose specific decisions. In places such as Brittany, the movement attracted the "regional autonomists" who perceived nuclear power as a vehicle for further political centralization. The national movement was also joined by feminist and other emancipatory groups attracted by the revolt against the political establishment.... [The] diverse agenda [of these groups] is underlaid ... by some powerful common themes: an apocalyptic image of the destructive potential of nuclear power; a pessimistic vision of an ecological, economic, and cultural crisis; and a critical analysis of its sociopolitical roots. Opposition to nuclear power in France has become a struggle for "human survival" and "political justice" against the powers of "nuclear technocracy" and "electro-fascism."[10]

Resistance groups were formed in response to reactor projects in a number of regions during the 1970s. Quite often local communities became sharply divided over a particular project. Reactors proposed for Fessenheim in Alsace, Port La Nouvelle near Bordeaux, and Creys-Malville in the Rhone Valley were all the focus of bitter opposition; mass demonstrations and site occupations were common and incidents of violence not infrequent. But the demonstration at Creys-Malville in July 1977 against the Superphenix project marked a turning point. Some 60,000 protesters showed up, the mood turned ugly after a police official referred to the many German demonstrators as a second "German occupation," and violence broke out. Several hundred people were injured, one fatally.[11] The backlash within the protest movement and in the general public hurt efforts to mobilize further mass demonstrations.

Anti-nuclear protest may have taken the form of turbulent site demonstrations at least in part because under the French system of governance the protesters had little chance to be heard effectively in any

[9]Irvin C. Bupp, "The French Nuclear Harvest: Abundant Energy or Bitter Fruit?" *Technology Review*, November–December 1980, pp. 30–39.

[10]Dorothy Nelkin and Michael Pollak, "The Antinuclear Movement in France," *Technology Review*, November–December 1980, pp. 36–37.

[11]Dorothy Nelkin and Michael Pollak, *The Atom Besieged* (Cambridge, Mass., MIT Press, 1981) pp. 72–75.

other way. The courts would hear complaints only on narrow, easily overcome procedural grounds, and there was little political leverage to be exercised at the local and regional levels. The French tradition of political and administrative centralization goes back through the Napoleonic era to the time of Louis XIV. The *préfets* (prefects) in charge of the local *départements* (regional administrative units) represent an extension of the authority of the government ministries in Paris.[12] They can countermand local actions and policies, and through their control of the police keep demonstrators from occupying nuclear construction sites without having to look to local elected officials for help. Projects that involve the taking of private property have to be justified in an *enquête de déclaration de utilité publique*, but these proceedings have in the past tended to be pro forma. In the *enquêtes* on the siting of reactors citizens could record their opposition to, or support for, a project by signing a register, but with neither the prefect nor the responsible officials in Paris likely to be moved in the slightest by the outcome. For instance, in the case of the nuclear station proposed for a site at Nogent-sur-Seine, in a heavily populated area sixty miles east of Paris, EDF persisted in its plans despite the fact that hardly a corporal's guard among the local citizenry favored the project, while more than 50,000 registered their opposition to it.[13]

All the important decisions were being made in Paris. There, at the national level, nuclear energy continued to enjoy support across a wide spectrum of French politics. The political parties making up Giscard d'Estaing's governing right-of-center coalition during the 1970s were all strongly behind nuclear development. On the left, the Communist party, together with the communist trade union, Confédération Généraie du Travail (CGT) was also strongly supportive, which was not surprising in view of the huge number of jobs (180,000 at the end of 1985[14]) that the nuclear industry provided. Part of the Socialist party and the Socialist trade union Confédération Française Démocratique du Travail (CFDT) took exception to the pace and direction of the nuclear program, but these dissident elements were not strong enough to bring about meaningful or sustained questioning of the ongoing program even on the floor of the Parliament, much less within the government.

[12]On the role of the courts and French centralized administrative control, see Nelkin and Pollak, *The Atom Besieged*, pp. 157–168; Bupp, "The French Nuclear Harvest," pp. 33–34.

[13]Brice Lalonde, "Letter from the Movement," *Technology Review*, November–December 1980, pp. 38–39.

[14]Interview by the author, October 21, 1986, with Eric Frémond, assistant to the nuclear attaché, French Embassy, Washington, D.C.

Beyond that, the Commissariat à l'Energie Atomique tended to withhold information to which nuclear critics or members of the press might
point with alarm. Several years ago a commissariat official explained his
agency's policy for the release of nuclear safety information this way:
"The publication of technical data concerning the precautions against
certain risks ... frequently has little other effect than to heighten feelings
of insecurity, the technically ignorant seldom retaining anything other
than the existence of the risks in question....There is nothing to be
gained for real information of the public by holding controversial debates."[15]

Jean-Paul Schapira, a theoretical physicist at the Institut de Physique
Nucléaire at Orsay and an early critic of the nuclear program, described
for me the situation in late 1980, about six months before the Socialist
Mitterrand government assumed office. "France is a country where there
is no institutional possibility to discuss technological options," he said.
"Everything is a pure political game. You are always put in a situation
where [if you are a critic] you are considered a political opponent. You
can't have any sound discussion the way you can in other countries."

Critics of the nuclear program, working on their own, were gradually
able to establish a technical basis for a sustained critique. In early 1975
four hundred scientists at prestigious institutions signed the *Appel des
Quatre Cents*, which warned that French nuclear development posed
unresolved safety problems; that these problems were being obscured
by secrecy and misinformation; and that the decision-making procedures
of the responsible ministries and agencies was a closed process, carried
forward without consultation of the public or even members of the
Parliament. Some of the scientists who initiated the *Appel des Quatre
Cents*, which ultimately had some 4,000 signers, founded the Groupe
de Scientifiques pour l'Information sur l'Energie Nucléaire (GSIEN), which
became France's most active public interest science group. Through its
publications GSIEN became a source of technical information for anti-
nuclear groups, especially Les Amis de la Terre, an offshoot of the
American Friends of the Earth and a leader of the fight against nuclear
power on the national level.[16] Although GSIEN was representative of
only a minority within the French scientific and technical community,
many of its members had better access to information than holders of
dissident views generally do in a technocratic system, for many were

[15]Public statement by the CEA information director, cited in Daniel S. Greenberg,
"Smooth Passage to Nuclear Power," *Washington Post*, July 5, 1979.
[16]The *Appel des Quatre Cents* and the formation of GSIEN are discussed by Nelkin and
Pollak, *The Atom Besieged*, pp. 89–90; and John Walsh, "Nuclear Power: France Forges
Ahead on Ambitious Plan Despite Critics," *Science*, July 23, 1976, pp. 305–306, 340.

employed by the CEA, COGEMA, and EDF (organizations in which the Socialist union CFDT was generally strongly represented). This gave them an insider's view of what was going on. The CFDT's *L'Electronu-cléaire en France*, published in 1975 as an inexpensive paperback, was the first widely accessible handbook for French nuclear critics and the general public. But the French nuclear program was still gathering momentum and proceeded virtually unchecked.

Nuclear Energy and the Mitterrand Government

The poor state of the French economy was the overriding issue in the 1981 elections that brought François Mitterrand and the Socialist party to power. Inflation, unemployment, and the status of foreign workers were the questions that dominated the campaign debates. Nuclear issues were only marginally important. Mitterrand's position on nuclear power was summed up by *Le Monde* two days before the runoff election: this energy source was "costly and uncertain," and a pause in the nuclear program—without however stopping projects already under construction—was needed to allow an extended nuclear debate and referendum.[17] "I do not reject nuclear power, I intend to master it," Mitterrand had said in the course of the campaign.[18]

Mitterrand and the new government were not long in office, however, before it became apparent that the nuclear "pause" might be only that, and that the great debate was going to consist of a few days' discussion in the National Assembly. Mitterrand had acknowledged during the campaign that a referendum on the nuclear program would be impossible without a constitutional amendment. Following the election, the idea of putting the nuclear program to a popular vote disappeared from the agenda. There was only one nuclear project actually cancelled—the proposed reactor station in Brittany at Plogoff, probably the most bitterly controversial project of them all and apparently the only one which Mitterrand as a candidate had definitely promised he would stop. Five reactor projects were temporarily halted, and a controversial project to establish a facility for the disposal of low- and intermediate-level waste at Saint-Priest-La-Prugne was suspended, then eventually abandoned.[19]

[17]*Le Monde*, May 8, 1981.

[18]*Agence France-Presse*, May 12, 1981. Translations from the French press, and from official French documents such as the *Journal Officiel* of the National Assembly, are by the author.

[19]John Walsh, "French Nuclear Policy Only Slightly Revised," *Science*, November 27, 1981; "The French Government Was Expected to Reactivate Five Nuclear Sites," *Nucleonics Week*, November 26, 1981; " 'I Am Not an Antinuclear Militant,' New French Energy Undersecretary," *Nucleonics Week*, April 14, 1983.

The expansion of reprocessing facilities at La Hague posed an especially delicate question for the new government because of COGEMA's contracts with foreign customers. The government announced that its commercial commitments to foreign countries would be respected, but that the "definitive calendar" for realization of the work at La Hague was one of the matters subject to the coming parliamentary debate.[20] Before that debate, which came in early October 1981, the Council of Ministers adopted a plan for "energy independence" which, though subject to the approval of the National Assembly, represented the new government's judgment as to what—if any—changes should be made in the nuclear program inherited from the Giscardist regime.[21]

Essentially, the Council of Ministers' program called for a continuation of the previous government's nuclear commitments, including those at La Hague. There would be only a modest reduction in the number of reactors to be ordered despite an expected decline in the rate of growth in energy demand. There was concern that "too brutal" a cutback from the previous government's plans would hurt Framatome, the reactor builder, and increase joblessness. Coupled with the continued commitment to nuclear energy, however, the program did promise greater emphasis on conservation and renewable energy sources. In addition, the Council of Ministers called for some decentralization of energy planning (especially with respect to conservation and renewable resources), a better flow of information to the public, and better observance of democratic practice in arriving at decisions. This was the Mitterrand government's energy program, worked out down to the last details. Prime Minister Pierre Mauroy, meeting with Socialist legislators (*deputés*) in a stormy party caucus before the parliamentary debate, warned that if the proposed energy program were rejected, the legislators would be voting down the Socialist government as then constituted.[22] As a result, the ensuing discussion in the National Assembly affirmed the French commitment to nuclear power. Minister of Research and Technology Jean-Pierre Chevènement quoted a bit of philosophy from François Mitterrand: "The denial of technical progress, the fear of the creative act are the mark of lost societies. The danger for humanity is not that man invents, but that he does not master what he has created."[23]

[20]Enerpresse no. 2875, government press release of August 3, 1981.

[21]The energy independence program as adopted by the Council of Ministers was described by Edmond Herve, Minister of Energy, at the start of the energy debate in the National Assembly on October 6, 1981. See the *Journal Officiel*, Assemblée Nationale, October 7, 1981, no. 36 A.N.(C.R.), pp. 1504–1508.

[22]Edward Cody, "French Ruling Party Shifts Nuclear Policy," *Washington Post*, December 7, 1981.

[23]*Journal Officiel*, October 6, 1981, p. 1510.

Overcommitment at La Hague?

While government support for the entire French nuclear program continued essentially uninterrupted into the new Socialist regime, there were clearly some major questions about the wisdom of the commitments at La Hague; these would eventually have to be addressed. The CFDT handbook, reissued in 1980 as *Le Dossier Electronucléaire*, had set forth the Socialist union's criticisms and complaints about this reprocessing center. It argued that the center had been incompetently designed and built, in part because in drawing up specifications for UP-2 and the oxide fuel head-end unit COGEMA had failed to consult with the workers and technicians who would have to maintain the plant: the management treated the workers with disdain, the CFDT complained.[24] (The Socialist union's rival for the loyalty of La Hague workers, the Communist union CGT, generally has been more a booster of La Hague than a critic.)

By early 1980 such criticism of La Hague was receiving attention from the nuclear trade press. A special report by the Paris correspondent for *Nucleonics Week* attributed to plant workers, environmentalists, and anti-nuclear scientists the view that the reprocessing facility there was "an inefficient, costly, and increasingly dangerous mess." The essence of their criticism—which stood unrebutted at the time, for COGEMA refused repeated requests by the *Nucleonics Week* reporter to answer them—was that La Hague had assumed a huge workload that it could not possibly meet in light of its demonstrated inability to reprocess high-burnup oxide fuel at better than a halting pace.[25]

Some 5,207 tons of foreign and domestic fuel was to be delivered to La Hague by 1986, with more to come later. Yet the amount of light water reactor fuel reprocessed the previous year—1979—was 65 tons, against the plant's nominal capacity of 400 tons. According to *Nucleonics Week*, the shortfall was attributed by the critics to two factors. One was that COGEMA had no choice but to give priority to reducing the backlog of magnesium-clad, gas-graphite reactor fuel at that time deteriorating in the storage pools; the bottom of one pool was already covered with a ton of radioactive sludge for which COGEMA had no disposal remedy. The other was that the greater amount of fission products (such as ruthenium and zirconium) in the high-burnup oxide fuel was degrading the extraction solvent and causing piping to be blocked by precipitates,

---

[24]Confédération Française Démocratique du Travail (CFDT), *Le Dossier Electronucléaire* (Paris, Editions du Seuil, 1980) pp. 336–337 (cited hereafter as CFDT, *Le Dossier*.)

[25]Vanya Walker-Leigh, "La Hague Critics Present Litany of Problems Facing Reprocessing Plant," *Nucleonics Week*, March 20, 1980.

which could be removed only by frequently shutting down the separations process and rinsing out the system. The report also cited other problems: the shears used to chop up the spent fuel needed to be replaced sooner than expected; and faulty hydraulic controls were sometimes causing packets of fuel rods to fall uncut into the acid dissolving solution, from which they then had to be retrieved.

Workers at La Hague were reported to see plant maintenance as a particular problem: "The plant was designed for highly automated functioning, with little thought given to how repairs might be undertaken. Pumps leak or break down, valves stick and pipes crack, and they are located in places difficult for workers to operate in. Increasing radiation levels are leading to a rapid turnover of repairmen." The report said that to enter some zones where no protective clothing had been required ten years earlier, workers were now having to don full protective suits. It indicated, too, that to spread out radiation exposures over a larger group COGEMA was relying heavily on temporary workers, and that these workers made up better than half of the La Hague work force of nearly 2,600.

Less than a month after the *Nucleonics Week* report appeared, a fire caused by a short-circuit destroyed the transformer unit at La Hague, knocking out both the plant's regular and emergency power supplies. The electrically operated forced cooling system for the high-level waste tanks was left useless for nearly an hour, and emergency power for this system became available only when mobile generators were brought in from the Cherbourg arsenal and elsewhere. Power for other safety systems, such as plant ventilation and atmosphere control, was not restored for about ten hours, which could have led to major contamination had there been spent fuel in the process of dissolution at the time of the power failure. By COGEMA's own reckoning this fire and power failure was the "gravest" accident ever to occur at La Hague, although the company has said that had it been necessary the high-level waste tanks—which begin to boil off after eighteen hours without cooling—could have been kept cool by water circulated by nonelectrically driven pumps. Nevertheless, workers regarded it as a scary episode.[26]

During the period from January 1980 to January 1981 a half-dozen other incidents occurred at La Hague, and while their importance was minimized by COGEMA, they aroused concern among workers and antinuclear activists.[27] Among them was another fire, this one in a silo for

[26]Paul André, "On a Frôle L'Accident Majeur à la Hague," *Le Matin*, April 18, 1980; Tim Keeley, "PuO Production Restarts at La Hague, Accident Now Seen Not So Serious," *Nucleonics Week*, April 24, 1980; Vanya Walker-Leigh, "La Hague Workers Demanding Several-Month Stoppage for Overhaul of Plant," *Nucleonics Week*, April 24, 1980.

[27]Michel Toussaint, "Incendie dans un Silo de Stockage à La Hague," *Le Matin*, January 8, 1981.

storage of pieces of magnesium and graphite from the decladding of gas-graphite fuel. According to one press account, two days after this incident a thousand workers met to protest the failure of the plant management to give them immediate notice of the fire and to take the precaution of evacuating the plant.[28] Another incident at this time involved leaks discovered in a pipe through which low-level waste is discharged to the English Channel.[29]

COGEMA Answers Its Critics

With Mitterrand's election, the Commissariat à l'Energie Atomique and COGEMA were under pressure to answer the critics of reprocessing and better justify their plans for La Hague. Not long before the parliamentary debate of October 1981 COGEMA leaked to three newspapers simultaneously a "confidential" report on the reprocessing program.[30] COGEMA argued that to consider irradiated uranium oxide a suitable material for containment of radionuclides was a mistake. A significant "fracturing" of this ceramic material could occur, the report said, with the release and diffusion of some fission products resulting. In COGEMA's view, from the standpoint of permanent waste isolation as well as conservation of energy resources, "reprocessing constitutes a wholly satisfactory solution" that should be adopted by all westernized nations having nuclear programs sufficiently large to justify a reprocessing industry. (As noted in chapter 1, U.S. experts in materials science do not credit vitrified high-level waste with greater durability than spent fuel.)

In its report COGEMA acknowledged that its first efforts to reprocess high-burnup fuel at the high-activity oxide unit at La Hague were overconfident:

One must remember that at the beginning of the 1970s a wind of optimism was blowing from the other side of the Atlantic where it had been announced that reprocessing light water reactor fuel, "a conventional industrial operation," would cost $25 a kilogram, that the cost of a plant of 1,500 tons a year capacity would be $200 million and that the plant could be run with 300 employees. Inasmuch as the training of French specialists in reprocessing had taken place in light of such expectations,... the effort which would have to be devoted to reprocessing was underestimated. One did not judge it necessary to conduct a

---

[28]Gilles Klein, "La Hague: Les Détecteurs, inutilisables, étaient eux-mêmes contaminés...," *Libération*, January 9, 1981.

[29]Paul André, "La Hague: L'Affaire des Fuites Rebondit," *Le Matin*, October 29, 1980.

[30]"COGEMA Has Underlined the Economic Benefits of Nuclear Fuel Reprocessing," *Nucleonics Week*, September 17, 1981.

program of development and experimentation of the critical equipment analo-
gous to that which had been carried out—and had produced such good results—
in the fields of isotopic enrichment and fast breeders. The HAO unit, conceived
too easily and put in use before being sufficiently tested, revealed itself to be a
difficult operation. The difficulties were of two types:

—Those linked to the use of equipment insufficiently tested: nothing fundamentally wrong, but an ordinary breakdown which could be repaired in an hour in
a conventional factory could stop production for a week in a nuclear plant
because of the difficulty of intervening. These breakdowns affect essentially the
mechanical apparatus for handling the fuel, the fuel chopping, the feeding of the
dissolvers and the transport of the wastes.

—Those having to do with the fact that certain problems linked directly to the
effects of radioactivity could not be discovered and dealt with at the time of the
preliminary trials.

... The experience with the HAO has been absolutely indispensable and irre-
placeable, for it is from experiencing difficulties that one gains savoir faire.
Serious mistakes in the conception of future plants have thus been avoided and
the French lead in reprocessing is fundamentally linked to that operational
experience.[31]

COGEMA also indicated that, despite the past difficulties in reprocessing
oxide fuel, the radiation exposure of individual workers actually had
declined from 1976 to 1980.

COGEMA cited a number of immediate and potential economic benefits offered by the planned expansion of La Hague. These included new
jobs for technically skilled workers at La Hague and in French industry,
and 53.5 billion francs (1981 value) in advance foreign payments for
reprocessing—money which would cover about half the cost of expansion. COGEMA also argued that expansion would permit France to
maintain a lead in nuclear research and engineering over other industrialized countries, and promote French dominance in the export of
nuclear technology. But COGEMA's economic arguments for La Hague's
expansion ignored the fundamental issue inherent in any plan for reprocessing: all things considered, how would the cost of fuel fabricated
with recovered fissile material compare with that of fuel made with fresh
uranium?

In the October 1981 parliamentary debate, Prime Minister Mauroy
declared that "no one can affirm today that there exists a better solution
for irradiated fuel than reprocessing." He added that the existing plant,
UP-2, would have to be modernized and enlarged if only to "better

[31]COGEMA, *Le Retraitment des Combustibles Irradiés en France et a l'Etranger* (Paris,
Editions MSA, September 1981) p. 13.

guarantee the safety of the workers," and that UP-3 would have to be built to honor COGEMA's foreign contracts. "This is a matter of the credibility of France at a moment when we can not, in this domain, permit ourselves the least faux pas," Mauroy said.[32] For his part, Chevènement, the research minister, spoke of a new emphasis on laboratory development of reprocessing technology looking to "entirely automated plants and teleoperated systems for maintenance and repair."[33]

But a Socialist deputy who had visited La Hague only the week before was deeply skeptical. "The working conditions in certain units were at the edge of acceptability," he said. "The plant could be put out of operation for an indeterminant time simply by failure of a single element." In regard to the waste problem this deputy added, "Until there is proof to the contrary, reprocessing does not reduce the volume of waste that must be disposed of and it favors the dispersion of plutonium."[34]

In light of the complaints that have since been heard in the British House of Commons about discharges of liquid waste from Sellafield to the Irish Sea, it should be noted that no such protestations were heard during the National Assembly debate about discharges from La Hague to the English Channel. Although discharges from La Hague have not been negligible, they have been much lower than those at Sellafield. In 1983, for example, discharges of plutonium and other alpha-emitters from La Hague were less than 2 percent of those from Sellafield.[35] Moreover, the powerful Raz-Blanchard currents that sweep around the Cotentin Peninsula are more favorable to the rapid dispersion of radionuclides than are the conditions in the Irish Sea. Present policy at La Hague is to accomplish the large planned expansion of plant operations without any increase in discharges above preexisting levels, as measured in total curies.

The Castaing Report on La Hague

The new government had made a concession to the critics of La Hague by going into the parliamentary debate promising that an objective evaluation of this center would be made by a special scientific commission. The commission was subsequently appointed, and its initial report

[32]*Journal Officiel*, October 7, 1981 (2 sess., October 6, 1981) p. 1562.
[33]Ibid., p. 1509.
[34]Deputé George Le Baill in *Journal Officiel*, October 7, 1981 (3 sess., October 6, 1981) p. 1541.
[35]United Kingdom, Radioactive Waste Management Advisory Committee, *Fifth Annual Report* to the Secretary of State for the Environment (London, Her Majesty's Stationery Office, June 1984) p. 19, table 2.

was issued about a year later, in January 1983. Although there were three Commissariat à l'Energie Atomique officials among its twelve members, there were also two scientists chosen from among La Hague's most severe critics. Raimond Castaing of the University of South Paris, a physicist and a member of the French Academy of Sciences, was named chairperson.

In its report the Castaing commission recognized that the existing UP-2 plant had been trouble-plagued, but it concluded that the two proposed new plants, UP-2-800 and UP-3A, should be able to perform to their announced capacities. The commission said that sweeping differences or improvements in the new plants would allow them to operate effectively.[36]

For the two scientists who had been critics of La Hague to have concurred in such a judgment was surprising. One of them was Jean-Paul Schapira, the theoretical physicist at Orsay; the other was Jean-Claude Zerbib, a health physicist at the Saclay Nuclear Research Center and at one time a Socialist union representative on a CEA committee investigating conditions at La Hague. In an interview in late 1980 I had found that neither Schapira nor Zerbib thought COGEMA was likely to deliver on its contracts to reprocess 6,000 tons of foreign fuel. Zerbib thought it possible that the cost of building and operating UP-3A would become high enough to lead foreign utilities to withdraw from their contracts at a penalty. Yet both Schapira and Zerbib signed the Castaing report, apparently accepting its conclusions, for no dissenting opinions were expressed.

That COGEMA will meet its commitments at La Hague has been something of an article of faith within the nuclear industry in Europe—a view held even by such technologically cautious individuals as Erik Svenke, the former head of the Swedish Nuclear Fuel Supply Company. "French pride is such that they will not allow La Hague to fail," Svenke told me.[37] He predicted that a substantial amount of fuel would be reprocessed, though at high cost and perhaps with some lengthy delays. La Hague plants UP-2-800 and UP-3A, whose maintenance will depend in part on radiation workers having to enter some of the process cells, are not to be built to the state-of-the-art technology that German reprocessors plan to incorporate in their Bavarian plant. Nonetheless, confidence in COGEMA's ability to deliver on its contractual commitments may be borne out: in 1985, for the first time, the existing UP-2 plant

---

[36]Council Supérieur de la Sûréte Nucléaire, "Rapport du Groupe de Travail sur la Gestion des Combustibles Irradiés," issued December 1982. The council appointed this working group on spent fuel management.

[37]Interview by the author with Svenke, November 17, 1982.

performed up to, and somewhat exceeded, its rated capacity to reprocess 400 tons of oxide fuel a year.[38]

Although confident of COGEMA's ability to reprocess at the planned rates, the Castaing commission noted the prospect for a large and continuing inventory of spent fuel. Some 9,000 tons would accumulate by 1992 and remain at least at that level thereafter, even barring accidents and delays. The 800 tons a year of reprocessing capacity to be reserved for French fuel would not be enough to accommodate fuel from reactors started up in 1992 and later years, as the commission pointed out. COGEMA at one time had plans for an additional 800 tons of reprocessing capacity at La Hague, but gave them up. Electricité de France, it seems, has long had reservations about the plunge into reprocessing. As Schapira told me not long before Mitterrand came to power, "The big argument in France—it is not actually a public argument because everything is hidden—is that the EDF takes the attitude that spending so much money in reprocessing is a burden because it already has to spend so much on reactors."[39]

According to the *London Financial Times*, Electricité de France today is keeping open the option of transferring some of its entitlement to reprocessing capacity at La Hague to foreign utilities.[40] Its large and fast-paced reactor construction program has entailed such large capital and debt retirement costs for the utility that reprocessing costs cannot be ignored. The lower-than-expected electricity consumption found in France (as elsewhere) reportedly makes retirement of the debt—put at $23 billion in 1984—difficult for EDF.[41] (The explanation for the debt problem lies more in politics than in economics: with EDF's total electricity sales in France and neighboring countries now at some 300 billion kilowatt-hours a year, the debt could be paid off easily if the government chose to permit a modest increase in the per-kWh price).[42]

Nuclear Waste Management

France's reputation as a leader in radioactive waste management has rested until now on a single major achievement: development of the

---

[38]Ann MacLachlan, "La Hague Plant Shows It Can Reprocess Over 400/Mt/yr of LWR Fuel," *Nuclear Fuel*, November 18, 1985.

[39]Interview by the author with Schapira, November 1980.

[40]David Marsh and David Fishlock, "French Nuclear Fuel Company's Battle for Supremacy," *London Financial Times*, January 23, 1985.

[41]"With Power to Spare, France Keeps Pushing Its Nuclear Program," *Business Week*, July 9, 1984.

[42]Personal communiction to the author from Bertrand Barré, June 21, 1985.

high-level waste vitrification facility at La Hague, together with the adjoining air-cooled surface facility where the canisters of vitrified waste are emplaced for temporary storage. Recently, however, there has been evidence of a commitment to find and develop in this century a site for permanent geologic disposal of high-level waste and long-lived transuranic or alpha-contaminated waste. While this effort faces significant technical uncertainties, as the largest country in western Europe France has an obvious advantage over the smaller countries in repository siting: it has more places to look and more candidate rock types to investigate.

The National Agency for Radioactive Waste Management (Agence Nationale pour la Gestion des Déchets Radioactif, or ANDRA) was created as part of the Commissariat à l'Energie Atomique in November 1979. ANDRA's director at the beginning was Jean-Marie Lavie, who previously had been the on-the-scene manager for French underground nuclear weapons testing in the South Pacific. His mission at ANDRA was to plan and develop the *centres de stockage* for permanent disposal of all categories of wastes. Although the vitrified waste can in principle be kept in surface storage indefinitely, Lavie approached the task of selecting a geologic disposal site with a certain urgency. As Lavie explained it to me, that urgency comes from the need to optimize the overall waste isolation system by taking into account the characteristics of a particular host rock and repository site in deciding how to treat and package the waste. "It is vain to perfect these [waste treatment and packaging] techniques without knowing the characteristics of the *centres de stockage*," he said.[43]

ANDRA's two initial undertakings to prepare for eventual establishment of a repository for high-level waste and a new burial facility for low-level waste got off to a clumsy and politically shaky start. They began in late 1979 or early 1980 in the highland region known as the *Massif Central*. One project was to drill two deep boreholes in a granite formation at Auriat, a village in the *département* of La Cruese. (Auriat was not considered necessarily a candidate repository site, for these were to be merely the first French boreholes in a geologic research program supported by the European Community.[44]) The other project was to convert to a low-level waste disposal site an open pit at a uranium mine, soon to be closed, at the village of Saint-Priest-La-Prugne.

At first the local villagers at Auriat believed that the big drilling rig was for prospecting for uranium, an activity long familiar to people in this part of France. But when it became clear that the drilling was unusually deep, descending some 30 meters a day, suspicions were

[43]Interview by the author with Lavie, November 1980.
[44]Ibid.

aroused. Both the departmental prefect, in replying to a letter of inquiry from Auriat's mayor, and the Minister of Industry, André Giraud, in replying to a question by a Socialist deputy in the National Assembly, equivocated and all but denied that the drilling figured in efforts to find a repository site.[45] Giraud, one of the powers in the Giscard government, provoked an outcry of protest on the Socialist benches by brushing aside the deputy's call for debate on nuclear policy and radioactive waste management. "The radioactive waste disposal problem has been solved," he said, observing that the high-level waste storage facility at Marcoule would suffice for many years.

But by now local officials had learned from the National Bureau of Geologic and Mining Research that the drilling was part of a long-term investigation looking to establishment of a repository. Incensed, Auriat's mayor later protested that "under a regime where the ministers do not hesitate to lie to members of parliament we can count only on ourselves."[46] The protests from Auriat finally faded away after the drilling stopped and the rig was dismantled and hauled away.

ANDRA's plan for Saint-Priest-La-Prugne was to construct a concrete pad on the floor of the large open pit created by past uranium mining operations, and to depose on this pad concrete containers of low- and medium-level wastes. The stacks of containers would then be covered with an impermeable layer of clay and topped with a layer of soil, which would be planted in grass. For the longer term, ANDRA might establish a disposal facility for transuranic wastes in the underground galleries of the old uranium mine, if geologic studies showed this to be safe.[47]

In September 1979 the Commissariat à l'Energie Atomique undertook to have the prefect of the Loire meet with the mayor of Saint-Priest-La-Prugne and officials of several other localities that would be affected by the project. In January 1980 this was followed up with a larger meeting to which all the mayors of the region were invited, but from then on things went from bad to worse, at least from the standpoint of the CEA, as opponents of the project made themselves felt.

An early move on the part of project opponents was to form a *comité de sauvegarde*, or committee of protection, established on the initiative

[45]*Journal Officiel*, Assemblée Nationale, April 9, 1980, pp. 148–149. The Socialist deputy was André Chandernagor, La Creuse.

[46]Agence Nationale pour la Gestion des Déchets Radioactif (ANDRA), *Defense des Monts d'Auriat*, quoting a letter from Mayor René Chambon to André Chandernagor, Socialist deputy and president of the Limousin Regional Council.

[47]ANDRA, "La Centre de Stockage de Saint-Priest-La-Prugne. Principaux Elements du Projet" (n.p., Commissariat à l'Energie Atomique, n.d.).

of teachers and farmers and made up in part of local elected officials and business and professional leaders. The committee, which deemed itself apolitical, propounded a list of twenty-two questions to be answered by the relevant ministries in Paris, including those for health and the environment and for industry. Pending answers to those questions, the committee decided that local officials should not lend themselves to further proceedings aimed at carrying the waste disposal project forward. In this way ANDRA was denied access to a dozen town halls for the deposition of official registers for the *enquête publique*.[48]

At Saint-Priest-La-Prugne ANDRA's promise that the project would mean 130 jobs counted for little,[49] and opposition was massive. The community resented the government's closing of the uranium mine and the offer of nuclear waste instead of direct economic assistance. For this reason and others, discussed below, the Saint-Priest-La-Prugne site was to become a lost cause.

In early June 1980 more than a thousand demonstrators took part in a protest march at Saint-Priest-La-Prugne, supported by the *comité de sauvegarde*, various agricultural and ecology groups, the CFDT, and the Socialist party. The demonstration ended with a country picnic at which the local politicians and Socialist officials mingled with the crowd and joined in complaining about ANDRA and the CEA.[50]

About this time, the *comité de sauvegarde* and some allied political bodies initiated what amounted to a counter-*enquête*. Technical experts from the Commissariat à l'Energie Atomique were brought face to face with hydrologist G. de Marsily of the School of Mines in Paris and physicist Jean-Paul Schapira at a meeting of the Loire *département* general council. De Marsily regarded ANDRA's report on the project as inadequate, and believed that the proposed site could not meet the criteria recommended by the International Agency for Atomic Energy. He told the council that the rock beneath the site was fractured and complex in its geologic structure (and hence difficult to study), and that its "absorption coefficient" for radionuclides was unacceptably low. He was especially concerned lest a waste facility there eventually be expanded to provide for disposal of transuranic waste in the mined galleries. And he was worried that given the wide disparity then existing between the French and American thresholds for classifying transuranic

[48]Ibid.

[49]Paul Chappel, "Les élus ont cédé à des pressions pour ne pas ouvrir leurs mairies, estime le préfet de la Loire," *Le Monde*, May 21, 1980.

[50]Paul Chappel, "Marche pacifique à Saint-Priest-La-Prugne (Loire) contre le stockage de déchets radioactifs," *Le Monde*, June 10, 1980.

waste, much of the waste that would be destined for the open pit facility properly belonged in a deep repository.[51]

Schapira accused the CEA and ANDRA of an outright misrepresentation of the wastes to be disposed of at Saint-Priest-La-Prugne. In a later statement before a special information commission on nuclear energy, Schapira testified that "ANDRA has carefully avoided telling the truth about the inventory of radionuclides that will contaminate the waste that it plans to dispose of at Saint-Priest-La-Prugne."[52] He noted that the Minister of Industry, replying to a question from the *comité de sauvegarde*, had said, "It must be underscored that in no case will irradiated fuel elements or wastes from their reprocessing be disposed of on the site." The minister failed to say, Schapira went on, that reprocessing wastes include, besides high-level waste (which would not be disposed of at this site), low- and medium-level wastes that can be significantly contaminated with plutonium. Schapira added that to exercise control over the plutonium content of waste on a routine and continuous basis was, as a practical matter, technically impossible. "It will be necessary [for ANDRA] to trust COGEMA," he said, by which he meant that when waste containers arrived at Saint-Priest-La-Prugne, if anyone knew the alpha content it would be COGEMA. Schapira also charged that an ANDRA information brochure was worded to imply that the site had been studied and found suitable by the Bureau of Geologic and Mining Research, when in fact no such finding had been made. The bureau's director had made this clear in a letter to ANDRA calling on the agency to correct its error.[53]

By this time, however, the official *enquête publique* on the Saint-Priest-La-Prugne project had long since run its course, arriving at the judgment that the project should be approved. But whether a *centre de*

[51]Ibid.; interview by the author with de Marsily, November 1980. In 1984 the French government lowered significantly the threshold for alpha waste accepted for shallow burial. Ann MacLachlan, "France Sets 'World's Strictest' Waste Safety Rules for Surface Storage," *Nucleonics Week*, July 5, 1984.

[52]Jean-Paul Schapira, "Le Projet d'un Stockage de Déchets Pouvant Contenir du Plutonium de Faible et Moyenne Activité à Saint-Priest-La-Prugne," statement submitted November 13, 1980, to La Commission de l'Information sur L'Energie Electronucléaire; Jean-Paul Schapira, "Comment Stocker des Déchets moyennement et faiblement radioactifs?" *Le Monde*, September 10, 1980. The special commission Schapira addressed was created by the Giscard government to placate critics who felt that the public was being denied objective information on nuclear issues. This body was abolished by the Mitterrand government.

[53]Letter of July 10, 1980, from the director, Bureau de Recherche Geologiques et Minières, to Jean-Marie Lavie, director of ANDRA, reprinted in *Libération*, November 12, 1980.

stockage actually would be established at Saint-Priest-La-Prugne depended on the issuance of a presidential decree. Giscard d'Estaing's
defeat in the May 1981 election meant that the decision would lie with
Mitterrand, who during the campaign had promised that the project
would be restudied, if not abandoned. Later, Jean Auroux, the Socialist
mayor of Roanne who had been a leading project opponent, turned up
as minister of energy in the Mitterrand government. Soon thereafter the
Saint-Priest-La-Prugne site was dropped from consideration altogether.[54]

A New Push for Geologic Disposal

In October 1982 ANDRA brought forth a long-range plan for radioactive
waste management and disposal, a plan subsequently reviewed and approved with some changes by the Industry and Research Ministry's High
Council on Nuclear Safety, and, ultimately, by the Mitterrand government in June 1984. The plan adopted contemplates a twenty-five-year
program, to cost some 60 billion francs (7 billion dollars), that will lead
to the establishment of three major facilities: two new *centres de stockage* for shallow burial of low- and medium-level waste, and one for deep
geologic disposal of high-level and transuranic wastes.[55]

The two low- and medium-level waste facilities, expected to be available before the end of the 1980s, would be similar to the existing *centre
de stockage* at La Hague. The limited amount of unused space at the
latter facility would henceforth be reserved for waste from the reprocessing plants. (Originally there was to have been only one new center
of this kind, but a second was added to the program out of considerations
of regional equity and convenience of transport.[56]) The sites would be
chosen from a list of a half-dozen candidate locations, to be publicly
announced.

The approved ANDRA program for the siting of the deep geologic
repository for high-level and transuranic waste has three phases. The
first consists of a review of existing information to identify four candidate
sites, one in each of four different kinds of rock: granite, schist (which,
like granite, is a crystalline rock), salt, and clay. The second phase
consists of investigations of the four sites from the surface, by geophysical surveys, shallow hydrologic test wells, and deep boreholes; in this
phase the aim is to determine which of the sites holds the greatest

[54]"'I Am Not an Anti-Nuclear Militant,'" *Nucleonics Week*, April 14, 1983.

[55]Interview by the author with Jean-Marie Lavie, December 1983; MacLachlan, "France
Sets 'World's Strictest' Waste Safety Rules."

[56]Interview with Lavie, December 1983.

promise for development of a repository. The third and last phase consists of developing a deep underground laboratory at the preferred site, extensive data-gathering and modeling, and, if all goes well, proceeding to licensing for repository development. The entire three-phase process is expected to last ten years or longer.[57]

By the fall of 1986 phase one had been completed and four candidate sites had been selected, though not yet announced.[58] But this announcement and the beginning of the second (field investigation) phase will have to come soon if ANDRA is to have any chance of meeting its goal of choosing a site for an underground laboratory by 1989.

Several things should be noted about the ANDRA site-selection process. One is that it makes not the slightest bow to the concept of public participation—a concept enshrined (at least in principle) in the United States and honored to one degree or another in the United Kingdom, Sweden, and even Germany (as in the International Gorleben Review). Another is that, from the standpoint of technical conservatism or redundancy, the French siting policy seems to fall somewhere between the German and the U.S. policies. In Germany all bets have been placed on the Gorleben salt dome site; in the United States the Department of Energy has been committed to the procedurally and politically demanding task of selecting three sites for deep-shaft exploration, or in effect, underground laboratory studies. The CEA decision to investigate only one site from an underground laboratory was quite deliberate. Some members of the Castaing commission argued that several sites should be so investigated; but others, including in particular the members from the commissariat, took the view that a suitable site could be obtained without comparative site studies.[59]

If ANDRA can avoid severe technical setbacks that embarrass or discredit the siting effort, the agency may have a good chance to overcome any political obstacles presented. Despite reforms sponsored by the Socialist government to promote greater decentralization, the power of the departmental prefects and the ministries in Paris remains strong. ANDRA might even have the good luck to find a technically suitable site where a repository would not be unwanted. In its search for low-level waste burial sites, ANDRA provoked no great uproar back in November

[57]M. A. Barthoux, "Deep Geological Disposal," in *French Industrial Experience in Radioactive Waste Management and Spent Fuel Storage* (Commissariat à l'Energie Atomique, 1985); interview by the author, October 22, 1986, with Eric Frémond, assistant nuclear attaché, French Embassy, Washington, D.C.

[58]Interview with Frémond, October 22, 1986.

[59]Conseil Supérieur de la Sûreté Nucléaire, "Rapport du Groupe de Travail sur les Recherches et Developpements en Matière de Gestion des Déchets Radioactifs (October 1983–October 1984)."

1984 when it announced several candidate areas, and in one or two instances a site was added at the suggestion of local officials.[60]

Like other industralized countries, France has communities where once-flourishing industries have declined and left pockets of unemployment and related ills. One such community is Cholet, a town in west-central France near Nantes. In November 1984 at a meeting of the municipal council, the mayor, a long-entrenched politician who had been a high official in the Giscard d'Estaing government, announced to the stupefaction of the Socialist and Communist members present that Cholet was prepared to host an ANDRA low-level waste facility. The community, which has known much joblessness, appears to have received the news calmly. "ANDRA is looking for a site, and we are looking for jobs," the mayor has explained.[61] Besides jobs, the community could also expect a one-time impact assistance grant of about $3.6 million, plus an annual bonus of roughly $150,000.[62] The challenge for ANDRA in this situation is to see that Cholet has an acceptable site as well as a receptive attitude, and to determine this in a technically unbiased manner.

If the Commissariat à l'Energie Atomique can meet France's domestic waste disposal needs, it may in time find it politically possible to offer relief to its foreign customers who have reprocessing wastes accumulating at La Hague. This might come to be seen as merely another opportunity to gain commercial advantage and demonstrate French mastery of another aspect of nuclear technology. It might also be an essential precondition for any initiatives by European waste generators or others looking to the establishment, inside or outside Europe, of international waste and spent fuel management and disposal facilities (see chapter 12). But progress of this kind may be impossible without a long period of political quiet for nuclear power in which the benefits of this energy source can be calmly weighed against the risks.

The Fate of Separated Plutonium

The nuclear enterprise in France and the rest of Europe certainly does not need a round of disturbing events and controversies associated with separated plutonium, especially now when uranium is so cheap that

---

[60]Ann MacLachlan, "French Agencies to Study Four Sites for Low Level Waste," *Nucleonics Week*, October 18, 1984; "French Commune Has Second Thoughts About Low Level Waste Storage," *Nucleonics Week*, January 17, 1985.

[61]Roger Cans, "Cholet Accepterait de Stocker des Déchets Nucléaire, la Poubelles aux Emplois," *Le Monde*, November 20, 1984.

[62]MacLachlan, "France Sets 'World's Strictest' Waste Safety Rules."

breeder reactors cannot be justified for anything other than experimental purposes. The present drift to the commercial use of plutonium fuel in light water reactors in France, Germany, Belgium, and Japan could bring on just that kind of trouble, with the potential for nuclear theft and terrorism perhaps being the most worrisome. The reprocessing at La Hague presents the safeguard problems inherent in a bulk separations facility where discrepancies of just 1 percent or less in the plutonium inventory suggest the possibility that enough material could be stolen to fabricate an atomic bomb (see chapter 3).

The introduction in France, in other European countries, and in Japan of new or enlarged plutonium fuel fabrication plants, together with increased shipments of plutonium oxide by COGEMA and other reprocessors to those plants, would raise significantly the risks of nuclear theft and terrorism. The risks could be reduced by building the plutonium fuel plants at the reprocessing centers and thus eliminating the transport of plutonium oxide, but this is not what the French are now planning. Substantial amounts of plutonium are to be moved from the separation plants at La Hague to fuel fabrication plants in Belgium and the south of France, with the plutonium fuel then to be shipped to EDF reactors. In late 1984 COGEMA and Belgium's Belgonucléaire set up Commox, a joint venture, to market plutonium fuel produced at Dessel, Belgium, and at Cadarache in the Rhone Valley, where COGEMA runs a pilot plant for the Commissariat à l'Energie Atomique.[63] This was made possible by the decision of Electricité de France to begin using plutonium fuel in some of its reactors.[64] (If this were not done, the plutonium separated at high cost to EDF at La Hague would bring no return whatever and would—every kilogram of it—impose on the utility significant further charges from storage fees and the eventual necessity of processing the plutonium again to remove the buildup of americium from radioactive decay. See chapter 12.)

A start toward this new regime will be made in late 1987 when one of EDF's 900-megawatt pressurized water reactors receives 8 tons of plutonium fuel. Use of such fuel is expected to increase progressively through 1995, when ten to twelve EDF reactors are loaded with 90 tons of it. Moreover, COGEMA and the Commissariat à l'Energie Atomique are considering building at Marcoule a plutonium fuel plant capable of an annual output of 100 to 150 tons by 1993,[65] which would presuppose another dozen or so reactors using such fuel, in France and possibly

[63]Ann MacLachlan, "French and Belgians Pool MOX Fuel Marketing, Prepare Recycle," *Nucleonics Week*, November 8, 1984.

[64]Daniel Chavardés, "Brief Update of the French Nuclear Program" (November 1985).

[65]Interview with Eric Frémond, October 22, 1986.

other countries. This emerging commitment to plutonium fuel is high-stakes poker. For instance, if terrorists managed to divert or forcibly seize a shipment of separated plutonium and then followed up with a bomb threat, the psychological impact—even if the threat were found to be a bluff or hoax—could be an unparalleled political disaster for nuclear power.

It is when separated plutonium is in transport that the danger of accidents or terrorist incidents is greatest. For this reason plutonium shipments, and even word that plutonium is to be moved, can bring on intense press attention, public apprehension, and political controversy. In fact, concern that something might go wrong can lead to security precautions so elaborate as to reinforce the public impression of a highly risky undertaking. The summer and fall of 1984 saw two separate but not unrelated shipping events which pointed up the large potential for notoriety and political trouble that plutonium recycling holds for the nuclear industry, not just in France, but around the world. One had to do with the shipment of a quarter of a ton of plutonium from La Hague to Japan for the experimental breeder Joyo, and offered a foretaste of what might be expected from the more numerous and routine shipments that will be associated with plutonium recycling in light water reactors. The other event was the sinking in the North Sea on August 25, 1984, of the French container ship *Mont Louis*, which was carrying thirty 15-ton cylinders of uranium hexaflouride (UF_6) when she was struck broadside in a fog by a German car ferry carrying nearly a thousand passengers.[66] The feed material for uranium enrichment plants, UF_6 is not a nuclear explosive material and is nothing like as toxic radiologically as plutonium. Nonetheless, the sinking of the *Mont Louis* was in the headlines of the French and European press for weeks. One can only wonder what the reaction of the press and the public would have been if the ship had been carrying plutonium. A closer look at these two happenings and the responses that they provoke offers further insight into the implications of a commitment to the use of plutonium fuel.

The UF_6 aboard the *Mont Louis* was being sent from France to the Soviet Union, where the uranium was to be enriched under a contract of long standing and then returned to western Europe for use in French and Belgian reactors. It was a routine commercial transaction involving nuclear materials at the front end of the fuel cycle. There was no need for the shipment to be regarded as particularly dangerous, and certainly no need to treat it as ominous or mysterious. But French authorities had

[66]See Ann MacLachlan, "Mont-Louis Accident Fuels International Debate on Nuclear Transport," *Nucleonics Week*, August 30, 1984; "Alerte aux toxiques en Mer du Nord," *Le Monde*, August 28, 1984.

foolishly and clumsily tried from the outset to hide the fact that radioactive material was aboard, and it was Greenpeace that first alerted the
press to the nature of the *Mont Louis*'s cargo. At one point some fifty
journalists were on hand to witness the difficult and prolonged operation
by divers and salvage vessels to extricate the UF_6 cylinders from the
wreck, which rested on its side in 15 meters of water off the Belgian
coast.[67]

Le Monde, in an editorial on August 28, 1984, called attention to the
extent to which the movement of radioactive substances had become
commonplace in France. Within the borders of France alone shipments
of such substances each year were running to a total of 5.4 million
kilometers and more than a million tons. Shipments were moving by
highway, rail, and, in the case of certain materials that are only barely
radioactive, even by mail. Added to these domestic shipments was a
growing international commerce in nuclear materials. *Le Monde* remarked ironically on the connection between the *Mont Louis* incident,
the shipment of plutonium to Japan, and the French government's habit
of secrecy:

Bought and sold and shunted about, the atom is thus reduced to a banality. For
all that, it retains its risks, and hence its emotional charge. For this reason
shipments are always surrounded by the greatest secrecy, not to say mystery.
The time of embarkation for the 250 kilos of plutonium to go from La Hague to
Japan is still a state secret. It took the revelations of the Greenpeace ecologists
for us to know what was aboard the *Mont Louis*.... So nuclear energy is made
a banality, except in the domain of information, where a mania for secrecy
reigns. Is it thus that one will learn to live with and master the atom?

French authorities had indeed shown poor judgment in not fully
revealing beforehand the nature of the UF_6 shipment and of the material
itself. Even had the UF_6 cylinders not been recovered there would have
been little hazard to the public, either radiologically or chemically. With
some qualification this statement could be made of any shipment of
nuclear material associated with a once-through light water reactor fuel
cycle. In particular, shipments of spent fuel present a certain risk, but
probably are less hazardous than shipments of many nonradioactive
materials ubiquitous in ordinary commerce (such as gasoline, propane,
and liquid chlorine) when the spent fuel is well-aged, robustly packaged,
and routed to avoid urban centers or to be escorted swiftly through
them. But shipments of separated plutonium are a different matter: that
material is a nuclear explosive, and too many precautions can hardly be
taken to deny to potential terrorists information about how and where

[67]Danièle Rouard, "Le premier fût a été retiré des cales du 'Mont-Louis,'" *Le Monde*,
September 4, 1984.

it is stored, when and by what routes it will be shipped, and by what means and at what strength it will be guarded.

In the case of the quarter ton of plutonium being returned from La Hague to Japan, the magnitude of the potential dangers perceived by the responsible officials was evident from the security precautions taken at the insistence of U.S. officials, who were involved because the plutonium had been separated from fuel of American origin. The shipment took place over about six weeks in October and November 1984. Approval of the security arrangements by the U.S. secretary of energy was given only after the original Japanese security plan had been substantially strengthened, for the Japanese had intended to ship the plutonium on the deck of a common and lightly guarded container ship.[68] This vessel was to be routed through the Suez Canal and the waters of the Middle East, where terrorism is rife and several governments are considered proliferation-prone. In the end, as widely reported in France and other countries, the shipment went via the Panama Canal on a heavily guarded vessel—*the Seishin Marie*—that was specially equipped for the mission. The canal transit was made secretly in the dead of night, with two U.S. naval vessels as escorts (on the arrival of the *Seishin Marie* in Japan, see chapter 11).[69]

Conclusion

To sum up, the moves led by COGEMA to use plutonium in light water reactors bring to the fore the problems and issues that surround plutonium as an explosive nuclear material, and carry a risk of lowering public trust and confidence in nuclear power in Europe and throughout the world. This seems all the more unwise at a time when popular acceptance of nuclear power may turn on whether the nuclear enterprise in France and other countries can catch up with such major problems as the disposal of radioactive waste and—in light of Chernobyl—convince the public that power reactors are indeed as safe, manageable, and economically beneficial as the leaders of the enterprise believe them to be.

[68]John J. Fialka, "U.S. Approves Shipment of Plutonium to Japan from France for New Reactor," *Wall Street Journal*, July 7, 1984; Kennedy Maize, "Democrats Plea to President: Hold the Plutonium, Monsieur," *The Energy Daily*, August 7, 1984.

[69]"Panama Protests Nighttime Movement of Plutonium Ship Through Canal," *Nuclear Fuel*, November 5, 1984. From start to finish the shipment was carried out in keeping with a security plan that DOE officials described to Congress and the U.S. Nuclear Regulatory Commission as going beyond national and international standards, with "U.S. military units [escorting the shipment] in designated areas to minimize response time in the event of an incident." Ibid.

11
Japan, the Pacific, and the Nuclear Allergy

Japan's ambitious nuclear program is still one of the world's largest, despite a slowdown in recent years.[1] In the mid 1980s the Japanese nuclear industry fashioned elaborate but problematical plans to establish facilities for a complete nuclear fuel cycle on the French model. Included would be commercial-scale facilities for waste disposal, uranium enrichment, and fuel reprocessing to separate large amounts of plutonium for eventual use in light water reactors.

If this major new development in Japanese nuclear energy actually materializes it will be played out against a political background in which two important attitudes diverge dramatically. One is the nuclear allergy of the Japanese population, an allergy almost pervasively present as a legacy from Hiroshima and Nagasaki. The other is the perception, strongly held by Japan's powerful governing bureaucracy and the long-dominant conservative Liberal Democratic party, that nuclear power is essential to this resource-poor nation's ability to reduce its heavy dependence on oil from the politically unstable Persian Gulf. Public opinion surveys over the past decade have shown that the greater part of the citizenry shares this perception, although since the Chernobyl accident there has been a growing uneasiness about the safety of nuclear power.[2]

---

[1] The Japanese program is the fourth largest, after the U.S., French, and Soviet programs.

[2] According to *Atoms in Japan* (August 1986), English-language magazine of the Japan Atomic Industrial Forum, the Japanese daily *Asahi Shimbun* has reported the following results of a poll conducted by the newspaper on August 6–7, 1986: 60 percent of the respondents favored maintaining Japan's nuclear power capacity at its present level; 13 percent would reduce the capacity; 10 percent would increase it; and 9 percent would abolish it. When asked whether they favored "promoting" nuclear power, only 24 percent said yes, while 41 percent said no—the first time since 1978, *Atoms in Japan* says, that more people have been found to oppose such promotion or expansion than to support it. Sixty-seven percent now feel concern about a major nuclear accident happening in Japan. *Asahi Shimbun* reported these results on August 29, 1986.

335

The progress made by Japanese commercial nuclear power since its beginning in the 1960s has on the whole been impressive, although at times painfully won. The siting of reactors has sometimes been long delayed by negotiations between industry representatives and farmers, fishermen's cooperatives, and local elected officials. Nonetheless, the sites needed have in the end been obtained. Lying ahead for the Japanese nuclear industry is the task of gaining local and prefectural consent to investigate and develop a geological disposal site for high-level and transuranic wastes. If this can be achieved, the overall record of the light water reactor program during its first twenty years will have been one of remarkable accomplishment and success, particularly in view of Japan's nuclear allergy. But the question remains unanswered whether public trust and confidence in nuclear power will be increased or diminished by what happens over the next ten to fifteen years.

Radioactive waste disposal is a problem on which the Japanese only now, after years of frustration, have made a degree of progress. In 1984 and 1985 the nuclear enterprise in Japan obtained a site for low-level waste disposal, and made at least a start—albeit one that is controversial and seems none too promising—at identifying an experimental site for high-level waste. Success now in coming to grips with the waste problem might allay some of the public's uneasiness about the atom. But a precondition for success will be a predictable and rational national nuclear waste policy that local and prefectural officials can depend on. As events in 1984 and 1985 pointed up, such a policy is still lacking. Beyond that, controversy and conflict over reprocessing and use of plutonium fuel, which might be provoked or heightened by nuclear accidents or terrorism in Japan or abroad, could aggravate the nuclear allergy and make public trust for the nuclear program harder to gain.

Japan's Nuclear Allergy

Japan is no doubt more politically sensitive to things nuclear than any other major country on earth. Given Hiroshima and Nagasaki and the continuing psychological hold those holocausts and their aftereffects exert on the population, every Japanese citizen might well feel uneasy about the atom. Moreover, since the end of World War II more than forty years ago there have been a variety of sensational scares at home and abroad: fallout from weapons tests showering the tuna boat *Fukuryu Maru* (Lucky Dragon); contaminated fish being found in the marketplace; radiation leaking from the nuclear ship *Mutsu*; the American accident at Three Mile Island; and a much-publicized leakage of radioactive waste from a Japanese nuclear power plant into a coastal bay. In most of these episodes the radioactivity present was at low levels and probably did no harm to those exposed. Despite this scientific reality, among many Japanese the psychological reality was one of alarm and anxiety.

Public support for Japan's long-standing "three non-nuclear principles"—no possession, no manufacture, and no introduction of nuclear weapons—is so strong that at present a nuclear-armed Japan seems unthinkable. Indeed, if the Japanese public could have its way, the superpowers would move rapidly toward nuclear disarmament and a weapon-free world. In 1982 an anti-nuclear weapons campaign collected an astounding 28 million signatures on a petition. Women's organizations, consumer groups, labor unions, youth groups, and the anti-nuclear groups of the Socialist and Communist parties (Gensuikin and Gensuikyo, respectively) all joined in this massive indication of opposition to nuclear weapons.[3]

The Japan Atomic Industrial Forum (JAIF), the most important industrywide group, identifies itself with the overwhelming popular antipathy to nuclear weapons. When several years ago it came to light that despite the three principles some American naval vessels calling at Japanese ports had been nuclear-armed, a contributor to the Forum's magazine, *Atoms in Japan*, expressed consternation and alarm. Later that year *Atoms in Japan* took U.S. Secretary of Energy James Edwards to task for suggesting that plutonium from the reprocessing of commercial spent fuel be used in nuclear weapons. An article by an official of the Japan Nuclear Fuel Service Company, which the Japanese utilities had established to undertake commercial reprocessing, warned that for the United States to do as Edwards suggested would create a "crisis in the NPT [Nonproliferation Treaty] regime," and declared that if abolishing all nuclear weapons immediately was impossible, it was "absolutely necessary that a definite line be drawn between peaceful and military uses."[4]

Contributors to *Atoms in Japan* clearly are frustrated by statements of anti-nuclear militants that link nuclear weapons and nuclear power. These authors have described, for example, the atmosphere of overwhelming opposition to nuclear power encountered at a ban-the-bomb

---

[3]Sunao Suzuki of the *Asahi Shimbun* editorial board reports that a total of 80 million signatures were gathered by the four ban-the-bomb petition campaigns active in Japan in 1982, but that as many of the same signatures appeared on more than one petition the number of signers was nearer 28 million. Sunao Suzuki, "Public Attitudes Toward Peace," *Bulletin of the Atomic Scientists*, February 1984.

[4]Koichi Kawakami, "Japan's Three Non-Nuclear Principles," *Atoms in Japan*, June 1981, pp. 4–6; Shigefumi Tamiya, "Commercial Nuclear Fuel Must Not Be Diverted to Weapons," *Atoms in Japan*, October 1981.

conference. Nuclear energy proponents warned the conference that if they were discouraged from joining the movement against nuclear weapons, the movement would be the weaker for it.[5]

Japan's nuclear allergy has also been much in evidence when even minor releases of radioactivity have occurred. Particularly striking was the reaction by the public and the Japanese nuclear industry itself to the radioactive spill at the Tsuruga-1 power plant in early 1981. In that incident liquid waste overflowed from a sludge storage tank, perhaps a ton of it escaping into a storm drain and flowing into Urazoko Bay. Subsequently, routine monitoring of bay waters detected cobalt-60 and manganese-54 present in seaweed at concentrations ten times higher than normal, though still at the very low level of less than one picocurie (or trillionth of a curie) per gram. The spill became a national scandal because, like several earlier spills at the Tsuruga plant, it had not been reported by the utility to regulatory authorities, and therefore smacked of a cover-up. By standards of American behavior, the Japanese response to this incident was amazing. The all-Japan council representing cities and towns where nuclear plants are found adopted a resolution expressing "irrepressible anger... at the shocking blows struck not only at Tsuruga City but also at all cities, towns, and villages ... where nuclear power plants are located." The response of the regulators and the Japan Atomic Power Company (JAPCO), the utility responsible, was similarly severe. The reactor was ordered shut down for six months, the president and board chairman of JAPCO resigned, and the entire Japanese nuclear industry put on the hair shirt.[6] Hiromi Arisawa, chair of the Japan Atomic Industrial Forum, said the Tsuruga spill had impaired public confidence in nuclear power when it was recovering from Three Mile Island, and that the "sensational press reports themselves caused serious damage to fisheries, tourism, the sale of local products and other industries in Tsuruga City and its environs."[7]

Japan's Nuclear Dependence

Alongside Japan's understandable anxiety about the atom is Japan's strongly perceived need for the atom. In economic terms, Japan can perhaps best be described as a dependent giant. Its economy, the noncommunist

[5]"World Conference Against A & H Bombs," *Atoms in Japan*, August 1981.

[6]The Tsuruga spill and the reaction to it have been described in detail by *Atoms in Japan* in its issues of April, May, June, and July of 1981.

[7]"JAIF 30th General Assembly, Chairman's Opening Remarks," *Atoms in Japan*, June 1981.

world's second largest, relies on imported fuels, mostly foreign oil.[8] The dependence on foreign oil persists even though considerable progress has been made in conserving energy and diversifying energy sources since the shock of the first oil embargo in 1973.

Nuclear power provided 26 percent of Japan's electricity in 1985, or on the order of 8 percent of the country's consumption of primary energy.[9] Despite the nuclear effort and the smaller but not insignificant effort to develop solar energy and other renewables, much of the primary energy necessary to sustain Japan's dynamic industrial economy will for many years come from imported fossil fuels. What such a dependence can mean was dramatically illustrated in 1979, the year of the second oil shock. At that time 41 percent of all Japanese spending for imports went for fuels, their total cost ($33 billion for oil alone) being equal to all of Japan's foreign exchange earnings from the sale of automobiles, steel, television sets, radios, cameras, and watches.[10]

Uranium too must be imported, but fuel costs are so small a part of the total cost of nuclear electricity that this relatively modest dependency is not to be compared with imports of oil, natural gas, or even coal for electric generating plants. It is also advantageous that Japan can tap the abundant uranium reserves in politically stable countries such as the United States, Canada, and Australia. Moreover, uranium is unique among major energy resources in the ease with which it can be stockpiled, and energy security is further enhanced by the ability of nuclear reactors to run for nearly a year without being refueled.

The foregoing advantages of nuclear power are all real and beyond dispute. Even so, leaders of the Japanese nuclear enterprise have gone further by making speculative claims for plutonium and the breeder reactor. As they see it, if breeders are not to be built on a large scale in the near term, this goal will be reached by an evolutionary path that will first include plutonium recycling in light water reactors.[11] Japan's

[8]By the size of its gross national product and national income, Japan ranks behind only the United States. In 1983 less than 10 percent of Japan's primary energy supply was domestically produced. Energy imports as a proportion of total primary energy were: oil, 60.8 percent; coal, 15.3 percent; and liquified natural gas, 7.2 percent. Nuclear power, derived from imported uranium, represented 7.4 percent of the primary energy supply. Statistics Bureau, *Statistical Handbook of Japan 1985* (Tokyo, Japan Statistical Association, 1985).

[9]"Nuclear Capacity Factor Sets Record of 76.0% in Fiscal 1985," *Atoms in Japan*, April 1986, p. 50.

[10]John W. Powell, "Nuclear Power in Japan," *Bulletin of the Atomic Scientists*, May 1983, p. 33.

[11]The Japan Atomic Industrial Forum, in a report to the Japan Atomic Energy Commission on July 18, 1984, called for use of plutonium fuel in light water reactors to be brought to a stage of "practical utilization" by the late 1990s. "JAIF Proposes Plutonium Recycle

nuclear program has included substantial research and development efforts for fuel cycle facilities, the breeder, and advanced thermal reactors, and a French-designed demonstration reprocessing plant at Tokaimura, north of Tokyo. French technology is also to be used in the commercial-scale reprocessing plant now planned.[12]

In sum, Japan is strongly committed to the nuclear option. This dependent giant's reasons for its commitment cannot be easily discounted or dismissed. But to go beyond the well-established light water reactors on the strength of uncertain and speculative claims for plutonium recycling can be a risky choice, especially in nuclear-allergic Japan.

The Siting Problem

As a result of the nuclear allergy and of other social, environmental, and geographic constraints, the siting of nuclear facilities presents a major problem in Japan, as it does in many countries. Reactor siting is difficult, for example, because reactors, numerous as they are, require many more sites than other nuclear facilities; because in Japan reactors are sited on the coast where cooling water is available, creating conflict with coastal farming and fishing communities; and because reactors must be sited in rural areas where vacant land is available, yet must also be near the urban centers where nuclear electricity is most needed. Finally, reactors must be built where the earthquake hazard—not entirely avoidable anywhere in Japan—is deemed acceptable, for given the enormous thermal energy present from the fission process and fission product decay, reactors present a greater risk of catastrophic containment failure than do nuclear installations of any other kind.

The siting of a radioactive waste repository in Japan appears easier than reactor siting in some respects, but harder in others. Only a few geologic disposal facilities for high-level waste will be needed in Japan for many decades to come, and only one will be needed early on. Repositories need not be near the urban centers; indeed, if a geologically suitable site is available in Japan, the most remote corner of the country will do. The problem of public acceptance aside, a major difficulty in

---

in Thermal Reactors," *Atoms in Japan*, August 1984. While no official decision has yet been made to do this, the large-scale program of commercial reprocessing to which the Japanese nuclear industry now seems committed carries its own logic. Once large amounts of plutonium begin to be separated the question becomes, "What is to be done with it?" The costs and other disadvantages associated with plutonium storage are such that recycling in light water reactors is the ready, convenient answer.

[12]"MITI Subsidizing Introduction of Large-Scale Reprocessing Technology," *Atoms in Japan*, January 1986, p. 34.

repository siting in Japan, as elsewhere, is the problem of geologic constraints.

Japan is situated at the edge of what scientists in the fields of vulcanology and plate tectonics call the Pacific Basin's ring of fire. Here, just to the east and to the south of Japan, two of the oceanic tectonic plates that make up part of the earth's crust are slowly sliding beneath the Eurasian continental plate, where they will melt and become part of the earth's mantle. Earthquakes, volcanoes, and hot springs are the surface manifestations of this subduction process, and no country is more noted for these phenomena than Japan. More than a dozen major earthquakes, and countless lesser ones, have visited Japan's main island of Honshu over the last century. There are two belts of volcanoes, the larger one stretching from the eastern tip of Japan's northern island of Hokkaido down through northern and central Honshu before reaching off shore south of Tokyo Bay to form the Izu, Ogasawara, and Kazan island chains. The active geothermal fields in the main Japanese islands number about one hundred, and are often marked by telltale fumaroles or steam cracks, as well as by hot springs with temperatures at or just below boiling.

Major earthquakes, with their violent ground shaking, faulting, landslides, slumping, and tsunamis, quite obviously can be a threat to nuclear facilities of any kind if precautions taken in siting and engineering design prove inadequate. Problems associated with vulcanism and geothermal fields are particularly relevant to the siting of geologic repositories, where containment must be for the very long term, beyond anyone's ability to confidently predict geologic phenomena. The hot water circulating upward from considerable depths in a geothermal field creates a mechanism for the transport of radionuclides to the earth's surface. A first consideration in building nuclear facilities in Japan must be to take account of hazards posed by the earthquakes, the vulcanism, and the other phenomena characteristic of the ring of fire.

Japan's generally mountainous terrain and high population density combine to pose another formidable siting constraint. Most of the population, now totaling almost 120 million, is crowded into the valleys and along the coastal margins, on the kind of terrain likely to be favored for both nuclear facilities and those non-nuclear industrial projects that compete for available sites.

The Players, Consensus, and Incentives

A close look at the circumstances of reactor siting in Japan reveals some of the technical, political, and geographic factors that apply as well to waste disposal siting. The siting, construction, and operation of reactors is the responsibility of Japan's nine private utilities, but the utilities are strongly supported in this endeavor by the government bureaucracy, especially the Ministry of International Trade and Industry (MITI).[13] Besides being responsible for nuclear safety regulation, licensing, and rate setting, MITI is heavily engaged in selecting reactor sites and promoting public acceptance of nuclear energy and the facilities necessary for its generation.

Backing up the utilities and the bureaucracy in their nuclear efforts has been the Liberal Democratic party (LDP), which as the party of big business, most civil servants and white-collar workers, and many farmers is one of Japan's most conservative institutions, despite its name. Its control of the Diet, by an absolute majority, goes back to the 1950s. Its leaders head the government ministries and share power with the permanent bureaucracy, which, in a somewhat circular fashion, has a strong influence on the Diet and the LDP. Just as it is dominant in the Diet, the party is also usually dominant at the prefectural (regional) and local government levels. This is especially true in rural Japan, where potential reactor sites are likely to be available. The Socialist and Communist parties find their greatest strength in Tokyo, Osaka, and the other urban agglomerations.

The Socialists and Communists have opposed nuclear power, although there is now apparent among moderate Socialist factions a tendency to move toward positions of at least qualified acceptance of this energy source.[14] Socialist and Communist politicians, many holding seats in prefectural, town, and village assemblies, have often led in questioning or resisting proposals for reactor siting.

Two specific local interests that figure among the key players in reactor and other nuclear facility siting are the farmers and the fishermen's cooperatives. The farmers usually own the land needed by the utilities, and inasmuch as condemnation is considered impolitic, siting is stymied unless they can be persuaded to sell. The fishermen's cooperatives hold a vested monopoly right to coastal waters, and plans to use these waters for reactor cooling or other purposes must have the fishermen's consent.

The distinctive and unusual feature of Japanese nuclear facility siting today is the emphasis on achieving consensus and not overriding any

[13]Richard P. Suttmeier, "The Japanese Nuclear Power Option: Technological Promise and Social Limitations," in Ronald A. Morse, ed., *The Politics of Japan's Energy Strategy* (Berkeley, University of California, 1981).

[14]"Swinging Nuclear Policy of Socialist Party," *Atoms in Japan*, February 1985; "Japan Socialist Party May Lean Toward Flexible Nuclear Power Policy," *Atoms in Japan*, September 1986.

significant local interest or group. This attitude has become so basic to the utilities' and MITI's siting strategy that it has taken on a time-honored quality. But while consensus-seeking is a tradition in Japan, the Japanese proverb "Bureaucrats revered, people despised" also expresses an important truth about Japanese tradition, as scholars point out. Japan's emergence as a rapidly evolving democratic industrial society is a phenomenon of only the last several decades, and in prewar imperial Japan consensus-seeking did not reach below the governing bureaucracies and elites. Bureaucrats in powerful ministries and in industry alike have discovered since the war—and during the past twenty years especially— that a greater effort must be made to include within the circle of consensus parties which in an earlier epoch would have been given short shrift.[15]

The democratizing influences have been manifold, but a particularly important, dramatic, and prolonged influence has been the struggle over the building of the new Tokyo (Narita) International Airport, arrestingly described by David Apter and Nagayo Sawa in their 1984 book, *Against the State: Politics and Social Protest in Japan*. The site is near Sanrizuka, formerly an imperial estate and a place of quiet villages some thirty miles from downtown Tokyo and east of Tokyo Bay, in Chiba prefecture. The conflict began in 1965 with small groups of farmers trying to save their land; their struggle soon rallied thousands of militants, including activists from New Left sects, student radicals from Tokyo University and other institutions, and advocates of a variety of peace, anti-nuclear, and anti-pollution causes. This Sanrizuka movement of social protest against the state continued to be led by farmers, adopting tactics of violent resistance. Not until 1978, thirteen years after the start of the project, did the airport finally open. The direct and indirect investment costs of the struggle approached a trillion yen, and the human costs included six dead and thousands of injured. The Sanrizuka movement has now faded, but some of the militants remain entrenched at the site to this day.

Why this violent and remarkably sustained spasm of political protest came about, how it might have been avoided, and what can and has been learned from it are questions not easily answered. But as Apter and Sawa point out, the Japanese government now recognizes that the political process is not neatly contained, that it reaches beyond organized parties, factions, bureaucracies, and interest groups, and that before large-scale ventures are undertaken different sectors of the public must be consulted and support mobilized.

[15]Interview by the author on August 15, 1985, with Susan J. Pharr, former holder of the Japan chair at the Georgetown University Center for Strategic and International Studies, now holder of the chair of Japanese politics at Harvard University.

In reactor siting, consensus-seeking took hold early, and has been reinforced by a generous policy of compensation, grants-in-aid, and other benefits which the nuclear industry clearly intends to apply to other nuclear facility siting problems, such as waste disposal. A lot of money and tireless persuasion went into an intense and largely successful effort at plant siting in the early to mid 1970s. The first Japanese commercial power reactor (the British-built, gas-cooled Tokai-1) had begun operating in 1966 without controversy, but in 1971 reactor safety became an issue in the general election. This was an echo of the American controversy over the adequacy of emergency cooling systems to prevent a reactor core meltdown, for by then Japanese utilities had embraced the American light water reactor technology.[16] Yet partly as a result of the generous compensation offered affected communities, reactor development was scarcely impeded. Over the next ten years or so more than a score of reactors, all built to Westinghouse and General Electric designs, were completed and put on line.

In 1974 the Diet enacted a per-kilowatt-hour "promotion tax" on electricity generation to fund a new Electric Power Resources Development Special Account. MITI was to use this account as a source of assistance grants—"cooperation money," as some would put it—to localities that accepted either a nuclear or non-nuclear electricity generating station. The funds for this purpose have been surprisingly large; in fiscal 1980 the total was more than $168 million.[17]

The rest of the siting strategy has called for an early start and an endlessly patient effort at persuading local officials and other interested parties to go along with the projects contemplated. The initial approach, coming some years in advance of the need for a site, has consisted of quiet talks by utility people with selected local leaders who could influence both local approval and negotiations with landowners and fishing cooperatives. The critical step of local acceptance is indicated when the local government invites the utility to make a formal siting survey.[18] The governor of the prefecture also has an important say, but typically defers to the local government on siting questions.

---

[16]Akira Matsui, "Gaining Public Support for Nuclear Power," *Nuclear Engineering International*, December 1979, pp. 70–72.

[17]Suttmeier, "Japanese Nuclear Power Option."

[18]Suttmeier, "Japanese Nuclear Power Option"; interviews by the author with Toru Namiki, director of the Japan Electric Power Information Center in Washington, D.C., and Tetsu Imai, his deputy, July 1985. The local official who fails to keep his political ear to the ground and gets out of step with his constituents will be in trouble, however. This was demonstrated in 1981 by the Liberal Democratic mayor of Kubokawa, a town on the island of Shikoku, who invited a reactor siting survey prematurely. The mayor only barely escaped with his political life: he was recalled by the voters in a special election, but

The Strategy Outcome So Far

Wherever reactor siting has been in controversy, there have been no violent disruptions by student radicals or New Left activists of the kind characteristic of the struggle at Sanrizuka. The utilities' consensual approach to siting has helped to avoid mass protests. The fact that most proposed reactor sites have been some distance from major cities and urban universities has meant less intervention by outside activists than might otherwise have been the case. Japan's authorized and ongoing commercial nuclear program includes fifty power reactors, and sites have been obtained or are available for every one of them. Thirty-three of the reactors have been built and put in operation; eleven were under construction in 1986; and six other planned projects have been approved by the Electric Power Development Coordinating Council, signifying that the prefectural governors have given their blessing and that local consensus in support of the projects has been achieved.[19] Although some reactor siting problems remain, the siting effort overall has been a considerable success.

Siting problems have not held up nuclear development in Japan in part because reactors have been concentrated at relatively few sites. Just how far the use of a particular site or area can be carried is evident in southwest Honshu on Wakasa Bay and the Sea of Japan. Kansai Electric Power Company and JAPCO will have thirteen reactors operating along this bay by the early 1990s. Five of these, together with two experimental reactors (including the Monju breeder) to be built by the government, will be on one small peninsula that juts out into the bay near the port of Tsuruga.[20] Once development of one or more nuclear stations in an area like this begins, the jobs and other direct and indirect economic benefits can be so important that the local communities want them sustained. But such clustering of reactors points up the conflicts among siting objectives inevitable in Japan. Having the Honshu reactors both convenient to population centers and far from major earthquake

---

reelected after promising that no reactor would be sited at Kubokawa without a local referendum. "Mayor of Kubokawa Recalled over Nuclear Power Plant Survey Decision," *Atoms in Japan*, March 1981; "The Recalled Mayor Re-Elected—Review of the Kubokawa Case," *Atoms in Japan*, April 1981.

[19]"Progress of Power Plants to Date" [March 31, 1986], *Atoms in Japan*, April 1986, pp. 52–53. Of the fifty reactors in the program, forty-seven are light water reactors, divided almost equally between the boiling water and pressurized water types developed by General Electric and Westinghouse, respectively. The other three reactors are the British gas-cooled reactor, Tokai-1; the experimental advanced thermal reactor, Fugen; and the experimental breeder, Monju (under construction).

[20]Ibid.

faults would not have been possible.[21] The risk of earthquake damage
to reactors was reduced by careful reactor design and siting—the latter
being of particular concern because foundation conditions may either
amplify or attenuate groundshaking.[22] But experience in designing and
building reactors or other nuclear facilities to withstand earthquakes has
been limited, and neither in Japan nor anywhere else has a reactor been
close enough (within ten miles) to a major earthquake to test the limits
of its design.[23]

Reactor siting in Japan is a highly complex art, and there is much
more at stake than whether or not the utilities get the sites needed. The
health and safety of the public are also at issue. Moreover, while in
principle the consensual approach to siting would allow the local and
prefectural authorities to refuse a project, this power can be illusory
because these authorities lack the expertise to make independent evaluations of project risks and environmental effects. That such weaknesses
exist in the siting program has been acknowledged in a study done by
the Japanese utility industry's own Central Research Institute.[24] These
problems are as relevant to the siting of nuclear waste and fuel cycle
facilities as they are to the siting of reactors.

Waste and Spent Fuel Management: A New Urgency

The nuclear enterprise in Japan as in other countries was slow to give
radioactive waste management the care and priority called for, but by
the early 1980s this problem was taking on a new urgency. Facts and
circumstances that generated this urgency could not be ignored.

[21]Four major earthquakes (magnitude 7.0 to 7.4) have occurred near Wakasa Bay since
the mid 1920s, three just to the west of the bay and one just to the east. The great
earthquake of 1891 (magnitude 8.4), Japan's largest since instrumented recording of such
events began a century ago, took place on faults 20 to 30 miles to the southeast, with
maximum displacements of up to 20 feet. Toshihiro Kakimi, "Quaternary Tectonic Movements," *Geology and Mineral Resources of Japan* (Hisamoto, Kawasaki-shi, Geological
Survey of Japan, 1977) table 19-1, p. 271, and fig. 19-4, p. 272.

[22]The magnitude 7.4 Miyagi-ken-Okai earthquake of June 12, 1978, that occurred about
70 miles from Tokyo Electric's Fukushima stations did no damage to any of the five
reactors then operating there, other than to crack one switchyard insulator. By contrast,
at the coastal town of Sendai, which was only about 12 miles closer to the earthquake's
epicenter than the Fukushima stations, many masonry structures not built to quake-
resistant standards were destroyed. Interview by the author with Gus V. Giesekoch,
geophysicist in the U.S. Nuclear Regulatory Commission's geosciences branch, July 24,
1985.

[23]Interview by the author with Walter W. Hays of the U.S. Geological Survey's Office of
Earthquakes, Volcanoes, and Engineering, August 13, 1985.

[24]Suttmeier, "Japanese Nuclear Power Option," pp. 126–127.

In keeping with Japanese regulations requiring that reactor licensing applications specify how the spent fuel generated is to be reprocessed, the Japanese utilities had moved early to obtain contracts with the British and French reprocessors. In addition, steps toward domestic reprocessing were taken. First, the government's Power Reactor and Nuclear Fuel Development Corporation started up a trouble-plagued demonstration reprocessing plant at its Tokai works north of Tokyo in 1977.[25] Second, the utilities and the nuclear equipment manufacturers were creating a new entity, Japan Nuclear Fuel Service Company, which was expected to build and operate by the mid 1990s Japan's first, full commercial-scale reprocessing facility. Meanwhile, spent fuel was—and is—kept in pools at the reactor sites, awaiting shipment to La Hague, Sellafield, or the Tokai works, or eventually to the commercial reprocessing center.

But these plans and arrangements left unanswered what was to be done with the high-level waste from reprocessing. Such waste resulting from the very small amount of Japanese fuel reprocessed abroad under the early foreign contracts would remain a responsibility of the British and French. Beginning in the 1990s, however, the greater part of the high-level waste generated abroad would be vitrified and returned to Japan, possibly accompanied by transuranic and other wastes from the reprocessing. These shipments would be received and kept temporarily in surface storage at the new domestic nuclear fuel cycle center, as would the vitrified waste produced by the new reprocessing plant at the center itself. It would not be until 1984 and 1985 that a site for this center would be chosen. Yet, pressing as it was, finding this site offered at best only a solution to the problem of temporary storage.

Clearly something would also have to be done about the ultimate disposal of high-level and other reprocessing wastes. The longer interim surface storage was continued, the more burdensome it would become, especially in the case of the large amounts of transuranic waste to be generated. The burdens could be political as well as economic. Any local community or prefecture agreeing to a fuel cycle center could very well want assurances that the storage of long-lived reprocessing wastes would be only temporary.

But of all the waste problems confronting the Japanese utilities, probably the most pressing and immediate has been what to do about the

[25]The Tokai plant, of French design, has a nominal capacity of 210 tons of fuel a year. The plant shut-downs, which have included two lasting for fourteen months each, have been much discussed in the trade press. See "Pinholes Develop in Dissolver at PNC Reprocessing Plant," *Atoms in Japan*, April 1982; Shota Ushio, "Reprocessing to Resume at Tokai Mura," *Nuclear Fuel*, November 26, 1984; "PNC Runs Tokai Reprocessing Plant," *Atoms in Japan*, February 1985; and "Tokai Plant Reprocesses 223 NTU of Spent Fuel," *Atoms in Japan*, September 1985.

low-level waste fast accumulating at the reactor stations. By 1980, there were already some 260,000 drums of waste in storage at these stations, and unless a means of permanent disposal is found there will be an estimated 1.2 million drums in temporary storage by 1990 and 3.8 million by the year 2000.[26] Local officials of the communities near the reactor stations were viewing this growing accumulation of waste with uneasiness. In fact, when the all-Japan council of nuclear plant host communities vented its anger in 1981 about the Tsuruga reactor spill and cover-up, opposition was also expressed to any permanent storage of waste at nuclear stations.[27]

The Search for Waste Disposal Solutions at Home

Japan's radioactive waste disposal problem seems to dictate that every plausible option be explored, at home and abroad. Progress of a sort in the search for solutions at home first became evident in 1984 and 1985. Local and prefectural officials agreed that Rokkashomura, in the largely rural Aomori prefecture at the northeastern tip of Honshu, should become the site of a major nuclear fuel cycle center, including a reprocessing plant and waste storage and disposal facilities.[28]

Initiatives were also made with respect to the disposal of high-level waste, but whether they will lead to anything more than conflict and protest is not clear. The town of Horonobe, in the far north of Japan on the island of Hokkaido, invited the government to consider its area for a waste storage and disposal facility. The Power Reactor and Nuclear Fuel Development Corporation (PNC), the government enterprise responsible for research on reactor development and the fuel cycle, and operator of the demonstration reprocessing plant at Tokai, responded eagerly. It proposed Horonobe as a site for a near-surface storage facility for PNC's own reprocessing wastes, both high- and low-level, and for a deep underground laboratory to investigate high-level waste disposal.[29] But while Horonobe welcomed this initiative, other communities nearby were opposed, and so was Hokkaido's prefectural governor, who was to

[26]Takehiko Ishihara, "Low Level Radioactive Waste Management," *Atoms in Japan*, August 1980.

[27]"Tsuruga Plant Incidents Evoke Strong Feelings at General Meeting of Council of Local Governments," *Atoms in Japan*, June 1981.

[28]"Aomori Governor Says Yes for Siting of Fuel Facilities," *Atoms in Japan*, February 1985; "Aomori Local Bodies Accept FEPC's Proposal of Siting Fuel Cycle Facilities," *Atoms in Japan*, April 1985.

[29]Interview by the author with Masami Katsuragawa, director of the PNC office in Washington, D.C., July 22, 1985.

spend a week in the United States in 1985 informing himself about the nuclear waste problem in this country.

Rokkashomura

The nuclear fuel cycle center proposed for Rokkashomura would be built and operated by companies established by the Federation of Electric Power Companies, the trade group of the nine Japanese utilities. To enter service during the first half of the 1990s, the center would consist of three major facilities:[30] a uranium enrichment plant that would start up at demonstration scale, but eventually expand production to meet about one-sixth of Japan's annual need for low-enriched uranium; a low-level waste storage and disposal facility with an eventual capacity of some 600,000 cubic meters (or 3 million drums), expected to start up about 1991;[31] and a reprocessing plant that would reprocess some 800 tons of fuel a year, equivalent to about half of Japan's annual spent fuel generation. Reprocessing would not begin until about 1995, but the plant's 3,000-ton spent fuel storage facility would begin to receive fuel about 1991, providing major relief to the utilities' reactor stations whose storage pools are filling up. Moreover, the plant would have a surface or near-surface vault storage facility to receive vitrified high-level waste from the plant itself and from France and the United Kingdom.[32]

Several years ago the nuclear industry seemed worried that finding a site for a reprocessing plant was going to be difficult, much less finding one that would accommodate a major waste disposal facility too. The early overtures made by the utility federation to various localities were in fact rebuffed, but at Rokkashomura on the Shimokita Peninsula the federation found an early base of support on what had once been terrain fiercely contested by the local partisans and opponents of industrial development.[33]

The Shimokita Peninsula is hatchet-shaped, the blade of the hatchet facing westward to enclose the upper part of Mutsu Bay. Rokkashomura itself consists of a string of six small settlements, and forms the handle of the hatchet, with the bay to the west and the Pacific to the east. It is

[30]Except where otherwise noted, the description of these facilities is from "Stepping up Industrialization of Back-End of Fuel Cycle," *Atoms in Japan*, April 1985.

[31]Ishihara, "Low Level Radioactive Waste Management."

[32]"PNC Rad-Waste Storage Eng. Center Proposed for Horonobe, Hokkaido," *Atoms in Japan*, August 1984. This article explains that the Horonobe facility would be only for PNC wastes. It noted that high-level waste from France and the United Kingdom would be stored at Rokkashomura.

[33]"FEPC Selects Site for Fuel Center in Rokkashomura," *Atoms in Japan*, July 1984.

a lightly populated, economically depressed rural community which good times have often passed by. The winters are long and hard, the growing season is short, the farming productivity is low, and the coastal fishing is not the best.

The utility federation's $4.4-billion fuel cycle center project has been embraced by the village and prefectural leaders as promising the progress that has long eluded Rokkashomura. The nineteen-fifties, sixties, and early seventies saw a succession of government-supported or -encouraged economic development undertakings that failed or were abandoned. In *Island of Dreams*, a book on Japan's environmental problems, three American authors have described at length the struggles of the Rokkashomura villagers to improve their lot without sacrificing their community's identity and rural environment.[34] But from these struggles Rokkashomura has emerged with a new mayor or headman, Isematsu Furukawa, who appears to have worked closely and shrewdly with representatives of the utility federation to win support for the new nuclear fuel cycle center, as has Governor Masaya Kitamura of Aomori prefecture.

The first public word that the federation was seeking a site in Rokkashomura for the proposed center came in April 1984. Governor Kitamura subsequently heard promises of financial aid and subsidies for the prefecture from government ministers, and traveled to France to visit La Hague. On July 18, 1984, the utility federation's board decided in favor of the Rokkashomura site. About a week later the board chairman called on Governor Kitamura and Mayor Furukawa and promised that the plant's design would be based on the best available technology from Japan and abroad.[35]

---

[34]Norie Huddle and Michael Reich with Nahum Stiskin, *Island of Dreams: Environmental Crises in Japan* (New York and Tokyo, Autumn Press, 1975) pp. 242–255. Efforts to encourage dairy farming, sugar beet production, and the growing of soybeans, corn, and rice all met with severe setbacks. The new dairy herds did poorly in the harsh winters, and the beet farmers and local sugar mill could not compete with cheaper foreign sugar. The last of the failed development initiatives was a controversial plan for a huge petrochemical complex. The prefectural governor wanted the complex and is said to have proposed it to government and industry officials, but Rokkashomura's mayor and many other village residents opposed it strongly, seeing a promise not of progress but of pollution. Not all of the villagers opposed the petrochemical project. Land prices had soared after the development plan was announced in 1969; those refusing to sell remained poor, while those selling were for the first time affluent enough to build new homes, buy new cars, and send their children to school well dressed. Opposition to the petrochemical project gradually lost strength, and by late 1973 the incumbent mayor had been voted from office. Yet just as the battle for public acceptance was being won, the first oil shock undermined the project's economic feasibility, and this project too was abandoned.

[35]"Japan May Build Fuel Cycle Facilities in Sparsely Populated Shimokita," *Nuclear Fuel*, April 23, 1984; "FEPC's Decision on Shinokita N-Fuel Site Nears," *Atoms in Japan*, June

A committee of experts impaneled by Aomori prefecture to review the safety of the project was to issue a favorable report at the end of October,[36] but long before then the prefectural government had initiated a series of explanatory meetings for prominent citizens and leaders, which were focused on safety, local development, and the importance of the proposed fuel cycle facilities to Japan. At the same time, the utility federation was conducting an intensive local public relations effort, establishing an office in an abandoned school and having some twenty-four utility staff people fan out through the village to visit each of its 3,000 homes.[37]

On January 16, 1985, a committee of the Rokkashomura Assembly decided to accept the proposed center, subject to certain conditions. There were three demands in particular: safety must be guaranteed; measures for village development must be taken; and establishment of a system for high-level waste management and disposal must be expedited. Approval by the town and prefecture was now assured, although signing of the formal agreement of acceptance by Governor Kitamura, Mayor Furukawa, and industry officials was not to come until April.[38] The prefectural assembly's overwhelmingly large Liberal Democratic majority was itself probably enough to predispose support of the project.

Despite all indications of support, substantial opposition to the project exists within Aomori prefecture. This became evident in early 1985 when about 88,000 people signed a petition demanding that the local assembly call a referendum on the project before approving it—a petition the assembly refused to act upon.[39] A year later plans to build the nuclear fuel center were still on, but no one should be surprised if in the end only the spent fuel storage pools and the low-level waste disposal facility are built. *Atoms in Japan* conceded in July 1985 that there were grave questions of economic feasibility with respect to both the proposed uranium-enrichment and fuel reprocessing plants. As an editorial indicated, the enrichment plant would use an already obsolescent centrifuge technology, and for this and other reasons would be a handicapped competitor in the international market for enrichment services.

1984; "FEPC Selects Site for Fuel Cycle Center in Rokkashomura," *Atoms in Japan*, July 1984; "FEPC Offers Plans of Fuel Cycle Facilities to Local Communities," *Atoms in Japan*, August 1984.

[36]"Aomori Prefecture Makes Efforts for Siting Fuel Facilities," *Atoms in Japan*, September 1984.

[37]"Japanese Utilities Expect to Win by December Approval to Build Fuel Cycle Facilities," *Nuclear Fuel*, September 24, 1984.

[38]"Rokkashomura Approves Siting of Nuclear Fuel Cycle Facilities," *Atoms in Japan*, January 1985, p. 36; "Aomori Local Bodies Accept FEPC's Proposal of Siting Fuel Cycle Facilities," *Atoms in Japan*, April 1985, p. 42.

[39]"A Country Without a Nuclear Toilet," *New Scientist*, October 3, 1985.

Similarly, "the economic prospect for reprocessing operations is not bright," the editorial observed, going on to comment that reprocessing costs are high, commercialization of the breeder has been postponed, and use of plutonium in light water reactors faces high fuel fabrication costs and public relations hurdles.

Horonobe

Horonobe, a town of about 4,200 inhabitants, has been languishing and losing population. Several years ago its leaders decided that the town would benefit from development of a nuclear waste center.[40] Overtures were made to PNC, the Science and Technology Agency's nuclear research and development corporation, which not surprisingly was receptive indeed. Besides its responsibility for taking the lead in solving the waste problem, PNC has its own growing inventory of wastes, especially at its reprocessing plant at the Tokai works.

Situated in the northern corner of Hokkaido, Horonobe is one of the most remote places in Japan, and is about 650 miles from Tokyo. Much of Hokkaido is mountainous and noted for heavy snows. But Horonobe, on the Teshio River less than ten miles inland from the Sea of Japan, is in a coastal region of modest relief, with a climate moderate enough to permit dairy farming—an economic mainstay of this none-too-prosperous area.

Settled only over the past century, Hokkaido is Japan's "frontier" province, and its people are known for a streak of independence. Hokkaido's prefectural governor, Takahiro Yokomichi, is himself becoming known for causing the Liberal Democratic government and the Tokyo bureaucracy considerable discomfort over the nuclear waste issue. As a legislator in the national Diet before his election as governor in 1983, he belonged to a moderate Socialist party faction similar in ideology to the Socialist parties of western Europe; as candidate for the governorship—a position the Liberal Democratic party had held for almost forty years—Yokomichi ran as an independent. He won only narrowly, and the Liberal Democrats have continued to control the prefectural assembly.

But Yokomichi is said to have since gained an increasingly strong popular following, presenting himself as a dynamic young governor vigorously pushing Hokkaido's economic development, and on most issues keeping on good terms with the leaders of the prefectural assembly's LDP majority. On the question of the nuclear waste center, however, the governor and the assembly majority have been in conflict, with

[40]"PNC Rad-Waste Storage Eng. Center Proposed for Horonobe, Hokkaido," *Atoms in Japan*, August 1984.

the latter, as a matter of party loyalty, tilting toward the national government's position favoring the center. Some have sought to explain Yokomichi's objections to the center as those of a "Socialist antinuke," but the merits of his position cannot be so easily discounted, as will shortly be evident. Moreover, while Yokomichi as a gubernatorial candidate reportedly declared that "no more nuclear power plants will be established in Hokkaido,"[41] he has not objected to Hokkaido's first two reactors, still under construction at the Tomari station fifty miles to the west of Sapporo, Hokkaido's prefectural capital.

It was August 1984 when Minoru Yoshida, president of PNC, first called on Governor Yokomichi to tell him about the nuclear facilities proposed for Horonobe and to seek his approval of such initial steps as conducting environmental surveys and land acquisition.[42] The Japan Atomic Energy Commission's nuclear waste policy advisory committee had in effect endorsed PNC's plans for a storage and R&D center as an important initiative in addressing the country's nuclear waste problem.[43] As laid before the governor, the plans included several major facilities. Among them were a near-surface interim storage facility for 2,000 canisters of vitrified high-level waste from PNC's Tokai reprocessing plant, and another near-surface facility for the storage of 30,000 drums of low-level waste in a bitumen (or asphalt-type) matrix. The facility for bitumen waste, which also would be coming from the Tokai plant and hence would no doubt be plutonium-contaminated, would be expandable to 200,000 drums by the year 2000. Only PNC-generated waste would be destined for these two interim storage facilities; no commercial waste would be received.[44] A third proposed facility was a deep geologic laboratory for the study of natural and man-made barriers to radionuclide migration and of interactions between waste and host rock. Construction of the laboratory (in American parlance, a test and evaluation facility) would not necessarily be a step toward designating Horonobe as the site for Japan's first commercial high-level waste repository, but PNC has not ruled out that possibility.[45] Governor Yokomichi and others were left to assume that should the site be found acceptable it might well be so designated, even though the PNC now points to a recommendation by the Japan Atomic Energy Commission for a nationwide search for geologic disposal sites.[46]

[41]Ibid.

[42]Ibid.; Shota Ushio, "The Japanese Government Plans to Complete a Radwaste Storage R & D Center," *Nucleonics Week*, September 13, 1984.

[43]Ushio, "Japanese Government Plans to Complete Center."

[44]Interview by the author with Masami Katsuragawa, July 22, 1985.

[45]Ibid.

[46]PNC's Washington representative, Masami Katsuragawa, emphasized to the author that the Japan Atomic Energy Commission advisory committee has called for a ten-year search

It should be noted that in any such national search north central Hokkaido, or the larger region of which Horonobe is a part, would likely figure as a prime area for candidate sites, perhaps one of few.[47] Besides being lightly populated and remote from the rest of Japan, this is not a region of high seismicity, intense volcanism, or active geothermal fields. Furthermore, in the center of the region a large body of ultramafic rock, running some thirty kilometers from north to south and several kilometers across, appears worthy of investigation. Ultramafic rock is a hard, coarse-grained material that resembles granite in its strength and low permeability and offers much the same advantages as granite for a repository site. Its heat conductivity is rather poor—New England soapstone foot warmers were made of ultramafic rock—but in principle this shortcoming could be compensated for in the repository design. Oddly enough, the presence of the ultramafic rock body did not enter into the PNC proposal in any way. The proposed facilities seem to have been confined to the boundaries of Horonobe; moreover, the PNC proposal appears to have been constrained to make the best of obvious drawbacks at the Horonobe site.

Governor Yokomichi was at first noncommittal toward the PNC proposal. While the prefectural assembly and the town of Horonobe were inclined to favor it, several other towns of north central Hokkaido expressed objection to it through formal resolutions. The governor soon found the PNC proposal wanting in two important respects. First, there was no overall national nuclear waste policy against which the undertaking at Horonobe could be evaluated. That is, there were no clearly formulated procedures and technical criteria for identifying, evaluating, and selecting sites, and no well-defined rights of participation for prefectural officials and other interested parties. Second, Yokomichi was taken aback by what he regarded as a disturbing paucity of information about either the specific project PNC was proposing (there was not yet in hand even a geologic map of Horonobe) or the more general problem of geologic disposal of nuclear waste in Japan. Yokomichi was also disturbed at how little information there was in Japan, and in Japanese, about the controversies he knew to be swirling about nuclear waste disposal in the United States and other foreign countries. His frustration on this latter score was compounded later when PNC failed to meet his

---

for disposal sites beginning in 1985, with no site selection to take place until sometime after 1990.

[47]Material on north central Hokkaido's plausibility as a site for a geologic repository is based on an interview by the author on June 25, 1985, with Maurice J. Terman, Asian and Pacific geology specialist with the U.S. Geological Survey's international program office; and an interview with Harry W. Smedes, consulting geologist, summer of 1985.

requests for even the most readily available American nuclear waste policy documents.[48]

In July 1985 Governor Yokomichi went abroad to inform himself directly about the waste problem and controversy in other countries, and particularly in the United States.[49] Preparations for the trip were made with remarkable thoroughness: American technical papers and reports were translated into Japanese, and Yokomichi studied them carefully, as those whom he called on in the United States discovered. The public announcement of his trip drew criticism in the prefectural assembly, Liberal Democratic leaders complaining that the trip was unnecessary and that there was more pressing business at home. But the Japanese press saw the trip as highly newsworthy, and traveling with the governor when he left Japan were reporters for four television stations and all of the country's major newspapers.

Stopping first in Washington state to visit the basalt waste isolation project at Hanford, Yokomichi saw the parallel to the situation at Horonobe: in each case the prospective host community was eager for a nuclear waste project, while the state or prefecture was either opposed outright or quite guarded. In a press interview, Yokomichi observed how both at Horonobe and Hanford the national governments seemed to have picked potential nuclear waste storage and disposal sites as far away from the national capitals as possible.

After a day at the Materials Research Laboratory at Pennsylvania State University, which has done extensive research into high-level waste forms, the governor went on to Washington, D.C. At the U.S. Geological Survey he displayed a map of the Horonobe area and pointed out the site for the proposed nuclear waste center; the half-dozen geologists and hydrologists present regarded the site as by no means favorable. According to Eugene Roseboom, a leader of the USGS team that advises the U.S. Department of Energy on nuclear waste isolation, the site was in or on the edge of the Teshio River flood plain—and a flood plain is typically a groundwater discharge area and a place where groundwater flows are upward.[50] At the Department of Energy itself Yokomichi met with Roger Gale, the waste program's director for policy integration, who found Yokomichi impressed by the U.S. Nuclear Waste Policy Act

[48]"PNC Rad-waste Storage Eng. Center Proposed for Horonobe, Hokkaido," *Atoms in Japan*; interview by the author with Governor Yokomichi, July 19, 1985; confidential interviews with Governor Yokomichi's Japanese consultants in the United States, July 1985.

[49]Information about the governor's visit to the United States is from the author's interview with him on July 19, 1985, except where otherwise noted.

[50]Interviews by the author with Eugene Roseboom and Isaac Winograd of the U.S. Geological Survey, July 26, 1985. Both were present at the USGS meeting with Governor Yokomichi.

and its procedures and criteria for siting repositories.[51] But Yokomichi also met with environmental lobbyists. They told him of the dissatisfaction among officials in the potential repository host states, who felt that the act's siting policy had been compromised by an excessively fast-paced siting schedule and by vague siting guidelines (see chapter 13).[52]

Governor Yokomichi also talked with Thomas A. Cotton, director of the congressional Office of Technology Assessment's nuclear waste study. Cotton later remarked to me, "He [Yokomichi] had picked up on the problem of getting sucked into a site selection process when there are no rules, when the rules are made up as you go along." Accordingly, in a statement to Japanese reporters before leaving the United States, the governor said that all of PNC's requests were going to be refused, even the request to conduct preliminary environmental surveys at Horonobe.[53]

Back in Japan, Yokomichi held hearings to get the views of leaders of thirteen towns and villages surrounding Horonobe, and found all of them against the project and opposed to the environmental surveys. On September 13, 1985, the governor presented to PNC a letter formally denying its request to do the surveys. The minister of the Science and Technology Agency asked the governor to reconsider, and the Hokkaido assembly, dominated by the Liberal Democratic party, passed a resolution in favor of expediting the environmental studies. When Yokomichi again refused, the minister said that the government intended to proceed with the studies anyway.[54]

An initial one-day survey was made by PNC that November despite the presence of local protesters, but nothing more was done over the winter. The Chernobyl accident and the Japanese national elections intervened during the spring and early summer of 1986, and all remained quiet at the Horonobe site. But in August 1986 PNC notified local officials that environmental surveys would begin again soon.[55]

---

[51]Gale was interviewed by the author on July 26, 1985.

[52]Environmental lobbyists who met with Governor Yokomichi included David Berick and Fred Millar of the Environmental Policy Institute, Brooks Yeager of the Sierra Club, and Thomas Cochran of the Natural Resources Defense Council.

[53]Interviews by the author on July 26, 1985, with Thomas Cotton and Masami Katsuragawa.

[54]"PNC's Horonobe Rad-Waste Site Survey Meets Difficulty," *Atoms in Japan*, August 1985; "Hokkaido Governor Rejects PNC's Environmental Assessment of HLW Center," *Atoms in Japan*, September 1985; "Horonobe Program: STA Minister Asks Governor to Reconsider," *Atoms in Japan*, October 1985, p. 26.

[55]"PNC Conducts Field Survey for Siting HLW Facility at Horonobe," *Atoms in Japan*, November 1985, p. 23; "PNC to Resume Site Survey at Horonobe," *Atoms in Japan*, August 1986, p. 30. Minoru Okuma and other members of a Hokkaido Socialist party delegation

In sum, in the 1980s there was indeed an urgency about Japan's waste problem. The siting of radioactive waste facilities in this densely inhabited island nation along the ring of fire was not necessarily impossible. But waste facility siting—particularly for a deep geologic repository for high-level waste—promised to be an undertaking whose success was neither technically nor politically assured. It would be only prudent for the Japanese nuclear industry to investigate the possibility of solutions outside of Japan as well as at home.

Looking Beyond Japan

There are essentially two possible ways for Japan to dispose of its radioactive waste outside of its borders but within east Asia or the larger western Pacific basin. One is to use the deep ocean floor for disposal of low-level waste and possibly the subseabed for disposal of high-level waste; the other is to seek disposal sites in another country, such as China. In either case an obvious advantage would be to move disposal of high-level waste well away from the ring of fire to land or seabed areas that are tectonically quiet.

The Chinese Land Mass

The nuclear technical cooperation agreement between Japan and China, signed in July 1985, calls broadly for cooperation in radioactive waste management, as well as in other fields.[56] The agreement conceivably could serve as a way to explore the possibility of China's accepting high-level waste and spent fuel from Japan in exchange for Japanese technical assistance in reactor design and construction and other nuclear fields.

There are parts of China, including extensive reaches of the Gobi Desert, that are both quiet tectonically and thinly inhabited. Moreover, according to U.S. Geological Survey Asian specialists, China has within its borders a wide variety of geologic environments, some being of the same kind that are under investigation for geologic disposal in the United States.[57] Pursuit of this option depends on both the willingness of the

---

investigating the nuclear waste issue have attributed PNC's long absence from the Horonobe site to political wariness. The author met with the delegation at Resources for the Future, Washington, D.C., in May 1986.

[56]"Japan-China Nuclear Cooperation Agreement Signed," *Atoms in Japan*, July 1985.

[57]Warren Hamilton, "Plate Tectonics and Man," in *U.S. Geological Survey Annual Report, Fiscal 1976* (Washington, D.C.) figs. 1 and 5; interview with Maurice J. Terman, June 25, 1985.

Chinese to entertain such a possibility and on the willingness of the Japanese—anxious as they are to achieve greater energy independence—to have vital waste disposal functions made subject to arrangements with a neighboring country.

That China has offered to receive spent fuel from European utilities suggests that the Japanese might find the Chinese receptive, particularly inasmuch as Japan is both splendidly qualified and conveniently situated to offer technical help to Chinese nuclear development. It should be noted, however, that even as a centrally directed communist state China is not entirely free of political constraints in the planning of nuclear facilities and activities. For example, a committee of Uyghur Turks has recently petitioned for their home province of Xingiang in northwest China to be made a nuclear-free zone, and for an end to all testing of nuclear weapons there.[58]

The nuclear cooperation agreement between Japan and China having entered into force in mid 1986,[59] it would be timely for either party to raise the waste disposal question. But if either has any plans to do so, there has not as yet been any public indication of it.

The Ocean Floor and the Subseabed

The possibility of using the deep ocean floor for disposal of low-level waste was initially of interest to the Japanese nuclear industry not so much as an alternative solution but as a key solution. In late 1979 a site at a depth of about 6,000 meters, some 900 kilometers southeast of Tokyo Bay, was selected for an experimental dumping operation. If this experimental dumping confirmed the suitability and safety of the site, full-scale dumping was to ensue, at a rate not to exceed 100,000 curies a year. Sixty percent of all the low-level waste generated in Japan was to be disposed of at this site, with the remainder (consisting of longer-lived waste not suitable for ocean dumping) to be disposed of on land.[60] It was calculated that the annual radiation dose to humans from full-scale ocean dumping would be equivalent to only one five-thousandth

---

[58]Dispatch from the Kyodo News Agency, May 19, 1986, reprinted in Foreign Broadcast Information Service, *Daily Report China*, May 19, 1986. In December 1985 Uyghur Turk students had demonstrated in Beijing, Shanghai, and Urungi against such testing. "Concern Grows in China over Safety of Nuclear Program," *London Financial Times*, May 23, 1986.

[59]"Japan-China Cooperation Agreement on N-Energy Comes into Effect," *Atoms in Japan*, July 1986, p. 39.

[60]Ishihara, "Low Level Radioactive Waste Management."

of the dose an individual receives from natural sources.[61] In the end, the proposed dumping operation, originally scheduled for the fall of 1981, was cancelled because of the opposition of South Pacific island nations.

The Japanese sea dumping proposal would have brought to the Pacific a practice that had long been going on in the Atlantic, where it was subject to the oversight of the Nuclear Energy Agency (NEA) of the Organization of Economic Cooperation and Development (OECD). A member of OECD, Japan has also been involved in an NEA-coordinated research and development effort aimed at the disposal of high-level and transuranic waste beneath the seabed. Apparently out of deference to the sensitivities of the South Pacific islanders, Japan has not played a large role in this effort, either with respect to funding or scientific participation.

The research on subseabed disposal has been under way since 1974, although it was dealt a serious blow in 1986 when the United States, provider of nearly half the funding, pulled out.[62] The undersea terrain of interest, whether in the Pacific or the Atlantic, is the gently rolling abyssal hill province found in the middle of the tectonic plates and toward the center of the wind-driven, surface-flowing circular currents (gyres). The abyssal hills typically lie at a depth of 5,000 to 6,000 meters, and are characterized by weak bottom currents, a high degree of seismic stability, very low biological activity, and an absence of mineral resources (apart from low-grade manganese nodules). A thick layer of reddish brown clay (up to 300 meters deep in the Pacific), accumulated

[61]Japan Radioactive Waste Management Center, "Safety on [sic] Sea Dumping of Low Level Radioactive Wastes" and "Dumping at [sic] the Pacific," brochures authorized by the Nuclear Safety Bureau of the Science and Technology Agency.

[62]Among the participants in this effort have been the United States, Canada, the United Kingdom, France, Germany, the European Community, and Japan. The overall multinational effort reached a peak level of funding at $25.3 million in fiscal 1985, when $11 million was provided by the United States. But the U.S. Department of Energy, citing competing demands for funds, asked for no money for this program in fiscal 1987, and Congress appropriated none. Continued financial support by the United Kingdom, where officials have become pessimistic about ever gaining international consent for use of the oceans for nuclear waste disposal, is also in doubt. William O. Forster, DOE's subseabed program manager, believes that the reasons for the U.S. withdrawal go beyond budgetary considerations. The principal motivation, he told me, is that the top DOE waste program leaders "see us as a thorn in the flesh" when subseabed disposal is suggested as an alternative at public hearings on geologic disposal sites. Scientific feasibility of the subseabed disposal concept was expected to have been tested by 1990, but now the program must move at a much slower pace. Forster was interviewed by the author on May 8, 1986. For British reservations about the subseabed program, see Peter Heywood, "Politics Snags Seabed Disposal Research Despite Findings of Technical Feasibility," *Nuclear Fuel*, October 7, 1985.

over the eons from a gentle rain of fine-grained volcanic ash and other material from the surface, overlies the basaltic basement rock.

The subseabed disposal concept is relatively straightforward.[63] Initially, the idea was to emplace canisters of high-level waste in holes drilled well into the basement rock. This implied an elaborate and costly disposal system, with much too-limited a capacity to accommodate both high-level waste and the large volumes of transuranic wastes generated by commercial reprocessing. But as later simplified, the concept envisioned canisters designed as free-fall projectiles that would penetrate some 30 meters into the plastic clay sediment. Laboratory experiments have indicated that because of its absorptive capacity and ability to retard radionuclide movement, 30 meters of this clay is more than enough to ensure containment for thousands of years.[64] An experiment conducted in 1984 showed that the projectile-shaped canisters (penetrometers) do indeed penetrate to the desired depth. Subsequent experiments will show whether the holes made by the penetrometers are self-closing, with the clay barrier left uncompromised. Other experiments are needed to show conclusively that the heat from the waste will not cause each canister and the clay around it to rise to the sea floor in a blob of fluid mud.[65]

In the Pacific, proposals for sea dumping have met with strong opposition from the South Pacific island nations from the outset, even though dumping would be carried out under international rules and supervision. This opposition warrants examination in detail because it illustrates clearly how blunders committed at the start of the nuclear era can leave a monumental legacy of distrust.

[63]The subseabed concept was first described in the scientific literature by W. P. Bishop and Charles S. Hollister, "Seabed Disposal—Where to Look," *Nuclear Technology*, December 1974.

[64]Interview by the author on September 15, 1982, with Glenn Boyer of the U.S. Department of Energy's Albuquerque operations office, and at that time in charge of the subseabed research program at the Sandia National Laboratory.

[65]In the final phase of these $5-million experiments, an automated "pipe rack" that has been likened to a moon lander would be lowered to the seabed to take anchor and to inject into the clay a slender canister, using the 8,000-pounds-per-square-inch hydraulic pressure of the deep sea as the propelling force. The canister would contain four 1-inch spheres of plutonium-238 (together generating some 400 watts of heat) plus an array of needlelike temperature gauges. The canister would be attached by a wire to the lander and would transmit data to the lander's memory system. After a year in place the canister would be pulled from the clay and withdrawn into the pipe rack, whereupon the device would trip its anchor release and float to the surface for retrieval by an oceanographic research vessel. Laboratory tests have indicated that no significant advection, or "burping," response by the clay to the heat would occur. This experiment, called ISHTE (in situ heat transfer experiment), was described to the author on August 8, 1985, by William O. Forster of the U.S. Department of Energy. ISHTE would also collect data relevant to the clay's ability to retard radionuclide migration.

Opposition by Pacific Islanders

In a world of vast continental land masses and more than four billion people, the tiny, isolated island communities of the South Pacific may seem at first to be capable of only token resistance to the plans of Japan and the western nations. Most of them are mere specks of land and humanity in an immense oceanic basin. For example, Micronesia, from the Marshall Islands to the Marianas and Palau, stretches across several million square miles of water, but contains less land than the State of Rhode Island and fewer than a quarter of a million people. Nonetheless, the island communities have been asserting their political identities, and in a world at least formally dedicated to the rule of law they have assumed an importance out of all proportion to their populations. Their claim to exclusive 200-mile economic zones around these small but multitudinous islands is in itself enough to make Japan and other nations take them seriously.

The ocean that separates these communities also binds them together in a single, intricate environmental system. Winds and currents serve as great conveyors of life forms across the vast reaches of the Pacific. Migratory ocean species such as the tunas and sea turtles visit the waters of one island nation, then pass on to the next. Marine life around one island may depend in part on breeding areas and nurseries for juvenile forms in waters around islands hundreds or thousands of miles away. The perception of the Pacific as environmentally unified is widely shared and contributes importantly to the intensity of opposition that proposals for radioactive waste storage and disposal inspire among the islanders. The report of the Conference on the Human Environment in the South Pacific, which took place at Rarotonga in the Cook Islands in March 1982, concluded with this observation: "It has long been accepted that 'no man is an island,' but today, no island is an island either."[66]

Central to the islanders' opposition to proposals for radioactive waste disposal in their region is the history of abuse associated with nuclear weapons testing in the Pacific. Governor Carlos S. Camacho of the Northern Marianas, testifying at a U.S. congressional hearing in 1979, called the unintended exposure of many people in the Marshall Islands to dangerous levels of radioactivity from nuclear weapons testing a "disastrous scientific miscalculation." He later spoke of "doubletalk delivered in the high language of science," and added that "the track record of scientists assessing radiological hazards, and especially radiological

---

[66]South Pacific Commission, *Report on the Conference on the Human Environment* (Noumea, New Guinea, March 1982) p. 91.

safety, is deplorable, and no responsible government, truly representing its people, can continue to accept their assurances without grave concern."[67]

The record shows that in a number of instances decisions affecting the health and safety of Marshall Islanders were made on the basis of grossly inadequate or just plain wrong information. A notable instance was the decision made in 1968 to allow the people of Bikini to return to their home atoll, from which they had been evacuated when the first weapons test was conducted in 1946. The Marshallese were assured by the U.S. Atomic Energy Commission that there was "virtually no radiation left on Bikini." AEC scientists accompanying the Bikinians on their return to the atoll in 1969 sought to show there was nothing to fear; according to one account, "when the Bikinians refused to eat any local foods, fearing radiation exposure, the scientists would consume coconuts, fish and other foods in front of the islanders to convince them."[68] Yet over the next several years the repatriated Bikinians were to show a steadily increasing body burden of radioactive cesium. It was now clear that the commission's scientists had underestimated the uptake of radionuclides from the soil by coconut trees and other food plants and the consumption of this locally produced food by the Bikinians. The American authorities next advised the Bikinians to stop eating the locally grown food and to eat instead only uncontaminated food shipped in from outside. But inasmuch as the people continued to eat the local foods, "especially during times when the food from outside was insufficient," the Bikinians were again evacuated in 1978.[69]

Another blunder remembered in the Marshalls was the failure to evacuate the inhabitants of the Rongelap and Utirik atolls before the 15-megaton BRAVO test shot in 1954. Many suffered heavy radiation exposure from fallout (up to 175 rems on Rongelap) and experienced nausea, skin burns, loss of hair, and other symptoms of acute radiation sickness. Nineteen of the twenty-two children exposed on Rongelap later had to undergo surgery for removal of thyroid nodules, and in 1972 one died of leukemia.[70]

[67]Testimony of Governor Carlos S. Camacho in *Nuclear Spent Fuel Storage in the Pacific*, Hearings before the Subcommittee on National Parks and Insular Affairs of the House Interior Committee, Honolulu, 96 Cong. 1 sess. (1967) p. 37.

[68]Giff Johnson, "Paradise Lost," *Bulletin of the Atomic Scientists*, December 1980, pp. 21, 25.

[69]U.S. Department of Energy, "Melelen Radiation Ilo Ailin in Bikini" (The meaning of radiation at Bikini atoll) (Washington, D.C., September 1980) pp. 28–29. This information booklet, in Marshallese and English, is intended to explain to the Bikinians why they were urged to return to their home island but then later reevacuated.

[70]Johnson, "Paradise Lost."

The islanders' sense of distrust from past U.S. nuclear weapons tests in the Marshalls has been intensified by the ongoing French weapons tests on Mururua atoll, east of Tahiti in French Polynesia. Following mounting protests by the islanders and by Australia and New Zealand against French atmospheric tests, the French in 1973 began testing underground. But in 1979 one of these tests is said to have caused a small tsunami; according to a report in *Le Matin* (which French officials denied), the nuclear device became stuck halfway down the emplacement shaft and was detonated in that position.[71] True or false, such stories have reinforced the islanders' belief that nuclear technology is fraught with hazards.

The legacy of abuse and distrust has been a palpable presence whenever representatives of the island communities have gathered in regional meetings. The Cook Islands Conference on the Human Environment in 1982 was attended by seventeen of the South Pacific's twenty-odd island communities, whose representatives made declarations and adopted resolutions that opposed nuclear weapons testing and radioactive waste disposal in the South Pacific Basin. One catchall resolution was meant to discourage three pending studies and proposals related to nuclear waste or spent fuel.[72] In addition to the Japanese plan for dumping low-level waste in the ocean and the multinational subseabed research, these included a United States and Japanese study of the possibilities for an island storage depot for spent fuel from such Pacific Basin countries as Taiwan and South Korea.[73] The motive behind the latter study, which the Carter administration had initiated and in which the Japanese had reluctantly joined, was to provide an alternative to reprocessing and thus to further U.S. nonproliferation policy.[74]

Significantly, the Cook Islands conference also set the broad outlines of an important regional political strategy by recommending that the

[71]Cited by Thomas O'Toole and Kevin Klose in "French N-Test Turning South Pacific Atoll into Swiss Cheese Isle," *Washington Post*, May 12, 1980.

[72]South Pacific Commission, *Report on the Conference on the Human Environment*.

[73]South Korea will have some 7.4 gigawatts of nuclear capacity by 1990, or the equivalent of seven large reactors. Taiwan will have a capacity of almost 5.0 gigawatts by then. Their combined nuclear generation will be of the same order as that of the United Kingdom. K. M. Harmon and coauthors, *International Nuclear Fuel Cycle Fact Book* (Pacific Northwest Laboratory, 1985). A nuclear program of this size can be expected to produce up to 3,000 tons of spent fuel over the course of a decade.

[74]The joint U.S.–Japanese study, completed by the end of 1982, was not much affected by resistance from the islanders because nothing was expected to come of it. The study named the American-owned Palmyra Island (900 miles south of Hawaii) as the depot site, but neither the United States nor Japan, or any other country, was actually proposing that such a depot be built. Taiwan and South Korea showed no interest in the island storage concept, and appeared able to handle their spent fuel storage problems. Interviews by the

South Pacific island nations become parties to the London Dumping Convention (on that convention, see chapter 12). They could seek to have the convention embrace—under Article VIII—the South Pacific Environment Program's policy against waste disposal or storage in the basin. The conference specifically recommended that the South Pacific communities be represented at the February 1983 meeting of parties to the convention.[75]

As it turned out, two of the new island nations—Kiribati, formerly a British possession, and Nauru, a former United Nations trust territory—were present at this London meeting of the convention, and played an influential role. Their proposal for an immediate global ban on ocean dumping had no chance of adoption, but it helped define a middle ground whereon a Spanish proposal for a two-year moratorium on ocean dumping, pending a scientific study, could be adopted. The vote by which the resolution passed gave evidence of a widening coalition of nations that either opposed ocean dumping or viewed it skeptically. It was a coalition that included, besides Spain and the two new island countries, three other Pacific Basin nations (New Zealand, Papua New Guinea, and the Philippines), several Atlantic fishing nations (including Ireland and all the Scandinavian countries), and several countries of Africa and Latin America. Although the United Kingdom and the other nations were not legally bound to observe the moratorium, it proved binding enough politically.

Japan was willing to respect the opposition of the South Pacific islanders to ocean dumping.[76] Japan's forebearance appears to have resulted from a mix of considerations: Japan has been cultivating an image of peace and conciliation in an effort to live down its prewar reputation as the Pacific Basin's strong-arm bully; it has made substantial investments in the island nations and territories, and wants to make more (as in Palau, where a large oil refinery and tanker terminal might be established); and it does not want to jeopardize its fishing privileges within the exclusive economic zones of existing or emerging island nations. Indeed, should these privileges be jeopardized by ocean dumping, Japan's politically powerful fishing industry would probably put a quick stop to it.

author with Michael Lawrence, deputy manager of the DOE waste program, June 15, 1982; Charles Van Doren, June 27, 1982; and Richard Scribner, a special assistant to the under-secretary of state for security assistance, science and technology who participated in the spent fuel island depot study, August 26, 1982.

[75]South Pacific Commission, *Report on the Conference on the Human Environment.*

[76]This Japanese policy was confirmed by Prime Minister Yasuhiro Nakasone during his tour of several South Pacific nations in January 1985. See "Nakasone Promises Not to Dump N-Waste Against Wishes of S-Pacific Countries," *Atoms in Japan*, January 1985.

In a recent development, thirteen South Pacific nations have negoti-ated a treaty to declare a nuclear-free zone, by which they hope to keep their oceanic region free of the use and testing of nuclear weapons and the disposal of nuclear waste.[77] Coincidentally, several days after the signing of this treaty the Greenpeace vessel *Rainbow Warrior* was sunk at her moorings in a sabotage incident for which France has since admitted responsibility. The *Rainbow Warrior* had been about to sail for the French nuclear test site at Mururua atoll.[78]

The status of the multinational subseabed disposal investigation as a project many years away from practical application has made it less vulnerable to the islanders' protests than ocean dumping. But the polit-ical obstacles to the use of the subseabed by Japan or any other country will continue to loom large. The London Dumping Convention prohibits ocean dumping of high-level waste; whether subseabed disposal consti-tutes dumping, the convention has still to decide. If the answer is no, and if such disposal is seen as promising effective containment, a new international regulatory regime will have to be worked out. However, the Pacific islanders have made no distinction between subseabed dis-posal and ocean dumping in the past, and they may make none in the future. The fact that dumping of low-level waste in the ocean has been sanctioned for years by the convention has not stayed the islanders' opposition, and they could oppose a convention-approved program of subseabed disposal just as doggedly.

Reprocessing, Plutonium, and the Nuclear Allergy

In light of Japan's nuclear allergy, the imperatives to contain radioactivity and safeguard plutonium may have particular force in Japan. Yet Japan is following the French and some other Europeans toward an unneces-sary, risky, and politically burdensome reprocessing venture, and doing so long before the commercial breeder that would make plutonium fuel economic becomes available. Fuel reprocessing that avoids unacceptable releases of radioactivity may be possible with the best available tech-nology, but as the British experience at Sellafield has shown, political distress may result from excessive routine releases, accidental spills, mismanagement of nuclear facilities, and obsolescence. In Japan, the

[77]Lena H. Sun, "South Pacific Nations Declare Nuclear-Free Zone," *Washington Post*, August 8, 1985; interview by the author on August 13, 1985, with Robert Gordon, counselor at the Australian Embassy, Washington, D.C.

[78]Michael Dobbs, "France Admits that Its Agents Sank Vessel," *Norfolk Virginian-Pilot*, September 23, 1985.

Tsuruga incident revealed the political cost of even relatively minor spills or failures of containment.

Even more than the need for containment, the imperative to avoid risks of nuclear proliferation and terrorism may weigh against reprocessing and the use of plutonium fuel in Japan. For if Japan itself is not a likely proliferation risk, the same may not be said of some of the nations of East Asia that might follow the Japanese example and venture into reprocessing. Countries such as Taiwan and South Korea face very real security problems, and under threat of overwhelming conventional attack or nuclear blackmail the presence of growing stocks of separated plutonium might prove an irresistable temptation, despite present adherence to the international nonproliferation regime.

But the possibility of terrorist incidents involving the threatened use of separated plutonium perhaps should be as great a worry in Japan as in any country. Even if such incidents do not materialize, the extraordinary precautions that must be taken to see that they do not inevitably attract press attention and heighten the public's sense of a connection between nuclear power and nuclear bombs. In this connection, one need only recall the predawn scene on November 15, 1984, when the *Seishin Marie*—the ship dedicated to the sensitive mission of transporting a quarter of a ton of plutonium to Japan from La Hague (see chapter 10)—arrived at dockside in the port of Tokyo. The security forces on hand included 18 harbor patrol vessels and 300 policemen, plus an escort of 15 vehicles and patrol cars to accompany the convoy of 6 tractor-trailers that transported the massive casks of plutonium to their destination at the Tokai works.[79] What were Japanese citizens to think when from television and their daily newspapers they learned of this display of force on behalf of the peaceful atom?

For the Japanese utilities, the alternative to reprocessing and plutonium recycling is extended spent fuel shortage, the technical possibilities of which have already been studied in Japan.[80] Such storage could be provided either at the reactor stations or at a central, away-from-reactor facility (like those being developed by the Germans and the Swedes), for which a site is already available at Rokkashomura. High-level waste returned to Japan from reprocessors in Europe could also be kept temporarily at one or more of the sites where spent fuel is stored. Interim surface or near-surface storage of spent fuel and high-level waste could continue until some means of permanent disposal becomes available

[79]"Plutonium Shipped from France Arrives PNC Tokai Works under Heavy Guard," *Atoms in Japan*, November 1984, pp. 21–22.

[80]Cindy Galvin, "Japan Eyes Interim Storage Alternatives," *Nuclear Fuel*, December 3, 1984.

either in Japan or abroad. The problem of what to do with stores of plutonium in excess of research and experimental reactor needs would remain, but at least the prospect of adding to the plutonium inventories would be sharply reduced.

Conclusion: Japan as a Nuclear Leader

The Japanese had no choice but to rely on foreign tutelage during the first few decades of their venture into fission energy technology. But today Japan is in a position to strike out independently and to assume an important leadership role in both nuclear technology and efforts to develop national and international regimes to reduce the actual and perceived risks of nuclear power and to increase public acceptance of this energy source.

If an effort to establish geologic disposal facilities for high-level waste in Japan is to be pursued, the government would do well to improve on the clumsy and apparently technically ill-advised initiative at Horonobe. But perhaps the most promising opportunity for Japan would lie in an arrangement with China, whereby the Chinese would provide for disposal of Japanese spent fuel and high-level waste in return for Japanese assistance in nuclear energy development or other endeavors. Both countries would have much to gain. Disposal systems could allow the option of retrieval, and Japan could retain the right to recover the uranium and plutonium contained in the fuel. For China, a cooperative arrangement with Japan surely would make more sense than any agreement to take spent fuel from Europe, on the other side of the world. Such spent fuel storage and disposal options should be made subject to International Atomic Energy Agency safeguards.

The immediate and urgent problem of low-level waste disposal facing the Japanese utilities must for now be dealt with at home in Japan. But the utilities should consider going beyond the kind of shallow-burial facilities planned for the fuel cycle center at Rokkashomura; these are to consist of reinforced concrete bunkers built in pits, which when filled with drums of waste solidified in concrete or asphalt would be capped with concrete and then covered with earth.[81] Although such a disposal method is consistent with methods now favored in the United States, it compares poorly with the much more robust solutions for low-level waste isolation adopted by the Germans and the Swedes. The Swedish

[81]"Land Disposal of LLW Managed by Four Stages, NSC Settles Guidelines," *Atoms in Japan*, October 1985.

concept of tunneling out from an on-shore terminal to create storage caverns beneath the seabed might be particularly applicable in Japan.

The Soviet accident at Chernobyl points up poignantly the need for Japan and other nations to design nuclear facilities that are free of the risk of major failures of containment. Well before Chernobyl, Japanese reactor manufacturers joined with General Electric and Westinghouse to develop safer light water reactors. But the Japanese could go further and undertake the building of more forgiving reactor types that are immune from loss-of-coolant accidents—as indeed they should in light of their country's propensity for large earthquakes. One Japanese re-search group has the concept of "intrinsically safe" reactors under study.[82] Elsewhere, the Japanese already are at the forefront of research to re-cover uranium from seawater; if successful, this effort could open up an inexhaustible source of fissionable material and eliminate all need for separated plutonium.[83]

The nuclear enterprise in Japan is presented with both major problems and major opportunities. Japan has the opportunity to make nuclear power safer and more environmentally benign over the long period of hazard, and, with neighboring China, to pioneer in seeking sensible solutions to radioactive waste problems. To accomplish only the latter would itself be a signal achievement, likely to have a positive influence far beyond East Asia and the Pacific Basin.

[82]"Trends in the Study of Intrinsically Safe Reactor in Japan," Atoms in Japan, February 1986.

[83]None of the various processes under development by the Japanese or others for recovery of uranium from seawater has been carried beyond the early pilot stage. But Michael J. Driscoll, professor of nuclear engineering at the Massachusetts Institute of Technology, believes that fission energy from uranium recovered from seawater for use in a once-through fuel cycle is likely to be competitive in cost with energy from breeders. Driscoll was interviewed by the author on February 21, 1984.

12
Transnational Problems and the Need for Multinational Solutions

The nuclear waste problem in Europe and Japan quite clearly calls for multinational or international solutions. But there has been little progress as yet toward achieving such solutions beyond cooperation in research and development, some international standards-setting (as in the International Atomic Energy Agency's voluntary safety guidelines pertaining to nuclear material transport), and the now-suspended ocean dumping program for low-level waste supervised by the Organization for Economic Cooperation and Development's Nuclear Energy Agency. No nation is going to take another nation's spent fuel or high-level waste for permanent disposal as long as the bureaucrats and politicians fear that they will be accused of allowing their country to become a nuclear dustbin.

Global or multinational solutions to the problems of spent fuel storage and disposal are needed, and needed now, as an alternative to reprocessing. The plans and commitments of European and Japanese utilities for reprocessing promise to create stocks of separated plutonium far in excess of what is needed for the early breeders or other experimental reactors. The plutonium surplus is leading to the introduction of plutonium fuel in light water reactors. This in turn will increase the risk of nuclear proliferation and terrorism for everyone and the risk of needless political troubles for the nuclear industry. In principle, reprocessing commitments could be cut back to correspond with the needs of experimental and demonstration breeders, with more spent fuel being placed in interim storage or (eventually) committed to retrievable disposal in deep geologic repositories. But the movement to large-scale commercial reprocessing appears to have gathered strong momentum, at least for the near term.

The urgency and multinational scope of the spent fuel storage and disposal problems are therefore evident. This chapter takes a broad look at the mix of spent fuel management, waste disposal, reprocessing, and

369

plutonium fuel issues that transcend national boundaries. It examines first the transnational problem of spent fuel and high-level waste disposal, paying particular attention to the geologic and economic constraints that make international solutions so essential, and the political constraints that make them so hard to reach. It then examines the forces that drive reprocessing and the introduction of plutonium fuel despite the lack of economic justification, and looks at the consequent increased political risk for the nuclear industry and the increased proliferation and terrorist risks for everyone.

Geologic Disposal and National Boundaries

Although suitable geologic disposal sites no doubt will be found in some West European countries and perhaps even in Japan, these countries are not, with few exceptions, the best places to be looking for them. Several years ago I talked with Rudolf Trumpy, professor at the Swiss Federal Polytechnical School's Institute of Geology, an adviser to the Swiss government on radioactive waste disposal problems, and a former president of the International Union of Geological Sciences. Trumpy drew for me a rough map to indicate those parts of Europe which, from a standpoint of tectonic stability, he regarded as "good," "mediocre," and "very bad." Most of the immediate Mediterranean basin, including all of Italy, the Balkans, Greece, and Turkey, he labeled as very bad because of the frequency of earthquakes and the presence of geothermal fields and volcanic activity. The greater part of France and most of the rest of Western Europe he rated as mediocre, or as an area once tectonically active which has not yet come entirely to rest. The only parts of Europe rated by Trumpy as good were Russia and most of Scandinavia, which like much of Canada are underlain by Precambrian crystalline rock, and have been "quiet" for many millions of years.[1]

The waste problems of the smaller European nations point up with particular clarity the severe geologic and political constraints that national boundaries place on the search for repository sites. On Trumpy's map the line dividing the "very bad" and the "mediocre" parts of Europe runs right through the middle of Switzerland. With five reactors now operating, Switzerland must eventually find a place to dispose of much of its spent fuel and perhaps some reprocessing wastes. But Trumpy thought prospects for finding a repository site in his country were poor. "If Western Europe were one country, we wouldn't be looking in Switzerland, of all places, for a place to put a repository," Trumpy told me.

[1]Trumpy was interviewed by the author in December 1980.

In his view, Switzerland's alpine region was clearly unsuitable for a repository for high-level waste because of its heavy faulting. If a geologically acceptable site could be found at all, it would be in the granite and gneiss that underlie densely populated northern Switzerland, south of the Rhine River. But in 1980 an advisory panel of leading Swiss geologists, which Trumpy chaired, reported to the Swiss government that even this region offered a doubtful prospect. "The presence of thermal springs implies water circulating at great depth, where it is heated up," he explained to me.

Since 1980 the Swiss nuclear waste agency, NAGRA, has nevertheless worked gamely at gathering geologic data, with nearly a dozen deep boreholes now having been drilled. The first three produced disappointing results, one revealing a major (and quite unsuspected) displacement and two encountering carboniferous layers where only competent rock was expected.[2] But NAGRA persevered, and on the basis of admittedly preliminary data announced in September 1984—just before a national referendum in which 54 percent of the voters said no to a nuclear phase-out—that safe geologic disposal of radioactive waste could be guaranteed.[3] Yet probably no specific site is to be chosen for another decade, and no repository is expected to begin operation before the year 2020. The granite formation of interest to NAGRA is deeply overlain by sedimentary rock, while just across the Rhine in southern Germany it rises to the surface, where technically it would be much more accessible;[4] but politically the German granite is beyond the reach of the Swiss.

In seeking a place to dispose of their own nuclear waste, the Belgians face geologic constraints which, if anything, are even greater than those confronting the Swiss. At their nuclear research center at Mol the Belgians have sunk a shaft some 220 meters deep to explore the properties of a thick clay layer, clay being Belgium's only geologic disposal medium. As such, clay has major limitations and presents major difficulties, as P. Dejonghe, the official in charge of nuclear waste management at Mol, explained to me. "It is a plastic formation, and you must freeze the clay" before sinking a shaft or tunneling through it, he said.[5] This however is

[2]Interviews by the author with Richard A. Robinson of the Office of Nuclear Waste Isolation, Battelle Memorial Institute, Columbus, Ohio, and Kent Harmon of the Battelle Pacific Northwest Laboratories, Richland, Wash., on December 18, 1984, and May 7, 1985, respectively.

[3]"Nuclear Waste Can Be Disposed of Safely, NAGRA Tells Swiss Voters," *Nuclear Fuel*, September 24, 1984.

[4]Interview by the author with Hermann Flury of the Swiss Federal Institute for Reactor Research, Wurenlingen, December 1980.

[5]Dejonghe was interviewed by the author in November 1980.

manageable, as the Belgians have demonstrated by creating an experimental chamber in the clay. The more fundamental problem is that "clay is a thermal shield and [its] conductivity is extremely low," Dejonghe told me. Unless the waste were allowed to cool over an extended period—seventy-five years or longer—the buildup of heat could be unacceptably high. "The clay would be hot, undergo structural changes, and lose its excellent absorptive capacity," Dejonghe added. Gilbert Eggermont, physicist and science adviser to the Fédération Générale du Travail de Belgique (Belgium's socialist-leaning trade union) and an articulate critic of his country's nuclear program, observed to me that if final disposal has to be so long delayed, Belgium has no solution to its nuclear waste problem.[6]

Austria's Initiative in Egypt

In the absence of adequate cooperative arrangements on a European scale, another small country—Austria—hemmed in by geographic and political circumstances tried to work out a waste disposal agreement with two Third World countries, Iran and Egypt. Although this effort was to come to naught, how it came about and why it failed is a remarkable, if little known, story. It took place before a November 1978 referendum in which Austrian voters narrowly decided not to start up the country's one power reactor, located at the village of Zwentendorf on the Danube River about twenty miles west of Vienna.

The Austrians expected to conclude a contract with France's COGEMA to have spent fuel from the Zwentendorf reactor reprocessed at La Hague, with the waste then to be returned to Austria for disposal. But where was this waste ultimately to go? Approximately the size of South Carolina, and for the most part covered by the Alps and hence heavily faulted geologically, Austria has few places to look for radioactive waste repository sites. Among the most promising areas available was the Waldviertel, a region of low mountains and evergreen forests adjacent to Czechoslovakia, sixty miles northwest of Vienna as the crow flies. In the early summer of 1977 the government was still hoping to build a repository there. But by then people of the region, including many farmers, were bringing strong pressure to bear on their local elected officials to refuse all overtures from Vienna to that end.

In late June, for instance, some 3,000 demonstrators, many of them driving farm tractors in a procession stretching as far as one could see, converged on a principal town of the region to urge officials to accept

[6]Eggermont was interviewed by the author in November 1980.

no inducements from the central government, no matter how presented. About ten days later the mayors of the region were in fact tested when Chancellor Bruno Kreisky, head of Austria's Socialist government, had them in attendance in his office and offered generous economic development assistance if they would cooperate in the siting of a repository. The mayors rejected his proposition, apparently in no uncertain terms, for after the meeting Kreisky reportedly said, "I don't want to start a civil war with the people of Waldviertel."[7]

By this time nuclear power had become such a hot political issue in Austria that none of the political parties wanted to take responsibility for the start-up of the Zwentendorf reactor.[8] Indeed, they were all committed to the proposition that this issue should be decided by the voters in the referendum set for November 5, 1978. Whatever the voters' decision, the Austrian Parliament was expected to see that it was faithfully implemented.

In early 1978 newspaper stories began to appear about Austrian officials having entered into negotiations with Iran and Egypt to establish a waste repository. A headline in the *Salzburger Nachrichten* read, "Atomic Waste: It Is Possible to Come to an Agreement with Iran."[9] The next day's *Oberösterreichische Nachrichten*, published in Linz, carried a cartoon showing Chancellor Kreisky, with a self-satisfied smile, sweeping atomic waste under a persian rug.[10] But 1978 was not a good year to be doing business with the Shah's regime about nuclear waste or anything else, and by summer speculation that Iran held the answer to Austria's problem had faded. In its place was speculation that the answer lay in the sands of Egypt.

The *Salzburger Nachrichten*, in its July 10, 1978, issue, carried a story headlined "Sadat Agrees to Atomic Waste Repository"; a subhead added,

[7]Initiative Österreichisher Atomkraftwerksgegner (IOAG), "Künstler Gegen Zwentendorf," campaign booklet issued by Austrian anti-nuclear initiative organization in October 1978. The booklet contains a detailed chronology of events described.

[8]Walter C. Patterson, "Austria's Nuclear Referendum," *Bulletin of the Atomic Scientists*, January 1979.

[9]This article in the *Salzburger Nachrichten* of March 3, 1978, attributed the hopeful news to Austria's foreign minister.

[10]Fritz Schmidt, an official in the federal chancellor's office, told me that Iranian officials were willing to provide a repository site if Austria could find one. Iran was then establishing a nuclear power program of its own; envisioned as part of it, according to Schmidt, was a repository large enough to accommodate the waste generated over the life of a 20,000-megawatt program. Iran was to take for itself 85 to 90 percent of the capacity and leave the rest for Austria, which was also to pay royalties and contribute to the construction of the repository (doing only the engineering and perhaps building and paying for the facility). The negotiations evidently were short-lived and the details were vague. Interview by the author with Schmidt, Vienna, October 1980.

"Contract Almost Completed—Waiting for the U.S.A." The fuel for the Zwentendorf reactor was of United States origin, and plans for the reprocessing of such fuel were subject to U.S. approval under its policies for preventing the spread of nuclear weapons. The Austrian article observed that Egypt was in a politically sensitive part of the world, and suggested that the United States might object to any fissionable material being sent there. (Austria was not, however, proposing to send spent fuel to Egypt, but rather the vitrified high-level waste from La Hague.) Kreisky was quoted as saying that he did not believe a repository site could be found in Austria. "I do not want to create any more possibilities for conflict," the chancellor added. The *Salzburger Nachrichten* was soon expressing further doubts about Austro–Egyptian plans and their underlying motivation. On July 25, 1978, the paper carried the headline "Egyptian Project Confused," observed that Chancellor Kreisky and the responsible minister were not releasing "concrete information" about the project, and suggested that the project might have been invented to delay until after the November referendum the question of the "real fate" of Austria's nuclear waste—including the possibility that disposal would have to take place in Austria after all.

How far negotiations went between Austria and Egypt is unclear; nor is it clear that the Egyptians had decided in what part of Egypt a repository might be built. Fritz Schmidt of the federal chancellor's office told me that a contract was negotiated whereby Austria would build, at its own expense, a repository in the Sahara several hundred kilometers west-northwest of Aswan. Austria was to give Egypt a billion shillings (approximately 60 million 1982 dollars) in advance as "good will" money, and provide certain infrastructure improvements. "Egyptian officials were readily agreeable [to the project], provided the International Atomic Energy Agency approved every step in the development of the repository," Schmidt said. "They had in mind getting a better technical infrastructure—transmission lines, roads, electronics, communication, and so forth." Schmidt added that the International Atomic Energy Agency did in fact assign a mission to review the proposed project, and that the site selected was found to be remarkably good. Chancellor Kreisky, he said, discussed this project personally with President Sadat.[11]

At the time of these negotiations President Sadat's science adviser was Farouk El-Baz, a naturalized American citizen who was then associate director for research at the Smithsonian's Air and Space Museum. El-Baz is a geologist who first made a name for himself as a scientific planner in the Apollo moon-landing program. As Sadat's science adviser—a role

[11]Ibid.

arranged through the White House at Sadat's request to President Carter—he visited Egypt several times a year for discussions with the Egyptian president.[12] As El-Baz has recounted for me, in one such session he explained to Sadat why the proposed repository project was a bad idea: first, no means of permanent disposal for high-level waste had yet been demonstrated anywhere in the world, hence there was no proven method available which could be applied in Egypt; second, the kind of detailed geographical and geologic information that would be necessary for the selection of a repository site was lacking; and third, Egypt lacked the scientific and technical infrastructure necessary to support the proper construction and safe operation of a radioactive waste repository. El-Baz told me that the principal area considered for a repository site was not in the Sahara west of the Nile, but in the Eastern Desert west of the Gulf of Suez and east of the Nile. To his mind, selection of a site there would have been ridiculous given the lack of information about this region's groundwater resources and its possibly large potential for future agricultural and oil and gas development.[13]

Once El-Baz made his views known to reporters following his meeting with Sadat, there was much discussion of the project in the press. According to El-Baz, the Egyptian Parliament eventually recommended that the project be put on hold. Losing ground in this policy debate was Egypt's minister of electricity at the time, Ahmad Sultan, who strongly favored the repository project. El-Baz says that Sultan viewed nuclear energy as a technological marvel in which Egypt was not going to be allowed to share unless the industrialized nations were given a compelling incentive, such as the promise of access to an Egyptian waste repository. President Sadat, whom El-Baz continued to serve as science adviser until Sadat's assassination in the fall of 1981, was well enough pleased to have the repository plan studied and publicly debated, but no binding commitments were ever made. The future of the project was rendered moot by the Austrian referendum.

Sigvard Eklund, former director general of the International Atomic Energy Agency, viewed the Austrian initiative in Egypt as the start of

---

[12]El-Baz, born in 1938 in the Nile delta town of Zagazig, first came to the United States in 1960 as a graduate student. He was keenly interested in the investigation and development of Egypt's natural resources and agricultural potential, including the possibility of converting desert regions to productive farmland. His work for the space program, his assignment as Sadat's science adviser, and his views on the desertification phenomenon are discussed by Constance Holden in "Egyptian Geologist Champions Desert Research," *Science*, September 28, 1979. El-Baz is currently vice-president for international development of ITEK Optical Systems.

[13]El-Baz was interviewed by the author on August 30, 1982. Material attributed to El-Baz below is from this interview, unless otherwise noted.

what might have become a significant international experiment. "One could think of some international area in Egypt where waste from different countries could be stored, [with] diplomatic immunity [established] for the area in question," Eklund told me.[14] Eklund was suggesting that a host country—in this case Egypt—would cede some degree of sovereign control for the sake of creating an international enclave in which the participating countries would have assured rights of use and access. But to persuade Egypt or any other country to agree to such an arrangement clearly would have been difficult.[15] In the negotiations between the Austrians and Egyptians there had been mention of the possibility that other countries besides Austria and Egypt (which had no power reactors but had ambitious plans for nuclear development) might use the repository;[16] but no other countries were actually brought into the discussions, and apart from the IAEA's role in reviewing plans for the project, no "internationalization" of the facility apparently was contemplated.

Multinational Initiatives in Europe

In truth, the effort to bring about international nuclear waste disposal cooperation is a matter of the greatest political sensitivity. In Europe there have been several multinational ventures to develop commercial nuclear services and technology, among them Eurochemic, a reprocessing project in Belgium; Eurodif, a huge uranium enrichment plant at Tricastin in the Rhone Valley; and the Superphenix breeder project at Creys-Malville in which France has been joined by several other countries, most notably Italy and Germany. But in the nuclear waste field multinational collaboration generally has not gone beyond the coordination and sponsorship of research and the development of safety guides and standards by the European Community, the Organization for Economic Cooperation and Development, and the International Atomic Energy Agency.

The European Community's involvement in waste management research began in 1975. The total five-year budget for its "direct" and "indirect" action programs in waste management research from 1980 to

[14]Eklund was interviewed by the author in Vienna, October 1980.

[15]"Establishing international enclaves is an idea that has been frequently talked about, but it has not gone anywhere," says Charles Van Doren, who was an assistant director of the U.S. Arms Control and Disarmament Agency and a special ambassador for nuclear nonproliferation during the Carter administration. Interview by the author with Van Doren, June 27, 1982.

[16]Interview with Schmidt, October 1980.

1984 was $130 million.[17] The indirect action program supported geo-logic and other research by individual community countries—the French looking at granite, for instance, the Germans at salt, and the Belgians and Italians at clay. The European Community, it should be noted, is the only existing multinational entity that would have both the money and the incentive to develop a "European" nuclear waste management sys-tem if the powerful political inhibitions now holding sway are ever overcome.

The one international effort involving the actual disposal of waste has been that undertaken by the OECD's Nuclear Energy Agency (NEA) in coordinating and monitoring the disposal of low-level waste in the North Atlantic. The Nuclear Energy Agency's role has been to oversee compli-ance with rules adopted by the International Atomic Energy Agency pursuant to the London Dumping Convention of 1972. Up through 1982 the United Kingdom, Belgium, the Netherlands, and Switzerland had been dumping low-level waste at a designated site some 500 miles southwest of England, the British disposing of far more waste by this means than the other three nations combined. Japan was eager to begin dumping in the Pacific under Nuclear Energy Agency supervision, but only if it could gain the concurrence of the Pacific island nations (see chapter 11).

There has been no sea dumping since 1982, however. In February 1983 the London Dumping Convention adopted a nonbinding resolution calling for a two-year moratorium on sea disposal, pending a scientific study of the risks involved. The Netherlands government announced that it would honor the moratorium, and later decided to give up sea dump-ing altogether. The other three governments intended to continue dump-ing despite the resolution, but found they could not, for the British National Union of Seaman refused to crew the *Atlantic Fisher*, the ship which had been carrying out the dumping for all four nations. Since then these nations have been in the difficult position of having to con-vince the skeptical London Dumping Convention majority that dumping does no harm to either humans or the environment.[18] The political plight of sea dumping makes this first multinational venture in waste disposal more of a symbol of trouble than of progress.

The lack of progress toward multinational solutions to nuclear waste problems is regretted by a number of people in the nuclear industry, as

[17]Serge Orlowski, "Development of the OEC's Activities Related to Radioactive Waste Management During the Year 1980" (Commission of the European Community, October 27, 1980).
[18]Stephanie Cooke, "London Convention Votes to Continue Ban on Radwaste Dumping at Sea," *Nucleonics Week*, October 3, 1985.

I found on a trip to Europe several years ago. "It's completely crazy to have each country trying to dispose of its own waste," Maurice Potemans, an official of Belgium's Union d'Exploitation Electrique, remarked to me.[19] P. Dejonghe of the nuclear center at Mol, Belgium, spoke similarly: "There is in this story something silly," he said. "There is the nationalistic approach. Many nations spend much money ... Yet the Rhine starts in Switzerland and flows between France and Germany into Holland. Suppose the Swiss adopt a solution to waste disposal that is less safe than the one developed by the Dutch. In such a case it would be better for the Dutch to dispose of Swiss wastes."[20]

A report by the Nuclear Energy Agency's Group of Experts had taken note of such situations and commented that "present national borders have no real significance given the long-term hazard of radioactive waste."[21] In its 1980 report the working group on waste management and disposal of the International Nuclear Fuel Cycle Evaluation, initiated by President Carter (see chapter 4), had pointed out that multinational repositories could offer greater economic efficiency for countries with small nuclear programs, and solve the disposal problem for countries lacking suitable geologic formations. With the number of disposal sites reduced, the International Atomic Energy Agency's job of safeguarding and maintaining accountability for spent fuel would be made easier. Some prominent industry people saw international collaboration and consensus on nuclear waste disposal as essential to public trust. By the same token, Erik Svenke, former president of the Swedish Nuclear Fuel Supply Company, has observed that opening the road to multinational repositories is not so much a "matter of rational optimization" as a "matter of confidence."[22]

Yet I also found that while there was a perception of need, there was much skepticism that multinational repositories would ever be a practical political possibility. André Finkelstein of the French Commissariat à L'Energie Atomique saw the very concept of such repositories as pie in the sky. "It's like an atomic airplane, it will never fly," he told me.[23] Hans W. Levi, then scientific director of the Hahn-Meitner Institute for Nuclear Research in Berlin, strongly favored the concept of multinational

---

[19]Interview by the author with Potemans, November 1980.

[20]Interview by the author with Dejonghe, November 1980.

[21]Nuclear Energy Agency, *Options, Concepts and Strategies for the Management of Radioactive Waste Arising from Nuclear Power Programs*, report by the NEA Group of Experts, September 1977 (Paris, Organization for Economic Cooperation and Development, 1977) p. 61.

[22]Erik Svenke, "Principles for Storage and Final Disposal of Used Nuclear Fuel and Reprocessing Wastes," paper prepared for the Spanish Atomic Forum Symposium, Madrid, November 7–8, 1983.

[23]Interview by the author with Finkelstein, November 1980.

solutions but saw enormous obstacles in the way. "If there is any public acceptance of waste disposal, it certainly ends where foreign waste is concerned," Levi said.[24]

The caution of the European Community in initiating studies and plans for a European waste disposal system has on occasion been publicly criticized. Jean-Pierre Hannequart of the Institute for European Governmental Policy, a private group based in Bonn, several years ago reproached the community's Council of Ministers for not acting on proposals made as far back as the early 1970s by the European Parliament, the European Community's Economic and Social Committee, and the Commission of European Communities itself.[25] These bodies had contemplated action plans to establish standardized waste management practices, a European network of disposal sites, and a European waste management agency. But Serge Orlowski, chief of the nuclear fuel cycle division in the European Community's Euratom Research and Development program, was impatient with such criticism. To do as the critics suggested, he said, "would put at risk not only the waste program but the entire nuclear program." By the end of the century, he suggested, it might be possible to have multinational waste disposal systems, but "you have to wait for the situation to cool down."[26]

One pioneering International Atomic Energy Agency effort in multinational waste management never really got around to the problem of finding a host state. Indeed, this initiative—the International Spent Fuel Storage (ISFS) project proposed by the Carter administration as a nonproliferation measure—never got anywhere because an ISFS working group set up by the IAEA could not overcome its sharp divisions on the reprocessing issue. The United States wanted fuel-vendor nations to insist that their foreign customers agree to commit their spent fuel to the international storage facility once it was established. But nations such as France, the United Kingdom, and Germany saw this as inimical to their ongoing or planned commercial reprocessing ventures.[27] Retrievable surface storage, not geologic disposal, was all that was contemplated, but the hang-up over reprocessing was paralyzing. The ISFS working group has been inactive for several years.

The present IAEA director general, Hans Blix, sees the prospects for multinational repositories as improving. At the agency's twenty-ninth general conference in Vienna in September 1985, Blix noted that a few

[24]Interview by the author with Levi, November 1980.
[25]Jean-Pierre Hannequart, "Les Communautés Européenes et le Problème des Déschets Radioactifs," *Amenagement de Territoire et Droit Foncier*, Series no. 5, April 1979.
[26]Interview by the author with Orlowski, November 1980.
[27]Interview by the author with Van Doren, April 13, 1986.

countries have actually offered to accept waste from abroad; he was alluding to offers extended by China to European utilities and by the Soviet Union to the Austrian owners of the Zwentendorf reactor (which had yet to split its first atom).

The Chinese and Soviet Initiatives

The first offer by the Chinese to accept spent fuel from European utilities came in early 1984. It was reported in the trade press to have been suggested initially by a principal in the German trading firm of Alfred Hempel. This firm, along with the nuclear supply company Nukem and Nukem's subsidiary Transnuclear, owns the consortium Inter-Nuklear Services, and was to be the agent for the China Nuclear Energy Industry Corporation. A letter of intent describing the proposed service was signed by representatives of the Chinese corporation and the German consortium's three parent companies on January 14, 1984.[28] The Chinese would charge $1,512 per kilogram of spent fuel received, whereas CO-GEMA was charging its foreign customers about $900 per kilogram for reprocessing; but the service provided by the Chinese would also include transportation and final waste disposal.[29] The Chinese fee would exceed more than fivefold the per-kilogram charge currently estimated by the Department of Energy for geologic disposal of spent fuel in the United States. If 4,000 tons of spent fuel were received by the turn of the century, China would earn as much as $6 billion in foreign exchange. According to the sketchy information available, much of it second hand, the waste disposal site would be in the Gobi Desert not far east of Lop Nor. It was at Lop Nor that the Chinese conducted their above-ground nuclear weapons testing before going to underground testing in the foothills of the Tian Mountains, farther to the west.

Utilities in more than a half-dozen countries—Germany, Austria, Switzerland, Belgium, the Netherlands, Spain, and Italy—were reported to be interested in the Chinese offer. Walter Fremuth, director of the Austrian national utility consortium, went to China for talks with Chinese Nuclear Energy Industry officials, and a commercial contract was expected to follow if Austria's policy paralysis over the Zwentendorf reactor and nuclear energy could be overcome.[30] Some German utilities

[28]Michael Weisskopf, "China Reportedly Agrees to Store Western European Nuclear Wastes," *Washington Post*, February 18, 1984.

[29]"West German Consortium Sends Offers to Take Spent Fuel to China," *Nuclear Fuel*, July 16, 1984.

[30]Padraic Sweeney, "Chinese and Austrians Reach Agreement in Principle on Spent Fuel Swap," *Nucleonics Week*, July 5, 1984.

were sufficiently interested in the Chinese offer to alarm proponents of German fuel reprocessing. "The Chinese offer, which is tempting to some utilities, could upset the financial basis for building our own reprocessing facility, and we would not allow this to happen," said Christian Lenzer, an important member of the Bundestag, whose comment reflected the views of Chancellor Helmut Kohl's government.[31] The government has since reportedly given its approval for German utilities to ship a modest amount of fuel (up to 150 tons) to China, but only under certain conditions: shipments could not be made if they would keep German reprocessing and spent fuel storage facilities from being fully utilized, the Chinese price would have to be competitive with alternative offers, and some of China's nuclear business would have to be reserved for the German vendor Kraftwerk Union.[32]

The Chinese offer gave rise to obvious questions about adherence to the international safeguards regime and U.S. nonproliferation policy. According to some estimates, more than half of the European spent fuel that might be sent to China would be of U.S. origin or contain uranium enriched in the United States; this would mean that more than half could not be sent to China without the consent of the United States.[33] The Austrians, in their discussions with the Chinese, found a way by which the problem of third-party consent could be avoided: Austria would buy its fuel from the Chinese, but Austria would insist that if the fuel were reprocessed by the Chinese, as expected, the plutonium recovered would not be reexported and would be used only in China's civilian nuclear program.[34] China had long refused to sign the Nuclear Nonproliferation Treaty, but was now joining the International Atomic Energy Agency and was later to announce that some of its nuclear installations would be placed under IAEA safeguards at an "appropriate time."[35]

The Chinese initiative seems to have encouraged a Soviet counteroffer, which was limited to Austrian fuel. In February 1985 Soviet foreign

---

[31]"The China Syndrome," *Radwaste News*, vol. 5, no. 6.

[32]Ann MacLachlan, "German Utilities Link KWU Work with Chinese Spent Fuel Deal," *Nuclear Fuel*, March 24, 1986.

[33]Senator Dan Quayle of Indiana and six of his colleagues felt that the United States should not allow fuel of U.S. origin to be sent to China without strong assurances against diversion or military use of the plutonium. In a letter of August 8, 1984, to the U.S. secretaries of state, defense, and energy, the senators said that title to all such fuel should remain with the utilities; that China should pledge not to reprocess it except under conditions acceptable to the United States; and that China should agree to IAEA safeguards against diversion and reprocessing of the fuel for military purposes.

[34]Sweeney, "Chinese and Austrians Reach Agreement in Principle."

[35]Ann MacLachlan, "China Says It Will Make Voluntary Offer to Accept IAEA Safeguards," *Nucleonics Week*, September 26, 1985.

minister Andrei Gromyko made a formal offer to Austrian chancellor Fred Sinowatz (who replaced Kreisky as Socialist party leader in 1983) to accept spent fuel from Zwentendorf if this reactor were started up.[36] The Soviet Union had been routinely taking back Soviet-supplied fuel from its East European allies, but this was the first time the Soviets had offered to take spent fuel of non-Soviet origin.[37] The offer was traceable to approaches made by the Austrians to the Russians in the summer of 1984, at the same time their widely publicized negotiations with the Chinese were going on. As matters have turned out, none of these offers have been of help to Austria's Socialist minority government in its desire to start up the Zwentendorf reactor.[38] The utility consortium that owns Zwentendorf has since taken steps to dismantle the reactor and sell off the parts.[39]

The Chinese and Soviet initiatives may be of limited relevance to the political problem faced by democratic nations in Europe that are attempting to establish multinational waste repositories. The political inhibitions discouraging such arrangements may not be easily dispelled despite the significant commercial advantages that could come to countries willing and able to accept spent fuel from foreign utilities. An example of such an advantage came to light in early 1985 when the Netherlands government asked the French and the Germans whether they would accept Dutch waste or spent fuel in return for reactor orders.[40] The inquiry was politely received, but no promises were made.

[36]Ann MacLachlan and Padraic Sweeney, "Soviet Union Offers to Take Austrian Spent Fuel for Storage," *Nucleonics Week*, February 21, 1985.

[37]For a summary of the little that is known about nuclear waste management and fuel cycle policies in the Soviet Union, see Barry D. Solomon, "Nuclear Waste Management in Japan and the Soviet Union," paper presented for Solomon at the International Conference on Energy Resources in Asia, Hong Kong, August 1986. According to Solomon, spent fuel from Soviet power reactors is being kept at the reactor stations, and liquid high-level waste from the reprocessing of fuel from military plutonium production is kept in tanks at the reprocessing centers. Reprocessing of civil power reactor fuel is planned for the future, but is not now being done. As is true of countries in the West, the Soviet Union has been developing methods to solidify high-level reprocessing wastes and dispose of them in geologic repositories. Pending development of reprocessing, solidification, and geologic disposal facilities, away-from-reactor spent fuel storage appears likely. Investigation of several potential geologic disposal media is under way, and there have been reports of deep excavations being carried out in salt domes just north of the Caspian Sea.

[38]Padraic Sweeney, "Austria's Governing Coalition Is Split over Zwentendorf's Fate," *Nucleonics Week*, November 1, 1984.

[39]Interview by the author on April 30, 1985, with Walter Greenert, press counselor at the Austrian Embassy in Washington, D.C.

[40]Linda Bernier, "Dutch Probing France, Germany on Return of Wastes from Future Units," *Nucleonics Week*, March 25, 1985.

For several years now the Nuclear Energy Agency of the OECD has been tip-toeing around a proposal for studying how an international waste repository might be established. First put forward in 1983 by the Netherlands representative to the NEA with the support of the United States representative, the resolution proposing this study was rejected in favor of another calling for a study of the "preconditions" to be met before an international repository could be considered. The adopted substitute called, in effect, for a study of what the study should address. Even then, nothing happened for fifteen months. In the NEA's Radioactive Waste Management Committee concern was expressed by some representatives, including Germany's, that the kind of study proposed by the Dutch might threaten national efforts to establish repositories. German critics of the exploration of the Gorleben salt dome, it was said, might argue that Germany should put aside the Gorleben project and look to an international solution to its waste problem. In February 1985, however, the NEA committee set up a "scoping" group to define the parameters of an international repository study. Allison M. Platt, an American and manager of nuclear technology at Battelle Northwest Laboratories, chairs the scoping group, which is expected to report its finding in late 1986.[41]

According to Platt, the group has found no legal, institutional, or other reasons why establishment of a multinational repository should not be feasible in principle. Platt believes that some potential host countries might eventually see major economic advantages in hosting a multinational repository. But he and his scoping group have concluded that establishment of a multinational repository will not be politically possible anywhere until one of the nuclear power countries has established its own national repository.[42] This latter finding, which is no doubt realistic, points up the international importance of current efforts in the United States and other countries to find and develop suitable repository sites. Availability of first national, then multinational or international repositories could offer all countries the alternative of direct geologic disposal of their spent fuel, which could be especially attractive if it included a retrieval capability that would preserve the reprocessing option for at least several decades. For utilities around the world to elect long-term retrievable storage or disposal of their spent fuel, part in surface facilities, part in geologic repositories, would all but put an end to the present premature development of commercial reprocessing—an activity now all the more untimely because it brings on the use

[41]Interviews by the author with Allison M. Platt on April 23, 1985, and April 16, 1986.
[42]Ibid.

of plutonium fuel in light water reactors, for which there is an abundance of cheap uranium fuel available.

Reprocessing and the Drift Toward Use of Plutonium Fuel

It is useful to summarize some of the assumptions and realities under-lying the interest of the European and Japanese nuclear industries in commercial reprocessing, and to look at the resulting economic and political burdens for the industry and the risks for the public. The risks become aggravated by what appears to be an inevitable drift toward use of plutonium fuel as reprocessing leads to the buildup of large stocks of plutonium.

Assumptions and Realities

The commitments of the Europeans and the Japanese to major programs of commercial reprocessing were made from assumptions which, for the most part, have corresponded poorly to the realities. First, in the very early years it was assumed that the reprocessors would keep the wastes from the reprocessing of foreign fuel—an assumption that was soon found to be quite wrong, although the reprocessors were to continue to get foreign business. Second, it was assumed that reprocessing is an essential step in waste management and disposal. And third—fundamen-tal to the entire endeavor—was the assumption that early use of the separated plutonium would be economically justified, if not in breeders, then in light water reactors as an intermediate step toward breeders. By the late 1970s, and certainly by the early 1980s, there was clear evidence that the latter two assumptions were wrong also. But this evidence brought no cutback in the contractual commitments between the French and British reprocessors and their customers; nor did it deter the Ger-mans and Japanese from proceeding with plans to establish major repro-cessing facilities of their own. These several faulty assumptions are each worthy of closer examination.

The reprocessor keeps the waste.   Before the mid 1970s nobody seems to have doubted that reprocessors would accept responsibility for all the radioactive wastes generated, whether the wastes came from the reprocessing of domestic or foreign fuel. In industry generally, nor-mal practice has always been for the responsibility for wastes to remain with the enterprise that produces them. Nations such as Austria, Swit-zerland, and the Netherlands, which had no plans for reprocessing on their own, saw this as a major advantage: their utilities could ship spent fuel to La Hague or Windscale without worrying about the wastes from the reprocessing of that fuel.[43] Similarly, the German and Japanese utilities, aware that it would be some years before reprocessing services could be available to them domestically on a large commercial scale, looked to the French and the British to take their spent fuel, reprocess it, and dispose of the wastes.

But by early 1976 it was clear that it was impossible politically for the reprocessors to keep and dispose of foreign waste, and would be for many years. The earliest complaints, from citizens fearful that their country was about to become a nuclear dustbin, seem to have been voiced in the United Kingdom in 1975. British Nuclear Fuels was entering into contracts to reprocess substantial amounts of Japanese fuel at the THORP facility planned for the 1980s. In May of 1975, British Friends of the Earth published an article entitled "Windscale to Be World Capital for Radioactive Waste?" Five months later the *London Daily Mirror*, on October 21, dropped the question mark and gave its own hyped-up version of the story, headlining its page-1 article "Plan to Make Britain World Nuclear Dustbin."

The story touched off an uproar. The secretary of state for energy, Wedgewood Benn, commended the *Mirror* for "sparking a public debate on Britain's nuclear dustbin problems," a debate which he said was "long overdue."[44] The upshot of the "dustbin debate" was that in March 1976 Benn announced in Parliament that British Nuclear Fuels could take on foreign business only if the contracts allowed the company the option of shipping back to its customers their share of the wastes.[45] This became

[43]Early contracts between COGEMA and BNFL and their foreign customers, signed between 1971 and 1974, made no provision for the customers to take back their share of the wastes generated. According to COGEMA president Georges Besse in a 1981 review of the evolution of COGEMA reprocessing contracts with foreign customers, "one still believed that certain of the wastes offered a potential value and that it could be to the reprocessor's advantage to keep them." "Les Contrats de Retraitment Etrangers de la COGEMA," Enerpresse document no. 2900, September 7, 1981.

[44]Denis Hayes, "Nuclear Power: The Fifth Horseman," Worldwatch paper 6 (Washington, D.C., Worldwatch Institute, May 1976).

[45]The British Royal Commission on Environmental Pollution, in its 1976 report *Nuclear Power and the Environment*, recommended against exercising the option to return reprocessing wastes to the countries of origin. It argued that additional transport risks would be involved, and that the wastes might be managed more safely in Britain than in an earthquake-prone country such as Japan. In the commission's view, a government prepared to sanction reprocessing of foreign fuel should be prepared to accept responsibility for the resulting foreign waste. Royal Commission on Environmental Pollution, *Nuclear Power and the Environment*, sixth report (Her Majesty's Stationery Office, September 1976) p. 161, paras. 422, 423, 424.

THE INTERNATIONAL WASTE PROBLEM

the policy in France as well, for COGEMA and BNFL, as partners in United Reprocessors (through which most of the existing reprocessing business was being divided up), insisted on similar terms in their contracts with foreign customers.

The new policy giving the reprocessor the option to return wastes to the customer did not harm COGEMA's and BNFL's foreign business. Reprocessing was still looked upon as a necessary and inevitable step in the nuclear fuel cycle, and even though the customer utilities could expect to have their pro rata share of the waste generated returned, they would gain a respite by sending their spent fuel to La Hague or Windscale. No waste or spent fuel (in the event that it was not reprocessed) would be returned before the mid 1990s. Moreover, in several countries—Sweden, Germany, and Japan in particular—the utilities had to have the means at hand for spent fuel reprocessing as a condition of reactor licensing and licensing renewal. In all countries, whatever the requirements of law, there was mounting political pressure to show that the waste problem was being effectively addressed; having reprocessing contracts with COGEMA and BNFL showed that something was being done.

For its part, the United States was doing nothing to relieve other countries of their growing accumulations of spent fuel, even though much of the fuel was of American manufacture and most of the uranium came from U.S. enrichment plants. Under United States law and nonproliferation policy, non-nuclear weapons states could not have fuel or uranium of American origin reprocessed without U.S. permission, but no meaningful offer was ever made to take back the spent fuel. (Permission for reprocessing, while never denied, has never been routinely granted either.) The offer made by President Carter in 1977 to provide storage for limited amounts of foreign spent fuel when to do so would support U.S. nonproliferation policy objectives soon became a dead letter (see chapter 4). In the Nuclear Nonproliferation Policy Act of 1978 Congress included an amendment, proposed by Senator James McClure of Idaho, which required elaborate justification in every instance where acceptance of foreign fuel was proposed. This was widely interpreted as discouraging implementation of the Carter policy. Furthermore, any proposal to accept foreign spent fuel in the absence of facilities for the long-term storage or permanent disposal of the United States' own fuel could hardly be politically realistic. Inasmuch as neither Japan nor any of the non-nuclear weapons states of Europe is regarded as proliferation prone, they might in any case have been unable to take advantage of the Carter policy.

Reprocessing is essential in waste disposal. This assumption has been widely held among people in the nuclear industry (some of whom hold it today), and, as noted above, enshrined in the laws and regulations of several countries. In 1980 and 1981, for instance, CO-GEMA and some high-ranking officials at Britain's Atomic Energy Authority insisted that whereas the feasibility of safe geologic disposal of reprocessing wastes was beyond doubt, the same could not be said of disposal of spent fuel.[46] That reprocessing is not in fact an aid to waste management is discussed in chapter 1, where it is pointed out that the containment imperative could be better met by keeping radionuclides locked up in fuel rods than by generating multiple streams of gaseous, aqueous, and solid wastes by chopping up the rods and dissolving the fuel. It is worth reemphasizing that the Carter administration's Interagency Review Group was not alone in concluding that spent fuel constitutes an acceptable form of waste. As noted in chapter 3, the International Nuclear Fuel Cycle Evaluation's working group on radioactive waste management reported in 1980 that wastes from any of the fuel cycles studied, including spent fuel from the once-through cycle, "could be managed and disposed of with a high degree of safety."[47] Since then there have also been generally favorable international reviews of the latest Swedish plan for direct disposal of spent fuel.[48] Furthermore, there is not in principle any reason why spent fuel cannot be emplaced in repositories in hard rock in a way that permits retrieval.

Beyond that, a U.S. Department of Energy study indicates that the cost of geologic disposal of spent fuel will not be more than about $300 per kilogram, whereas the reprocessing costs to be borne by COGEMA's foreign customers have been variously reported at $800 or $900 per kilogram.[49] To these reprocessing costs must be added the cost of disposal of the high-level and low-level reprocessing wastes—and DOE analyses indicate that this cost is not going to be less than that of disposing of spent fuel.[50]

[46]See, for example, Walter Marshall, "The Use of Plutonium," *Atom*, no. 282, April 1980; see also COGEMA, *Le Retraitement des Combustibles Irradiés en France et a L'Etranger* (Paris, Editions MSA, September 1981) pp. 8–10.

[47]International Nuclear Fuel Cycle Evaluation, *INFCE Summary Volume*, INFCE/PC/2/9 (Vienna, International Atomic Energy Agency, 1980) p. 21.

[48]Swedish Ministry of Industry, *Review of the KBS-III Plan for Handling and Final Storage of Unprocessed Spent Nuclear Fuel*, DsI 1984:17 (Stockholm, 1984).

[49]U.S. Department of Energy, "Nuclear Waste Fund Fee Adequacy: An Assessment," DOE/RW-0020 (February 1985); "COGEMA Dangles Prospect of Cheaper Reprocessing," *Nuclear Fuel*, July 16, 1984. The latter source says that COGEMA had put its reprocessing cost at $650 per kilogram; by adding to that the 25 percent profit COGEMA charges its foreign customers (according to *Nuclear Fuel*), one arrives at $812.50. All contracts are not identical, however. See also "West German Consortium Sends Offers to Take Spent Fuel to China," *Nuclear Fuel*, July 16, 1984.

[50]Office of Nuclear Waste Policy Project Office, "Report on Financing the Disposal of Commercial Spent Nuclear Fuel and Processed High-Level Radioactive Waste," DOE/5-0020/1 (July 1983) table 3-5, p. 14. According to this table, disposal of spent fuel would

Reprocessing under way and planned is justified by pluto-nium recycling. The rationale for the reprocessing now under way and planned for the 1990s in France, Germany, Belgium, and Japan has varied somewhat from country to country, but a common element has been an expectation that plutonium fuel will find an early and econom-ically defensible use. Until recently the French emphasis has been on early deployment of a series of commercial breeders. The near-term emphasis by the nuclear enterprise in Germany and several other coun-tries has been on the use of plutonium fuel in light water reactors, commonly referred to in the industry as thermal recycling. The British recognize that economic conditions are not yet right for commercial breeders, and certain leaders of the British nuclear industry have taken exception to thermal recycling in principle. Walter Marshall, formerly deputy chairman of Britain's Atomic Energy Authority and now chair of the Central Electric Generating Board, has said that use of plutonium fuel in light water reactors by countries with advanced nuclear programs could lead other countries, including non-nuclear weapons states, to undertake reprocessing and plutonium fuel fabrication themselves, with a resulting increase in the risk of weapons proliferation.[51] Nonetheless, much of the plutonium separated from foreign fuel by British Nuclear Fuels may well wind up in thermal reactors after its return to the customer, whether BNFL fabricates the plutonium fuel for that purpose or not.

The reality is that plutonium recycling in either breeders or thermal reactors will not pay off economically for many, many years. The pros-pects for an economically supportable breeder program now seem par-ticularly remote, even though as late as the end of the 1970s the French Commissiariat à l'Energie Atomique entertained the hope of a major and relatively early commitment to breeders.[52] Besides the fact that uranium has become abundant and cheap at the same time that the prospect for rapid growth of nuclear energy in the United States and elsewhere has

cost from 6 to 7 percent more than disposal of high-level waste from reprocessing, but the cost of solidifying the latter waste was not taken into account. Also to be considered is the cost of disposing of the transuranic and other wastes associated with reprocessing.

[51]Marshall, "Use of Plutonium," p. 92.

[52]In an article in *Science* in early 1979, C. Pierre L. Zaleski, nuclear attaché at the French embassy in Washington, D.C., and senior vice-president of the European Nuclear Society, foresaw a rising demand for uranium around the world and the possibility of a growing French dependency on uranium imports. The way to overcome this problem and stabilize the uranium demand, he wrote, was to introduce commercial breeders on a large scale before the year 2000. This would serve also "to get rid of the large amount of plutonium that will be produced by the end of the century." Zaleski, "Energy Choices for the Next 15 Years: A View from Europe," *Science*, March 2, 1979.

disappeared, the breeder cycle can now be seen to present peculiarities that will militate strongly against its commercial introduction in all but the most favorable circumstances. The breeder reactor itself, with its special construction requirements (for instance, the liquid sodium used to cool the core is a hazardous material that cannot be confined to a single circuit[53]), is so expensive to build that the cost of electricity from the French Superphenix reactor is now expected to be at least twice that of electricity from a light water reactor.

To the problems associated with building individual breeder reactors must be added the complexity and high cost of establishing an overall breeder system. Breeder fuel cannot be routinely and conveniently reprocessed and fabricated in the plants built for light water reactor fuel and thermal recycling. Separate facilities must be built. Five times more plutonium will be present in the breeder cycle than in the light water reactor cycle, and because the burnup of breeder fuel will be several times higher than that of light water reactor fuel, the radioactivity present will be correspondingly greater. (It is instructive that if a dependable technology for thermal recycling is now within reach, it is partly because the exigencies of the light water reactor fuel cycle are being relieved somewhat by allowing much more time for fission-product decay before fuel is reprocessed than was originally envisioned.[54]) The cost of an overall breeder system will surely not be affordable unless spread over a considerable number of reactors, but here again one runs into the breeder reactor's exceptionally great capital cost—high even by comparison with that of light water reactors.

Although thermal recycling is less demanding than the breeder in its technology and capital requirements, its own economic prospects are poor, at least for the next few decades. A major advantage of thermal recycling, compared to the breeder cycle, is that no new reactor is required. The light water reactor that operates on uranium fuel can also be made to operate on plutonium fuel. Nonetheless, predictions made almost ten years ago that the economics of thermal recycling would be

---

[53]Liquid sodium offers advantages that have made it the preferred coolant for breeder reactors, but it also presents major hazards: if exposed to air, it ignites; if it comes in contact with water, a violent chemical reaction occurs. Because the primary liquid sodium circuit that removes heat from the reactor core is radioactive, every precaution must be taken to prevent leaks to either air or water. Accordingly, the heat from the primary circuit is transferred to the nonradioactive secondary liquid sodium circuit, which in turn transfers the heat to the water-steam generator circuit that drives the turbines generating the electricity.

[54]COGEMA has found that the optimal time for reprocessing is four years after the fuel is removed from the reactor. Bertrand Barré, "The French Nuclear Experience," *EPRI Journal*, July–August 1984.

unfavorable have been amply borne out by experience.[55] Bertram Wolfe, general manager of General Electric's Nuclear Fuel and Special Projects Division, and Burton Judson, manager of advanced engineering in GE's Uranium Management Corporation, have put the cost of a bundle of plutonium fuel at from three to four times that of a comparable bundle of enriched uranium fuel. Wolfe and Judson have concluded that the "cost of plutonium recycle has clearly gotten out of hand, and attempts should be made to bring the situation back to rationality."[56] By this they mean that the great bulk of spent fuel should be kept in retrievable storage to await the breeder. Ray Sandberg of Bechtel National and Chaim Braun of the Electric Power Research Institute also have concluded that reprocessing and thermal recycling are economically marginal at best, and that direct disposal of spent fuel is preferable if economic considerations are considered dominant.[57]

Ideological Momentum—and the Risks

Despite the questionable present economics involved, France, as discussed in chapter 10, is now moving toward use of plutonium fuel in light water reactors to make up for the delay in breeder commercialization. That similar developments are afoot or contemplated in other countries, such as Belgium, Germany, and Japan, has been noted. While in principle a retrenchment in reprocessing would be a sensible alternative to the use of plutonium fuel, so much has been set in motion by past commitments that for the moment this option is not likely to be taken, certainly not in France. La Hague is one of the busiest construction sites to be found anywhere, and the foreign fuel to be reprocessed is arriving. Riding on the fulfillment of this undertaking are the professional

[55]See Ford Foundation, *Nuclear Power Issues and Choices* (Cambridge, Mass., Ballinger, 1977) p. 333.

[56]Bertram Wolfe and Burton F. Judson, "Closing the Fuel Cycle," *Nuclear News*, January 1984, pp. 84–89. According to the authors, an enriched uranium light water reactor fuel bundle represents an investment today of about $1,100 per kilogram of uranium. For an equivalent kilogram of plutonium fuel, at least 4.5 kilograms of spent fuel would have to be reprocessed. In figure 3 the authors show reprocessing cost estimates ranging between $500 and $800 per kilogram (in constant 1982 dollars); figure 4 shows the range of estimates for plutonium fuel fabrication at between about $700 and $1,000. Ibid., pp. 85, 86.

[57]Ray O. Sandberg and Chaim Braun, "Economics of Reprocessing—U.S. Context," paper presented at the American Nuclear Society meeting, Washington, D.C., April 8–11, 1984. The authors concluded that given present conditions the economies of reprocessing and plutonium recycling in light water reactors are "marginal, at best, and once through cycles [with direct disposal of spent fuel] are economically preferable," though in their view reprocessing offers noneconomic advantages that should be considered.

pride and the reputations of scores of government and nuclear industry officials, engineers, and technologists. For the French now to entertain the thought of initiating cutbacks in the commitment to reprocessing clearly would pose a painful dilemma. Retrenchment would also be painful for the British, German, and Japanese industry and government officials who have lent themselves to plans for commercial reprocessing. Moreover, the German and Japanese laws and regulations that have made reprocessing a condition of reactor licensing are still on the books. On top of everything are ideological momentum and the desire to keep up pursuit of the old dream of abundant fission energy.

Given the commitment to reprocessing so evident in France and a few other countries, the move toward thermal recycling seems inevitable, for two reasons: first, without thermal recycling there is no foreseeable use for the reprocessors' principal product, the separated plutonium; second, reactor-grade plutonium is expensive to store and while in storage it undergoes progressive degradation, with plutonium-241 (one of the higher isotopes of plutonium) decaying to americium-241. This degradation represents a loss of fissionable material, and because americium is a gamma-emitter it also creates a radiation exposure problem for fuel fabrication plant workers. Once the americium buildup occurs, the plutonium must be purified in another expensive reprocessing step, which may be necessary after the plutonium oxide has been in storage for as little as two to three years.

By the turn of the century, total reprocessing capacity in Europe and Japan will be about 3,500 tons of spent fuel a year if present plans are carried out. Annual plutonium separation at that scale would be about 20,000 to 25,000 kilograms, barring major breakdowns and losses of plant availability. The amount now expected to be needed for breeder and other advanced reactor programs appears to be at most about 7,000 kilograms a year, with the actual amount probably not exceeding 5,000 kilograms.[58] The surplus would present a formidable problem without thermal recycling. The manager of Belgonucléaire's thermal reactor fuels program, H. Bairiot, has put the cost of storing plutonium oxide powder

[58]Michel Rapin, director of the Institute for Technological Research and Industrial Development of France's Commissariat à l'Energie Atomique, speaking at an American Nuclear Society meeting in late 1984, put the total amount of plutonium needed by the year 2000 for early breeders and advanced thermal reactors in France, the United Kingdom, and Italy at 6,330 kilograms a year, to which I have added another 650 kilograms for the fast breeder program planned by the Germans at Kalkar. Rapin included 1,100 kilograms a year for a Superphenix-2 and a like amount for a Superphenix-3, follow-on breeders to Superphenix-1; neither of these has yet been authorized. See Cindy Galvin, "Growing Supply of Plutonium Envisioned for Use in LWRs Toward End of Century," *Nuclear Fuel*, December 3, 1984.

for ten years at from $570 to $750 (1983 dollars) a kilogram, and the cost of subsequent purification at up to $450 a kilogram.[59]

According to Bairiot and another industry source whom he cites, by looking to thermal recycling to provide an alternative to extended storage and to reduce the need for enriched uranium, the savings for each 1,000–megawatt reactor loaded with plutonium fuel could range from $3 million to $6 million a year, depending on the price of uranium, enrichment services, and plutonium fuel fabrication.[60] Bairiot concludes that to store plutonium for twenty to forty years until significant demand for fast breeder fuel materializes cannot be economically justified. On the other hand, use of plutonium fuel in light water reactors can be economically justified only if the separated plutonium is available in any case—that is, if it can be regarded as a "free" resource—and if it would be a substantial liability if kept in extended storage. As others have aptly put it, the present move toward use of plutonium fuel in light water reactors "appears to be a result rather than a cause of the commercial reprocessing which is now taking place and promises to take place in the future."[61]

Set against the combination of factors driving the expansion of reprocessing and preparations for thermal recycling are risks which most leaders of the nuclear enterprise have either ignored or found too speculative to be compelling. As Walter Marshall of the United Kingdom candidly has acknowledged, thermal recycling could, under certain circumstances, further encourage the spread or proliferation of nuclear weapons.[62] The existing safeguard and plutonium accountability regimes are admittedly not adequate to ensure timely warning of plutonium diversion from bulk reprocessing and plutonium fuel fabrication plants, especially inasmuch as some of the countries believed to be proliferation-prone have chosen not to adhere to the Nuclear Nonproliferation Treaty or to permit the International Atomic Energy Agency to inspect all their nuclear facilities. But at least with respect to proliferation there exists a relevant safeguards regime, however imperfect. That much cannot be said of the risk of terrorism, to which the existing safeguards regime has little relevance. Yet if terrorists stole or forcibly seized a shipment of plutonium, the possibility that they might be able to fabricate a crude fission bomb could not be precluded, especially if they

---

[59]H. Bairiot, "Laying the Foundations for Plutonium Recycle in Light Water Reactors," *Nuclear Engineering International*, January 1984, table 5, p. 28.

[60]Bairiot, "Laying the Foundation for Plutonium Recycle." Bairiot cited R. Cayron, "Economics of Plutonium Recycle in LWRs," paper presented in Tokyo, March 1983.

[61]David Albright and Harold Feiveson, "The Deferral of Reprocessing," *F.A.S. Public Interest Report, Journal of the Federation of American Scientists*, February 1985.

[62]Marshall, "Use of Plutonium," p. 92.

were supported by a Qaddafi or a Khomeini. Government leaders would be confronted with an appalling dilemma: either to give in to the terrorists' demands or to run the risk of having the terrorists detonate a bomb, perhaps in the heart of a major city.

Nothing of that magnitude has occurred during the first three decades of commercial nuclear power, but there have been numerous disturbing incidents of one kind or another at nuclear facilities around the world. Some have been alarming, as when fifteen armed men overran Argentina's Atucha-1 reactor in 1973, possibly only to gain attention for their group. According to a U.S. General Accounting Office survey, more than a score of incidents occurred at foreign nuclear facilities from 1966 to 1979.[63] Two involved thefts of uranium fuel, more than a dozen involved bombings at reactors or other facilities (with extensive damage in several cases), and four involved armed attacks or intrusions. In one of the latter, the intruder, intent on demonstrating security weaknesses, entered the German Biblis reactor with a Panzer-Faust bazooka and presented it to the plant director. On January 19, 1982, terrorists fired five Soviet-made antitank rockets at the French Superphenix breeder reactor; damage was light, but four direct hits were made on the reactor's containment shell. All of these incidents occurred when and where there was no widespread use of plutonium fuel. Once such use begins and the circulation of this nuclear explosive material becomes increasingly routine, the risk of an incident of truly grave magnitude can only be greater— how much greater depending on capacities for human mischief and inventiveness that lend themselves poorly to probability analysis.

David Albright of the Federation of American Scientists and Harold Feiveson of the Center for Energy and Environmental Studies at Princeton University have estimated the number of shipments of separated plutonium and plutonium fuel that can be expected in the late 1990s. Within Europe alone there may be about 270 shipments of plutonium each year from reprocessing centers to fuel fabrication plants. This plutonium would be shipped as an oxide, most of it going by truck, about 100 kilograms per shipment. About 50 shipments a year would be bound for Japan, with the first leg of the journey being made by truck to a harbor or airport, from which the plutonium would go by ship or plane.[64]

When the plutonium is fabricated into fuel, as a mixed oxide of plutonium and uranium, it still must be regarded as a potential target

[63]U.S. General Accounting Office, "Obstacles to U.S. Ability to Control and Track Weapons-Grade Uranium Supplied Abroad," GAO/ID-82-21 (August 21, 1982).

[64]David Albright and Harold Feiveson, "Plutonium Recycle and the Problem of Nuclear Proliferation," Center for Energy and Environmental Studies, PU/CEES report no. 206 (Princeton, Princeton University, April 1986).

of terrorists. The plutonium would be recoverable by a chemical separation process which is easy by comparison with the reprocessing of irradiated fuel. According to Albright and Feiveson, if fifty light water reactors in Europe and Japan were operating on plutonium fuel by the late 1990s as many as 400 shipments of mixed oxide fuel (assuming two fuel assemblies per shipment) would be necessary.

All shipments of separated plutonium and plutonium fuel must be attended by elaborate security. This is not to say that terrorists who stole or forcibly seized a few kilograms of plutonium oxide powder might fashion a powerful "coffee-can" bomb by wrapping high explosive around it. To the contrary, an international task force on nuclear terrorism has found that experts now agree that fabricating even a crude bomb is considerably more difficult than some of them once supposed. But the task force also concluded that the fabrication of such a device is "within reach of terrorists having sufficient resources to recruit a team of three or four technically qualified specialists."[65]

The task force concluded as well that while the probability of nuclear terrorism remains low, it is increasing because of a "confluence of factors," among them the "growing incidence, sophistication and lethality of conventional forms of terrorism, often to increase shock value"; "apparent evidence of state support, even sponsorship of terrorist groups"; the "increasing number of potential targets in civil nuclear programs," including facilities and shipments in which plutonium is present in forms suitable for use in weapons; and "potential black and gray markets in nuclear equipment and materials." The task force said that it saw no signs yet of any terrorists having "the essential combination of capability and will to engage in an act of nuclear violence," but it then observed: "A plausible threat or hoax involving a nuclear device or sabotage could have enormous coercive and disruptive results without mass killing or destruction; indeed, we believe this may be the most likely form of nuclear terrorism at this time." The task force warned specifically that if present reprocessing plans are carried out abroad, by the mid to late 1990s "tons of plutonium will be in commercial transit, posing increased opportunities for theft and diversion by terrorists." The storage and disposal of spent fuel was recommended "until such time as [the need for plutonium fuels] is clearly established, the threat of

[65]"Report of the International Task Force on Prevention of Nuclear Terrorism" (Washington, D.C., Nuclear Control Institute, June 25, 1986). The task force's twenty-five members included Stansfield Turner, former director of the Central Intelligence Agency; Harold Agnew, a former director of the Los Alamos National Laboratory; Bernard O'Keefe, chairman of the executive committee of EG&G, Inc., the DOE contractor in charge of weapons testing at the Nevada Test Site; and Bertram Wolfe of General Electric's Nuclear Fuel Division.

terrorism has lessened, and the adequacy of safeguards and physical-protection systems has improved."[66]

One certain consequence of thermal recycling is that new controversies will be generated, possibly doing serious harm to the nuclear industry by intensifying public fears of the link between nuclear power and nuclear explosives. By refusing to promise categorically that plutonium from the Superphenix will not be used in their nuclear weapons program, French officials already have shown remarkable indifference to public concern about this linkage.[67] Controversies over the use of plutonium fuel could be especially damaging to nuclear power in Germany, where political support for the atom seems to have become increasingly precarious since Chernobyl. All in all, the risks posed by thermal recycling for the public and the nuclear industry appear too great to justify the distinctly marginal—if not illusory—benefit such recycling affords. Walter Marshall concluded as much in weighing the proliferation risk alone.[68]

The Growing Accumulation of Spent Fuel

Important as the commitment by the Europeans and Japanese to reprocessing has been, by far the greater part of the spent fuel they will generate by the year 2000 will not have been processed by that time. Recent official estimates of spent oxide fuel accumulation through the end of the century put the total for Western Europe and Japan at about 76,000 metric tons.[69] Much of it would be generated in the four countries—France, Germany, the United Kingdom, and Japan—which expect to have reprocessing plants working. Only a few countries have deliberately chosen to have none, or very little, of their spent fuel reprocessed; Sweden is a notable case in point. Nonetheless, the reality is that all of the nuclear power countries will see large accumulations of unreprocessed fuel, with the total almost certainly exceeding 50,000 tons, or more than two-thirds of the total generated. An even larger backlog

[66]Ibid.

[67]David Albright, "French Military Plans for Superphenix," *Bulletin of the Atomic Scientists*, November 1984.

[68]Marshall, "Use of Plutonium." Marshall has favored reserving the use of plutonium fuel for the breeder cycle, with its promise of a thirtyfold extension of the uranium resource and its possibilities (as he sees it) of being confined to a few technologically advanced countries, which would start up new breeders with the plutonium recovered from spent light water reactor fuel purchased from other countries.

[69]Kent Harmon, L. T. Lakey, I. W. Leigh, and A. G. Jeffs, *International Nuclear Fuel Cycle Fact Book*, PNL document 3594 (5th rev., Richland, Wash., Battelle Pacific Northwest Laboratories, January 1985).

will develop unless all of the 3,500 tons of annual reprocessing capacity now planned is in service by the early 1990s, and unless operating efficiency is at much higher levels than that yet achieved by commercial reprocessing plants.

Furthermore, with an annual generation of about 6,500 tons beyond the year 2000 (assuming that total installed nuclear capacity remains steady at about 192 gigawatts), the backlog will grow steadily larger unless reprocessing capacity is greatly expanded. But even if reprocessing capacity were doubled to 7,000 tons a year by the turn of the century, and even if reprocessing operated at 75 percent efficiency over the following two decades, the backlog by the year 2020 would be on the order of 75,000 tons.[70] The actual accumulation, barring an abrupt reversal in the fortunes of nuclear power, is likely to be larger than that. The reprocessors seem destined to be players in a game of catch-up in which they remain hopelessly behind. Moreover, the production facilities that are supposed to be completed at La Hague, Windscale, and elsewhere over the next decade may turn out to be one-time commitments to undertakings that are not economically sustainable.

For the nuclear industry the growing backlog of spent fuel could well become both an economic nuisance and a political embarrassment. By the turn of the century the nuclear power industry, in the interest of gaining greater public trust, may have to begin the deep geologic disposal of some spent fuel and high-level waste to demonstrate that custodial care will not be forever necessary. At hard-rock sites the fuel could be emplaced retrievably for a period of 50 to 100 years, if not longer, thus leaving the reprocessing and plutonium recovery option open as a hedge against the possibility that the breeder will some day be both needed and practicable. At the same time, the fuel could be removed should the repository site in the end be found unacceptable. But this brings us back to our starting point, for no geologic disposal site has yet been found and certified as clearly acceptable. And for a number of nations the best solution, if not the only solution, will lie in multinational or international repositories, of which today there are none in sight.

Conclusion

To sum up, an international system of spent fuel and waste management is still very much needed, but despite a few encouraging signs is still

[70]Ibid.; Albright and Feiveson, "Plutonium Recycle and the Problem of Nuclear Proliferation," table 1.

beyond the horizon. An important but unanswered question for several countries is what form most of the waste should ultimately take. Reprocessing capacity in the foreseeable future will not suffice to prevent the buildup of large backlogs of spent light water reactor fuel. The Swedes alone have definitely opted for direct disposal of their fuel. The Germans have not entirely ruled out direct disposal, but remain officially committed to reprocessing. The French are the most firmly committed to reprocessing, but their backlog of unreprocessed fuel will be the largest of any country in Europe. If the economics of reprocessing and the recycling of plutonium in light water reactors remain as unfavorable as they now appear to be, the collapse of reprocessing as a commercial enterprise is likely by the turn of the century. Furthermore, the premature move to the use of plutonium fuel will almost certainly do more to increase the political controversy surrounding reprocessing than to improve reprocessing economics.

The reprocessing services offered to other nations by companies such as COGEMA and BNFL are only a stopgap solution that is on shaky economic footing. Yet the European Community, the OECD's Nuclear Energy Agency, and the International Atomic Energy Agency have felt severely constrained by the extreme political sensitivity of proposals for international or multinational management of spent fuel or radioactive waste. The few bilateral or multilateral initiatives seen to date, including Austria's approach to Egypt and the Chinese and Soviet initiatives, have either suffered from political or commercial opportunism or are of limited relevance to democratic societies. Nevertheless, if one or more of the democratic countries that are leaders in nuclear technology were to demonstrate safe geologic disposal within their own boundaries, multinational initiatives would be encouraged.

Beyond that, the retrievable disposal of spent fuel in hard-rock formations such as granite would help prevent the buildup of backlogs of fuel in surface storage, while leaving a hedge against the possible need for plutonium and breeders some decades hence. There appears to be little interest yet in such a compromise solution.

If direct, retrievable disposal should show early promise of success in the United States, the example might have a major influence in Europe, particularly among cost-conscious utilities. But as the concluding chapter makes clear, the U.S. program is caught in a procedural, legal, and political morass from which some way of escape must be found.

Part 4
A Time to Act

13
Common Ground

Two Paths to Political Failure

By January 1987 four years had passed since President Reagan hailed passage of the Nuclear Waste Policy Act (NWPA) as the long-awaited answer to the radioactive waste problem. But the act is proving not to be the answer. As now pursued, the repository siting effort in the United States appears to be headed into a political cul-de-sac and will not come close to its goal of establishing the first geologic repository by 1998. The very concept of geologic disposal might even be discredited and lost as a practical political possibility.

On May 28, 1986, the Department of Energy, after close consultation with the White House and with the approval of the president as required by NWPA, announced that the list of candidate sites for the first repository had been narrowed to the three which had been tentatively nominated earlier in the states of Texas, Nevada, and Washington. These sites would each undergo a characterization process that would include exploration from deep shafts; total costs at each site could run as high as $1 billion. The disclosure represented a major step toward the selection in 1991 of a site in the West for the first repository. At the same time, the department also announced that the site screening for a second repository which had been going on in the upper Midwest and in the East was to be suspended indefinitely.[1] Taken together, these announcements were the latest developments in what has been called in department parlance the "First Round" and the "Second Round" of repository siting efforts.

Key decisions in the Second Round had been expected to follow those of the First Round by five years or more. But now the Second Round,

---

[1]Secretary of Energy John S. Herrington, prepared statement presented at a press conference, May 28, 1986, Washington, D.C.

subject to unexpectedly intense political pressures, had been essentially abandoned, although DOE was maintaining a pretense of keeping it alive by some generic studies of crystalline rocks and by participation in research in Canada and other countries.[2] As for the First Round, despite representations by department officials of how confident they now were of progress, it was in deep legal and political trouble. For one thing, shutting down the Second Round in the Midwest and East had upset the balance Congress had sought to write into the waste act to satisfy the West.

Remarkably, the two rounds had each moved toward political failure by paths which were very different, except that in both cases the Department of Energy was insensitive to land-use and environmental conflicts that greatly alarm the public, intensify host-state resistance, and make geologic uncertainties all the more important.

The First Round

How the inventory of First Round sites came about has been described in chapters 4 and 5 in the accounts of the evolution of nuclear waste policy and the search for sites from the mid 1970s to the early 1980s. The reader will recall that interest in salt as a disposal medium remained high. Salt formations being where you find them, the emphasis was entirely on technical considerations in identifying the candidate salt sites, of which there were seven when the waste act was passed in 1982. Yet all of these sites presented major land-use or environmental conflicts, real or perceived: for instance, the Richton, Mississippi salt dome, deemed by DOE to be the best of the domes investigated, was next to a town; the best of the Utah bedded salt sites was next to a national park; and the Texas bedded salt formation of interest was beneath the Ogallala aquifer, on which the area's rich farming economy depends for irrigation water. In identifying these sites, political feasibility had essentially been ignored.

But politics was definitely a major consideration in arriving at candidate sites on the Hanford reservation, in basalt, and on the Nevada Test Site, in tuff. Looking for sites on these large, remote, desertlike reservations where the neighboring populations were accustomed to nuclear activities had seemed politically convenient indeed. But whether the sites found there would be technically suitable and free of conflicts was somewhat a matter of chance. The Hanford site was to present major technical difficulties and large conflicts, especially those arising from this

---

[2]Ibid.; interview by the author with Ben C. Rusche, June 12, 1986.

site's complex geohydrologic regime and its nearness to the Columbia River. The Nevada site, by contrast, was to show probably the greatest technical promise of any site. While this site was not free of conflicts, these were quite different from those present at the other sites, and probably more easily resolvable.

So the inventory of sites available when Congress passed the waste act consisted of seven sites in salt, of which the three mentioned above in Mississippi, Utah, and Texas were to be preferred; the tuff site in Nevada; and the basalt site in Washington—five principal candidate sites altogether. The tight siting schedule mandated by the act gave the Department of Energy no choice but to confine its siting activities, for better or for worse, to this limited universe of sites, at least for the First Round.

Although the waste act called for "consultation and cooperation" between the Department of Energy and the potential host states, the most fundamental issue had to do with making up the short list of candidate sites, and that question had been essentially decided before the act was passed. The states were consulted in the preparation of the siting guidelines, but as finally issued with Nuclear Regulatory Commission approval in December 1984, the guidelines eliminated none of the existing sites. They could serve only to help the department pick and choose among sites which the host states regarded as unsuitable. The states brought suit to have the guidelines invalidated.[3]

In the spring of 1984, Ben C. Rusche, a former director of regulation at the Nuclear Regulatory Commission, was confirmed as the first presidentially appointed director of the Office of Civilian Radioactive Waste Management. He had been chosen with the strong support of utility industry officials, who regarded him as a forceful, decisive administrator. Rusche took the view that it is in the nature of things for the relationship between the Department of Energy and the states to be at best uneasy. He was determined not to allow state objections to deflect him from the goals mandated by the waste act.[4]

In late 1984 the Department of Energy issued in draft the "environmental assessments" evaluating and ranking all the sites in the First Round inventory. It was then that the Hanford basalt site in Washington,

[3]Department of Energy, "Nuclear Waste Policy Act of 1982; General Guidelines for the Recommendation of Sites for the Nuclear Waste Repositories; Final Siting Guidelines," 10 CFR, pt. 960, *Federal Register* vol. 29 (December 6, 1984) pp. 47714–770; Mary Louise Wagner, "States and Environmentalists Sue DOE over Repository Selection Process," *Nuclear Fuel*, December 31, 1984. At this writing in late 1986 the guidelines litigation has still to be decided.

[4]John Graham, "Ben Rusche Talks About the Waste Programs," *Nuclear News*, November 1984, pp. 180–181.

the tuff site in Nevada at Yucca Mountain, and the salt site in Texas in Deaf Smith County were first tentatively designated as the three preferred for characterization. Nominated as the fourth and fifth choices, but not preferred for characterization, were the salt sites in Mississippi and Utah.[5]

The response from the host states was predictable. There was an outpouring of critical comments from host-state officials and citizens in letters to DOE and in often angry statements at public hearings. One particularly telling complaint came from Governor Booth Gardner of Washington, who testified before a congressional subcommittee that state analysts had found the methodology used to rank the sites to be seriously flawed.[6] He recommended that independent experts be empaneled to reevaluate thoroughly the existing sites and the past assessments. Here Governor Gardner was raising a critically important issue. The confidence of state officials and others with a serious interest in the siting effort was not going to be gained unless all the data, assumptions, and analytical methods bearing on the department's decisions were made available to independent experts for their evaluation and comment in a timely way, well before the final decisions were made.

The upshot of Governor Gardner's complaint was that Ben Rusche asked the National Academy of Sciences' Board on Radioactive Waste Management to review the site-ranking methodology.[7] But the Department of Energy never accepted the board's recommendation that its assumptions and conclusions be reviewed by an independent panel of experts. The board had cautioned the department that reliance on in-house experts alone could introduce bias and "mask the degree of uncertainty" involved.[8] This was important advice. By failing to take it, DOE was increasing the likelihood of its site-screening decisions turning out to be at least politically indefensible.

The DOE announcement in May 1986 that the Hanford, Nevada, and Texas sites had been selected as the three to be characterized occasioned a mild surprise among outside observers. There had been speculation that the Hanford site might be eliminated in favor of either the Mississippi or the Utah salt site. But the department had chosen to stick with

[5]Office of Civilian Radioactive Waste Management, Department of Energy, *Draft Environmental Assessment*, 9 vols. (Washington, D.C., December 1984). The volumes for the five sites are DOE/RW-0010 (Davis Canyon, Utah), DOE/RW-0012 (Yucca Mountain, Nev.), DOE/RW-0013 (Richton, Miss.), DOE/RW-0014 (Deaf Smith, Tex.), and DOE/RW-0017 (Hanford, Wash.).

[6]Testimony of Governor Gardner before the House Energy and Commerce Committee's Subcommittee on Energy and Power, August 1, 1985.

[7]Letter of Ben C. Rusche, August 29, 1985, to Frank L. Parker, Chairman, Board on Radioactive Waste Management. The methodology used is known in the professional literature as "multiattribute utility analysis."

[8]Letter of Frank L. Parker to Ben C. Rusche, October 10, 1985.

the Hanford site even though it had placed last among the five sites according to the site-ranking methodology.[9] The department's strained attempts to justify this choice despite the exceptional technical uncertainties and high costs associated with the Hanford site were an extreme measure of the political—and quite possibly the legal—vulnerabilities which the First Round siting effort was taking on. The vulnerabilities were being made worse by the department's decisions, but to one degree or another they were inherent in that effort.

All five of the principal candidate sites had been ranked with respect to the advantages and disadvantages expected of them during both the "preclosure" phase, when the repository would be built and operated, and the "postclosure" phase, which would follow the closing and sealing up of the repository. In both preclosure and postclosure ranking, Hanford had come in last, but the department was ready to explain why this could be deemed unimportant. Preclosure, said DOE, the Hanford site actually had ranked tops among the five sites if cost of repository construction and operation and of spent fuel transportation were not taken into account; DOE also observed that the siting guidelines "place cost among the least important category of considerations." Postclosure, the expected performance of the Hanford site in containing radioactivity over 10,000 years was, by DOE estimates, five hundred times better than the Environmental Protection Agency standard.[10] In favoring the Hanford site over the Mississippi and Utah sites, the department was giving substantial weight to the advantages of its location on a remote federal reservation and to the greater geologic diversity that the basalt site would provide.[11]

But these arguments for including Hanford were scarcely convincing. The differences in repository construction and operating costs at the various sites were very large. Between the Hanford and the Mississippi sites they ran to nearly $4.0 billion (1985 dollars), and between the Hanford and Nevada sites to $5.4 billion. The total estimated cost at Hanford was $12 billion, with the cost uncertainty for this site as well as for the others put at 35 percent either way, high or low.[12] At Hanford the estimate might well prove low; indeed, to build a repository at this site might prove to be impossible at any price.

[9]Office of Civilian Radioactive Waste Management, Department of Energy, *A Multiattribute Utility Analysis of Sites Nominated for Characterization for the First Radioactive-Waste Repository—A Decision-Aiding Methodology*, DOE/RW-0074 (Washington, D.C., May 1986) (hereafter cited as DOE, *Multiattribute Utility Analysis Report*).

[10]Department of Energy, "Recommendation by the Secretary of Candidate Sites [sic] for Site Characterization for the First Radioactive Waste Repository," DOE/5-0048, May 1986.

[11]Ibid. The Mississippi salt dome site would have also offered a degree of diversity, since a dome salt formation has characteristics somewhat different from bedded salt.

[12]DOE, *Multiattribute Utility Analysis Report*, p. F-54.

Besides the fact that the drilling of the seven 15-foot-diameter shafts that are necessary at Hanford is beyond the demonstrated state of the art, miners would face difficult and potentially hazardous conditions in excavating the 97 miles of waste emplacement tunnels. David L. Siefkin, hydrologist and a program leader at Roy F. Weston, Incorporated, a Department of Energy technical support contractor for repository siting, acknowledges that it is by no means certain a repository can be built at Hanford. "I can tell you, it's a close call," Siefkin says, adding that numerous discussions took place between DOE and the Hanford contractor before it was decided that construction of the repository would be within "reasonably available technology."[13] That a Hanford repository might be impossible to construct was not, however, taken into account in the site rankings except as indirectly reflected in construction cost.[14] Inasmuch as the department chose to treat cost as of relatively little importance, the "constructability" factor, though critical to the usefulness of the site, was largely ignored in the selection of the three sites to be characterized.

Department of Energy assertions that a Hanford repository would far exceed the Environmental Protection Agency performance standard are regarded by the U.S. Geological Survey as without scientific foundation. "Available data are insufficient to conclude much of anything with regard to groundwater travel time or direction," the Geological Survey has said. According to Siefkin, the DOE assertions about site performance are based on sensitivity studies that use a wide range of assumptions with respect to velocity and volume of groundwater flow to make up for the lack of data. "We have found [in] using these standard sensitivity analyses ... that we have plenty of grace," says Siefkin. But William Meyer, a Geological Survey hydrologist in Tacoma, asks, "If you go through a sensitivity analysis with bad data, what have you got?"[15] The National Academy of Sciences' Radioactive Waste Board as a whole has remained neutral on the site selections, its chairman has observed, although some board members have been highly critical of the choice of the Hanford site.[16]

[13]Siefkin was interviewed by the author on July 1, 1986.

[14]Interview by the author, June 26, 1986, with Thomas Longo, a DOE staffer in charge of applying the ranking methodology.

[15]U.S. Geological Survey, "Review Comments on Environmental Assessments," March 6, 1985, p. 28; Siefkin interview, July 1, 1986; interview by the author with William Meyer, July 1, 1986.

[16]Spencer Heinz, "Federal Site-selection Process for Nuclear Dump Defended," *Portland Oregonian*, September 24, 1986, reporting observation of Frank L. Parker. The critical members were Ross Heath, dean of the College of Ocean Sciences and Fisheries at the University of Washington, and Kai Lee, a member of the Northwest Power Planning Council in Seattle. See Louise Schumacher, "Scientists Call Hanford Choice $1 Billion Farce," *Seattle Times*, June 6, 1986.

The alternatives to Hanford were distinctly uninviting politically and may go far to explain why the Department of Energy chose to stick doggedly with the basalt site. To have dropped Hanford in favor of the Utah salt site next to Canyonlands National Park would have meant a battle not merely with the State of Utah but also with the National Park Service and the national environmental community.[17] To have dropped Hanford for the Mississippi salt dome site would have presented two kinds of political difficulties. First, it would have undercut in some measure the DOE proposal to establish a monitored retrievable storage facility at Oak Ridge, Tennessee, which was now being justified primarily as a spent fuel consolidation and packaging facility and a staging area for shipments to a repository.[18] An Oak Ridge site for an MRS facility would make the most sense in conjunction with a repository in the West. How large this loomed in DOE thinking is not clear, but it definitely was considered.[19] Second, Rusche and his associates at the Department of Energy could not have been unaware that Mississippi is a Deep South state supersensitive to federal intervention. In resisting the exploration of a site for a nuclear waste repository next to a small town, Mississippi officials would have held strong political cards. In sum, by sticking with Hanford the Department of Energy chose badly, but there was no way for it to have chosen well.[20]

Immediately after the selection of sites for characterization was announced, the states of Washington, Nevada, and Texas filed lawsuits challenging the administration's decisions and how they were reached.

[17]Both the National Park Service and its parent agency, the Department of the Interior, are on record as being strongly opposed to the siting of a repository adjacent to the Canyonlands park. See letter of March 28, 1985, from Bruce Blau of the Department of the Interior's Environmental Project Review to the Department of Energy, and attachments on the Davis Canyon site environmental assessment.

[18]Office of Civilian Radioactive Waste Management, Department of Energy, *Monitored Retrievable Storage Submission to Congress*, DOE/RW-0035 (Washington, D.C., December 1985).

[19]Confidential interview with a participant in the DOE site-ranking process, July 1986.

[20]More light was shed on how the Department of Energy went about comparing and ranking the sites when two House subcommittees (the Energy and Commerce Subcommittee on Energy Conservation and Power, and the Interior and Insular Affairs Subcommittee on General Oversight, Northwest Power, Forest Management) obtained several documents from DOE files, including drafts of the DOE methodology report (the "decision-aiding" document intended to inform the department's choices) and the energy secretary's site recommendation report to the president. After reviewing the documents obtained, the subcommittee leaders wrote Secretary of Energy Herrington accusing DOE of manipulating data and analytic techniques to arrive at a "predetermined set of sites," and of having "tailored the methodology report to justify the final decision," as in the deletion of a passage stating that there are no "realistic assumptions [about site performance] that can result in Hanford being anything but the last ranked site." Letter of October 20, 1986, from Congressmen Jim Weaver, Edward Markey, Al Swift, and Ron Wyden to Secretary of

Suits were also brought challenging the secretary of energy's formal declaration of a "preliminary determination" that the three sites are suitable for repository development. The states argued that this determination, which is required by the waste act, should follow, not precede, characterization. The litigation on First Round issues was accompanied by suits disputing DOE's authority to suspend the search for Second Round sites. At a minimum, the legal attacks could be expected to delay the siting effort, and they might well derail it. In Texas, the Department of Energy also faced the problem of obtaining the shaft-excavation permit required by state law to protect aquifers. But in the fall of 1986, when the Ninety-ninth Congress inflicted sharp funding cuts on the siting effort, it appeared that the clearest and most imminent danger to the siting program was likely to be a severe loss—even a collapse—of congressional support.

The Second Round

The Second Round's path to political failure was different but just as certain as that of the First Round. Because site nominations for the second repository were not due under the waste act until mid 1989, the Department of Energy had time to undertake a new search and to give potential host states a voice in the site screening from the start.

The department had been developing plans for a siting effort in granite and similar rocks, and this crystalline rock program became the vehicle for the Second Round search. Crystalline rocks occur in all major regions of the United States, but the search was to be limited to a seventeen-state area that included the Precambrian shield region of northern Minnesota, Wisconsin, and Michigan in the Midwest, plus most of the states along the Atlantic seaboard from Maine to Georgia. The search began with a survey of geologic literature, and from this reports listing 235 rock formations or "rock bodies" were prepared and issued after review by the affected states. These rock bodies typically are large and some

---

Energy John S. Herrington, with subcommittees' staff memorandum of the same date. According to the subcommittees' analyses, the tailoring was done when drafts of the two DOE reports, being prepared simultaneously, were coming to sharply divergent conclusions—the methodology report recommending the Richton salt dome over Hanford, the secretary's report recommending Hanford over Richton. Thomas Isaacs, director of repository coordination, recorded in a memorandum that a certain draft of the secretary's report "blatantly manipulates the methodology." Isaacs also found that the draft "arbitrarily throws out cost" as a consideration, arbitrarily limits the weight given postclosure performance, and wrongly suggests that it takes three rock types to meet geologic diversity requirements. Yet in these respects the secretary's recommendation report as finally issued was similar to the draft that Isaacs found so manipulative. Ibid., staff memorandum, containing Isaacs memorandum.

are very large, as in the case of the Wolf River batholith in Wisconsin, which extends beneath an area bigger than the State of Rhode Island. An elaborate regional site-screening methodology was then designed, with participation by state representatives. The aim was to screen out all but 20 or fewer of the rock bodies as candidates for field investigations, still using only information already available.[21]

The intent behind the screening methodology was to apply the Department of Energy siting guidelines in an explicit, systematic, quantitative, and objective manner that was above suspicion of bias.[22] Each rock body was mapped on 1-square-mile grid cells. Each grid cell was then reviewed against a short list of disqualifiers, such as those that would eliminate cells where deep mines or quarries are present or where there would be an encroachment upon highly populated areas or upon protected federal and state lands such as national parks.

Next, sixteen regional variables were applied, having to do with such "potentially adverse" and "potentially favorable" conditions as whether or not the rock bodies were near points of groundwater discharge or exploitable mineral resources. The geologic data available for the seventeen-state region were limited, however, so at this stage it was necessary to give land-use and environmental concerns an important place in the screening, even though these and other data relevant to public concerns about a repository were also limited. The variables had to be limited to those for which a consistent data base was available, and for which no special analytical work would be required. For example, the demand by some states to have convenience of transportation included as a variable was rejected as too complex and demanding.[23]

The Department of Energy, recognizing that some variables were more important than others and that there was disagreement as to their relative importance, held workshops for federal and state representatives to develop weighted sets of variables that would reflect public concerns. In the end, the 235 rock bodies were ranked against each of the nine sets of weighted variables, four of which had been prepared by state

[21]See Office of Crystalline Repository Development (OCRD), *A National Survey of Crystalline Rocks and Recommendations of Regions to Be Explored for High-Level Radioactive Waste Repositories*, OCRD-1 (Columbus, Ohio, April 1983); and the following reports prepared for the Crystalline Repository Project Office by the OCRD: Department of Energy (DOE), *North Central Regional Geologic Characterization Report*, DOE/CH-8 (final) vols. 1 and 2 (Argonne, Ill., 1985); DOE, *Northeastern Regional Geologic Characterization Report*, DOE/CH-7 (final) vols. 1 through 3 (1985); DOE, *Southeastern Regional Geologic Characterization Report*, DOE/CH-6 (final) vols. 1 through 3 (1985); DOE, *Region-to-Area Screening Methodology for the Crystalline Repository Project*, DOE/CH-1 (Argonne, Ill., April 1985).

[22]Interview by the author with William J. Madia of the OCRD, October 30, 1985.

[23]DOE, *Region-to-Area Screening Methodology*. For a discussion of the limitations of the regional screening methodology, see especially appendix A.

representatives. As it turned out, 9 of the 12 rock bodies tentatively chosen as potentially acceptable sites were among those ranking highest against all of the sets of variables.[24]

Nonetheless, the department's announcement in January 1986 of its tentative screening choices provoked a thunderous protest from the seven states where the 12 preferred rock bodies were located: Minnesota and Wisconsin in the Midwest; Maine and New Hampshire in New England; and Virginia, North Carolina, and Georgia in the Southeast. Nowhere was the protest louder than in Maine. The choice of the Sebago Lake batholith, only six miles north of Portland and not far west of the Lewiston–Auburn metropolitan area, was an extreme provocation. Not only was the site near the two largest communities in Maine, but above the rock body were numerous sizable lakes and immediately adjoining it was Sebago Lake itself, the second-largest lake in Maine and the source of Portland's drinking water.[25] When department officials went to Maine to brief the public on how the screening had been done, some 3,000 worried, upset, and angry people showed up, including the governor, and the meeting lasted until 3:30 in the morning.[26]

How could the Sebago Lake batholith have been chosen? The answer seems to lie in the nature of the weighting process: federal and state technocrats had pondered the relative importance of different screening variables and come up with weighted sets which, as it turned out, allowed this site's positive features to offset the negative ones that the public was going to find emotionally and politically salient. For the department, the Sebago Lake batholith presented a number of advantages, including its large size and the presumed absence of earthquake or other tectonic phenomena that might impair waste isolation. But what mattered to the people in Portland was their fear that the drinking water might be poisoned with radioactivity. Across the border in New Hampshire, the overriding concern of the people of Hillsboro, a rural town of about 3,000 residents sitting atop another potential site, was that their whole community and way of life were threatened. A repository project, if it came, would result in the government's purchase of thousands of acres of their land.[27]

[24]Department of Energy, "Area Recommendation Report for the Crystalline Repository Project," DOE/CH-15(1) (Argonne, Ill., Crystalline Repository Project Office, 1986), draft. Here "highest ranking" refers to the top 20 among the total of 235 rock bodies, as ranked against a particular set of weighted screening variables.

[25]The batholith itself underlies nearly all of Sebago Lake, but that part of the batholith designated as a potential site underlies only some of the lake's northern fringes.

[26]Phil Gutekunst, "Explosive Situation Handled Calmly," *Portland Press Herald*, February 12, 1986.

[27]Joyce Maynard, "The Story of a Town," *New York Times Magazine*, May 11, 1986.

Department of Energy and White House officials were hearing pleas to stop the Second Round from citizens, governors, and members of Congress from all three affected regions—the Midwest, New England, and the Southeast. Their protests came at a time when the Reagan administration was already worried that the Republicans might lose control of the U.S. Senate in the 1986 election. Four Republican-held Senate seats were at stake in the Second Round states of Wisconsin, New Hampshire, Georgia, and North Carolina. In all of these states the Republican incumbents or candidates could be hurt by the nuclear waste issue because it was their party that was running things in Washington. North Carolina congressman James T. Broyhill, running for the Senate, had been particularly embarrassed by the selection of two candidate sites in his state because he had been one of the principal sponsors of the waste act. He was now insisting that lowered projections for spent fuel generation meant that the siting effort for the second repository could be terminated.[28]

New Hampshire's Republican governor John Sununu was attempting to mobilize the utility industry against the Second Round siting. For Sununu, facing reelection in November, there was little doubt that the public controversies over the proposed Hillsboro nuclear waste site and the Seabrook reactor would make nuclear power an issue in the gubernatorial race. In mid May, Governor Sununu addressed a joint meeting of the Atomic Industrial Forum and the American Nuclear Energy Council (ANEC), the latter group being the industry's lobbying arm. According to Thomas J. Price, an ANEC vice-president, Sununu warned that the Second Round siting effort was mobilizing massive opposition that was dangerous to the entire nuclear industry.[29] He urged that the industry work hard to kill it. In the end, ANEC had no influence on the administration's decision on Second Round siting because its support of the suspension came after the fact. But the general drift of industry thinking seems to have been known to members of Congress from Second Round states—as well as to Ben Rusche and Secretary Herrington—at least a week or two before the decision was reached.[30]

---

[28]Congressman James T. Broyhill, "Washington Report," April 21, 1986, press release.

[29]Price was interviewed by the author on July 15 and August 5, 1986.

[30]By the last week of May there had been reports from Capitol Hill of a "dialog between members of Congress, the Executive Branch, and the nuclear industry," and Governor Sununu was predicting that the Second Round siting was a "dead duck." "Sununu: Nuke Dump 'Dead Duck,' " *Manchester Union Leader*, May 22, 1986. Senator Gordon Humphrey of New Hampshire, also a Republican, was saying that the Second Round search was a "public relations nightmare" for the nuclear industry and was predicting that once the administration's choice of First Round sites was announced, "the industry will be satisfied and will feel comfortable backing off from a second site." Ibid.

When Secretary Herrington announced the Second Round suspension he justified it principally on the grounds that Congressman Broyhill and others had argued, namely, that the decline in spent fuel generation meant the second repository could be deferred. But Herrington said the decision was also influenced by progress made toward selecting a first repository site—a reference to President Reagan's approval of the three First Round sites for characterization—and by the hope that Congress would authorize the monitored retrievable storage facility at Oak Ridge. According to Herrington, the need for a second repository would not have to be considered again until the mid 1990s, or even later. "To go ahead and spend hundreds of millions of dollars on site identification now would be both premature and unsound fiscal management," the secretary said.[31]

Herrington dismissed suggestions that the Second Round suspension was in response to political pressures: "Politics is not in this decision," he remarked. Ben Rusche has since said that the White House never told DOE that the Second Round siting must be stopped.[32] But internal Department of Energy policy option papers that have been made available in response to congressional demands were quite explicit in recognizing the "immediate political relief" that would come from terminating the Second Round search.[33]

The decline in spent fuel projections, while significant, had not been so great as to make a second repository clearly unnecessary. Indeed, on April 23, 1986, little more than a month before the suspension of the Second Round was announced, Rusche told a congressional subcommittee that it appeared that a second repository would be needed.[34] As for the possibility that the monitored retrievable storage facility proposed for Oak Ridge would be available and delay the need for another repository, there was no reason to be confident that this controversial project would be going forward. The Oak Ridge area, long comfortable with

[31]Prepared statement presented by Herrington at a press conference, May 28, 1986, Washington, D.C.

[32]Herrington response to questions, press conference of May 28, 1986; interview by the author with Rusche, July 3, 1986.

[33]See Department of Energy, Crystalline Rock Program Chicago office, "Crystalline Options," May 13, 1986. This paper was among those obtained by Congressman Edward J. Markey, chairman of the House Subcommittee on Energy Conservation and Power, in response to a letter of July 15, 1986, from Markey to Secretary Herrington.

[34]Ben C. Rusche, prepared statement presented to the Subcommittee on Energy Conservation and Power of the House Committee on Energy and Commerce, April 23, 1986. Rusche also noted that military high-level waste as well as spent fuel would have to be disposed of, and that Congress had put a 70,000-ton limit on the amount of spent fuel and equivalent defense wastes to be received by the first repository pending the opening of the second.

nuclear activities, welcomed the MRS project (on certain conditions) but the State of Tennessee did not. The Department of Energy would be stymied again unless Congress should override the state's veto, and this it might be unwilling to do, especially if not convinced that the MRS proposal is linked to a geologic disposal program that will succeed.

The Political Fallout in the West

The irony was that while stopping the Second Round was probably politically inevitable, this decision made the already bad problems of the First Round even worse. Several key sponsors of the waste act and all the First Round host-state senators from Texas, Nevada, and Washington immediately denounced the suspension. They saw it as a clear violation of the act's explicit requirements for a Second Round siting process and as an upsetting of the "delicate balance" of compromises that made the act possible.[35]

Senator Slade Gorton of Washington, a Republican, had been a prime mover behind efforts in 1981 and 1982 to have the waste act provide for a second repository and limit the spent fuel accepted by the first repository until the second is in operation. Now, in testifying before a Senate energy subcommittee, Gorton emphasized that the various elements of the act were "inseparable" and that "the siting of a second repository is a key element that cannot be removed without jeopardizing the entire Act." He insisted that if the department was going to disregard the act's requirements for the Second Round, then the First Round site selection process should be reopened too. "The department should conduct a nationwide search which culminates with the selection of a single site," Gorton said. Governor Gardner of Washington, appearing before the same subcommittee, warned: "If the federal government won't play by the rules, we will see you in court. The future of a repository will be tangled in the nation's court system for years to come."[36]

[35]Letter of June 11, 1986, to Secretary of Energy John S. Herrington from James A. McClure and J. Bennett Johnston, chairman and ranking minority member of the Senate Committee on Energy and Natural Resources, respectively; Alan K. Simpson, chair of the Senate Environment and Public Works Committee's Nuclear Regulation Subcommittee; Morris K. Udall, chair of the House Interior and Insular Affairs Committee; Senators Pete V. Domenici of New Mexico, Steven D. Symms of Idaho, Lloyd Bentsen and Phil Gramm of Texas, Paul Laxalt and Chic Hecht of Nevada, Slade Gorton and Dan Evans of Washington; and Congressman Sid Morrison of Washington, whose district takes in Hanford and the tri-cities.

[36]Testimony of Senator Slade Gorton, Hearings before the Subcommittee on Energy Research and Development of the Senate Energy and Natural Resources Committee, 99

Another effect of the Second Round suspension was to point up contradictions in the Department of Energy's repository siting philosophy. As Senator Dan Evans of Washington tartly observed following the DOE announcement of its First and Second Round decisions, arguments made to justify the one did not square with arguments to justify the other. "On the one hand," Evans said, "you reach outside the bounds of your formal rankings to pick [the Hanford basalt] site in order to increase geologic diversity. On the other hand, you postpone indefinitely the search in crystalline rocks." Evans also alluded to the further contradiction that whereas saving some "hundreds of millions" by stopping the Second Round siting was prudent management, the several billions of dollars more that it would cost to build a repository at Hanford compared to other sites was treated as of little importance.[37]

Trouble was also brewing for the Department of Energy in the Senate Appropriations Committee, chaired by Senator Mark Hatfield of Oregon, where, as in Washington state, public concern about the possible siting of a repository near the Columbia River at Hanford has been growing. In Hatfield's committee funding for the waste program was cut by at least a third and the department was enjoined not to begin deep-shaft exploration at any of the three candidate sites during the 1987 budget year.[38]

In sum, by the fall of 1986 the repository siting effort was in a crisis from which it might never recover.

Lessons

An overridingly important lesson from the First and Second Round siting experience is that insuperable difficulties have been created by not excluding areas where a repository project would present major land-use and environmental conflicts. Time after time DOE's choice of sites for study has created such conflicts, from the worries in Texas over the Ogallala aquifer to those in Maine over Sebago Lake. In some cases the conflicts have been indisputably real, as at Hillsboro, New Hampshire, where many citizens would face the loss of their homes. In other cases

Cong. 2 sess., June 16, 1986; Governor Booth Gardner, prepared statement presented to the Subcommittee on Energy Research and Development of the Senate Energy and Natural Resources Committee, June 16, 1986.

[37]Senator Dan Evans in an exchange with Secretary Herrington and Ben Rusche at hearings conducted by the Subcommittee on Energy Research and Development of the Senate Energy and Natural Resources Committee, 99 Cong. 2 sess., June 16, 1986.

[38]*Making Continuing Appropriation for Fiscal Year 1987*, H. Rept. 99-1005, 99 Cong. 2 sess. (1986) pp. 215, 651 (Nuclear Waste Fund).

the conflicts are arguably not real but are merely strongly perceived. The fact that a repository would be close to a town or beneath an aquifer does not necessarily mean that it would be unsafe. But perceived risks pose real problems. In many instances the problem is that people not merely are afraid, but that they fear the fear of others, worrying for example about declining property values or the possibility that sale of their farm products may be hurt by rumors of contamination.

Where siting choices present major conflicts, close questions bearing on containment or technical feasibility are often present too, or will certainly appear to many people to be present. For instance, at Hanford, where the volume, velocity, and direction of groundwater flow is a major technical issue, experts argue endlessly over whether there is or is not a threat of contaminating the Columbia River. The lesson should be never to propose a site anywhere near the Columbia or any other major river. At the Deaf Smith County site in Texas, the Department of Energy must show that a repository beneath the Ogallala aquifer would not present the unacceptable risk of having water rush through or around the shaft to flood the mined openings. It also must show that there is no credible way that the aquifer could ever be contaminated. However confident the department may be that fully satisfactory answers are in hand, these questions would never arise if it was policy to stay away from sites beneath prolific aquifers.

Still another lesson is to keep the geographic scope and the procedural complexities of the site search within modest limits. To seek a distribution of potential repository sites over several regions is more likely to spread the misery than to promote a sense of equity and fairness. The elaborate, long-drawn-out site screening process which the waste act explicitly and implicitly prescribes amounts to a cruelly demanding political marathon. For instance, each of the several stages of the Second Round—regional screening, the nomination and selection of sites for characterization, then selection of one site for licensing—would require voluminous documentation, information briefings and public hearings in the host states, and responses to state comments and lawsuits. As controversy heightened in the host states, the potential for trouble would increase back in Washington, where the Department of Energy is dependent on Congress for annual appropriations to the Nuclear Waste Fund and on the White House for steady political support. The rigors of the marathon would be the greater by virtue of the political importance in the federal-state system of even the smallest state, the special case of New Hampshire as an early presidential primary state being a case in point. It is also important to recognize that in the effort to weigh evenhandedly the pluses and minuses of a multitude of sites, issues certain to be emotionally and politically salient become diluted and obscured—but only for the Department of Energy decision makers, not for the host-state politicians. Furthermore, the First Round effort has made it clear that the task of comparing even a half-dozen or fewer sites is formidable and controversial, and that much information relevant to the sites is not available on a regional scale.

To be sure, the First Round site evaluations, comparative rankings, and ultimate choices could have been done better and without such disturbing lapses of integrity as seem to have been involved in the selection of Hanford as one of the top sites. But whether they could have been done well enough to promote a technical consensus as to the fairness and soundness of the choices made is another question. Part of the difficulty lies in the paradoxical nature of the problem: to choose the sites with the greatest promise in the absence of the geohydrological data that only characterization can provide. The mistrust engendered by such an exercise means that attempts at "consultation and cooperation" between the Department of Energy and the host states are doomed to frustration.

Not to be forgotten here is public distrust of the government's ability to cope competently and fairly with radioactive wastes and other hazards of nuclear technology. As I write, the postman brings a plea from the Atomic Industrial Forum to all journalists, imploring them to stop referring to repositories as "dumps." The request is warranted, for as the forum points out, a repository is intended to be a carefully engineered facility and not a valley of the drums. But the persistence of the dump image comes principally from the still-remembered record of waste mismanagement at places like Hanford, Maxey Flats, and West Valley.

Furthermore, assurances that a repository and not a dump is being sited will not be widely believed so long as the Department of Energy, by such behavior as the stubborn preference accorded the Hanford site, reveals signs of a closed system of bureaucratic groupthink. Trust will be gained by building a record of sure, competent, open performance that gets good marks from independent technical peer reviewers and that shows a decent respect for the public's sensibilities and common sense. An overly ambitious attempt at screening and investigating widely scattered sites is likely to make such a performance impossible. Credible performance may also be impossible unless independent experts are allowed to exercise a critical voice *before* key decisions are made, rather than afterwards when the bureaucracy has become entrenched in its positions.

But while the foregoing points up why repository siting is at an impasse, there are some positive lessons to be mentioned. One is that NIMBY, or "not in my backyard," is a cliche that does not always comport with reality. When local communities perceive significant net benefits and an absence of major conflicts, they are willing to look for something other than the dark side of repository siting. To establish a local base of cooperation and support is not a sufficient condition for host-state support, but it is surely a necessary condition.

In some instances, as at Hanford, there can be willing hosts in the absence of a defensible site. But there are instances where the local host community has been willing and the site has been adequate for the limited geologic disposal or spent fuel storage project proposed. One case of this kind centers on the WIPP site near Carlsbad, New Mexico, and another on the proposed monitored retrievable storage facility site at Oak Ridge, Tennessee. In the New Mexico case, local support clearly helped to keep the project alive. In Tennessee, where the state has promised a veto, the project may not survive, but the case is nonetheless instructive. Local leaders at Oak Ridge found incentives of two kinds for supporting the project: first, the possibility of economic benefits such as substantial payments in lieu of taxes, land for a new industrial park, and commitments for the development of project-related activities (centers for spent fuel transportation management and research, for example); second, a chance to gain commitments necessary not only to ensure a safe MRS operation, but also to effect an earlier and more complete cleanup of environmental problems from past AEC and DOE operations.[39]

The lessons from past nuclear waste facility siting in the United States, together with insights gained from foreign experience, illuminate the possibilities for new policy choices.

A Way Out

After nearly thirty years of fits and starts by the Department of Energy and its predecessors in grappling with the waste problem, a way out of the present impasse must be found with some urgency. Unless a confident show of progress is made soon, the geologic disposal effort will take on the appearance, if indeed it has not done so already, of an interminable trek toward an ever-receding mirage.

[39]Clinch River MRS Task Force, *Recommendations on the Proposed Monitored Retrievable Storage Facility* (Roane County/City of Oak Ridge, October 1985). See also Elizabeth Peelle, "Innovative Process and Inventive Solutions: A Case Study of Local Public Acceptance of a Proposed Nuclear Waste Packaging Facility," paper presented to the National Forum on Managing Land Use, René Dubos Center for Human Environment, New York, April 3–4, 1986. For an insightful general discussion of the place of incentives in waste facility siting, see S. A. Carnes and coauthors, "Incentives and Nuclear Waste Siting: Prospects and Constraints," *Energy Systems and Policy* vol. 7, no. 4 (1983).

To undertake a new national search for sites now, as officials of some First Round states have suggested, would try everyone's patience and be a thoroughly impractical endeavor. A far more promising approach would be to focus the search on a few areas, and, indeed, to place the emphasis on a single primary candidate site. The siting of a second repository should in fact be postponed; the problem of siting the first one is quite enough to take on for now. The insistence by First Round states that the search for a second repository site not be abandoned reflects a strong concern for regional equity, but there are better ways of addressing that concern than lighting political fires over much of the eastern half of the United States.

Leaving aside for the moment the political and equity questions and considering only the need to find a technically excellent site, several points should be made. One is that the present U.S. strategy of identifying several primary candidate sites and then exploring each of them through billion-dollar characterization projects is something no other country plans to do—or would feel that it could afford to do. The aim of the American strategy is to lend redundancy and an important element of technical robustness to the program of geologic isolation. But while the thorough exploration of multiple sites should in principle permit the selection of the best of those studied, doing this confidently and convincingly is highly problematic, as the First Round experience indicates. No site, upon characterization, will be found free of technical uncertainties.

Conscious of this, the Swedes are planning literally to overwhelm the uncertainty by placing the spent fuel elements in four-inch-thick copper canisters that are expected to last hundreds of thousands of years.[40] Swedish plans for site screening and characterization, on the other hand, are simplicity itself compared to those afoot in this country.[41] For the United States, given the geologic siting program's present predicament, something akin to the Swedish approach seems indicated. The appropriate strategy would place greatly increased emphasis on creating a multibarrier containment system, with the combination of natural and

[40]Swedish Nuclear Fuel Supply Co./Division Kärn-Bränsle Säkerhet (SKBF/KBS), *Final Storage of Spent Nuclear Fuel-KBS-III*, vol. 3, *Barriers* (Stockholm, May 1983) table 10-3, p. 10:15.

[41]Ibid., vol. 2, *Geology*. As described in this volume, Swedish site screening and characterization will proceed as follows: by 1990 two or three sites will have been selected from about ten candidate areas that have been investigated from the surface by geophysical methods, boreholes, and other means; a more detailed investigation of these few sites will follow, still from the surface, and by the year 2000 the site for the repository is to be selected, subject to licensing; between 2000 and 2010 the sinking of a shaft and construction of a pilot plant are a possibility; issuance of the final building permit is expected in 2010, and issuance of the operating permit in 2020.

artificial barriers optimized to ensure containment of radionuclides for a far longer time than would be possible for a repository relying principally on geologic barriers.

A strategy of this kind was at the heart of the recommendations of President Carter's Interagency Review Group in 1979, and it is this strategy that has contributed to the broad consensus among American earth scientists and engineering geologists that geologic isolation of nuclear waste is feasible. Yet as the U.S. geologic disposal effort has proceeded, the role of engineered or artificial barriers has assumed nothing like the significance it holds in the Swedish program and in fact has been treated as of distinctly secondary importance. Indeed, whereas present regulations prescribe that the waste package shall contain the radioactivity after repository closure for "not less than 300 years nor more than 1,000 years,"[42] the plutonium and some of the other radioactive species in the spent fuel will remain dangerous for tens of thousands of years.

When Congress revisits the Nuclear Waste Policy Act, as it surely will have to do, the systems approach to geologic isolation should be emphasized as a key to simplifying repository siting and improving its effectiveness. The policy adopted should aim for early identification of a site that is technically suitable and relatively free of conflicts; and it should avoid vain and far-flung site-screening attempts that commit the Department of Energy (or possibly a new waste agency) to a punishing procedural marathon that leads nowhere. The policy should seek to avoid close questions and to overwhelm uncertainty. Further, it should seek to overcome distrust by emphasizing a new openness, including a voice for independent experts before decisions are reached. Also, as I shall be emphasizing again, state and local officials and citizens should be assured that hosting a repository offers significant benefits, thus giving them the incentive (and political room) to examine what is proposed on its merits.

A Role for the National Academy

If such a policy is to be instigated, at the outset Congress might commission the National Academy of Sciences to do a study addressing

---

[42]Nuclear Regulatory Commission, "Disposal of High-level Radioactive Wastes in Geologic Repositories Technical Criteria," 10 CFR, pt. 60, para. 60.113, *Federal Register* vol. 48, no. 120 (June 21, 1983) p. 28224. As noted in chapter 1 of this book, these regulations also provide that 1,000 years after repository closure the release of radionuclides from the "engineered barrier system" (meaning the waste packages and the labyrinth of waste emplacement tunnels) shall not, with certain exceptions, exceed 1 part in 100,000. In order to satisfy this requirement under some conditions, a waste package promising containment beyond 1,000 years might be called for.

several key questions. First, are there among the sites in DOE's First Round inventory any that are both technically promising and relatively free from land-use and environmental conflicts? In essence, the academy's task in this instance would be to see if there is not already at hand a site worthy of designation as the primary candidate for a repository. Formal site rankings need not be contemplated. Enough is known about the various sites to permit knowledgeable experts, simply by an exercise of careful judgment, to recommend one for immediate characterization.

Second, what specific strategies and technologies can best be followed or applied in developing this robust system of containment? In particular, what types of waste canisters or casks can be used as part of a strategy to overwhelm uncertainty? How might foreign and domestic waste packaging and other technologies be used in a test and evaluation facility (T&E) to be developed at the chosen site with the aim of optimizing the overall system of containment?[43]

Third, how can an independent and credible process of peer review be established to increase public trust in repository siting and development? Other important questions related to public trust are: By what means can state and local governments best be given a voice in siting investigations, and perhaps later in oversight of actual repository operations? And should the Office of Waste Management be separated from the Department of Energy and made an independent agency?[44]

---

[43]The nuclear waste act gives DOE the option of establishing a test and evaluation facility, but DOE officials have shown no interest in pursuing this possibility. "I haven't heard that mentioned since the first few months I was here," Ben Rusche remarked when asked about a T&E facility in an interview with the author on June 12, 1986. But the Office of Technology Assessment's 1985 report, *Managing the Nation's Commercial High-Level Waste* (Washington, D.C., 1985) points to a T&E facility which would provide for temporary emplacement of modest amounts of spent fuel as a possibly useful and necessary means to test and verify handling and emplacement procedures. The OTA report also sees an advantage in going to a waste package with a design life exceeding regulatory requirements, and using it as a "fully redundant barrier." The report notes that there are possibly attractive alternative concepts of waste emplacement to be considered, such as the flexibility that might be offered by the "universal cask" concept which contemplates use of a massive 70-to-100-ton cask for spent fuel surface storage, transport, and final geologic emplacement. See ibid., pp. 145–146 and 149.

[44]Proposals to establish an independent agency or federal corporation to manage the nuclear waste program go back at least to 1977, when Mason Willrich and Richard K. Lester recommended it in their book *Radioactive Waste: Management and Regulation* (Macmillan). In 1984 the DOE Advisory Panel on Alternative Means of Financing and Managing Radioactive Waste Facilities also recommended such an institutional innovation. OTA's 1985 report, *Managing the Nation's Commercial High-Level Radioactive Waste*, pp. 162–164, suggested that to make an entity of this kind more accountable the members of its board might be appointed by several different authorities (such as the Congress, the president, and the secretary of energy), with some members possibly coming from state

The National Academy of Sciences study suggested here would be an ambitious undertaking requiring a coordinated approach by several separate panels. The academy's record in the nuclear waste field shows some conspicuous blemishes, but there probably is no other organization better able to assemble the relevant expertise and experience and produce a report that is credible and influential. The reports, books, and professional papers that have been published about the radioactive waste issue run to many, many thousands of titles; the problem is not that nothing is known, but that nobody seems to know what to do with what is known. The point of an academy study would be to bring coherence, direction, and authority to efforts to get something practical done.

To ask the academy to see whether there is a suitable primary candidate site at hand is not fanciful. There may be one in Nevada at Yucca Mountain. Similarly, there is no reason why the academy could not set forth a strategy for a systems approach at Yucca Mountain (or elsewhere) to overwhelm uncertainties about containment. The academy has already reviewed Swedish plans for encapsulating spent fuel in thick copper canisters, and in 1984 an academy panel concluded that the Swedes' claims for the canisters' durability appear to be well supported.[45]

Nor is there any reason why an academy study could not come to grips with the question of how to make timely and effective independent outside review a part of the siting process. The academy's Radioactive Waste Board is familiar, for example, with the critical role that New Mexico's Environmental Evaluation Group has played in the WIPP project. Other possibilities should suggest themselves. From its own recent experience in critiquing the First Round site-ranking methodology for the Department of Energy, the academy should be well aware how relatively ineffective outside peer review can be when its focus is narrowly limited and constrained. Consideration could be given to the creation of an independent peer review board with the authority and the staff to look into any technical nuclear waste issue at any time, with no information to be withheld from it. Such a board, which might include scientists nominated by the host state, would be expected to challenge foolish and contrived assumptions of the kind that went into the evaluation of the Hanford site. And just as the New Mexico Environmental Evaluation Group has done, the review board could insist that DOE (or

---

and local government and other nonfederal entities. Underlying the independent agency concept have been the objectives of increased efficiency and greater insulation from political pressures.

[45]National Academy of Sciences, Board on Radioactive Waste Management, "A Review of the Swedish KBS III Plan for Final Storage of Spent Nuclear Fuel" (Washington, D.C., National Academy Press, 1984) p. xii.

any successor agency) carry out any studies believed by the board to have been neglected.

The Promise of Yucca Mountain

Any hope of an early start on site characterization and development of a test and evaluation facility rests on identification of a primary candidate site from the existing First Round inventory. The Yucca Mountain site in Nevada seems by far the most promising; a key question is whether it is promising enough to justify immediate characterization of this one site alone. Technically, Yucca Mountain offers the important advantages of being high above the water table in a desert region of little rainfall. The Department of Energy and the U.S. Geological Survey believe, but must now confirm, that little or no water would infiltrate downward from the surface to the repository. If no water comes in contact with the waste canisters or casks, there would be no corrosion and no mechanism for radionuclide transport.

A repository about 1,000 feet beneath the top of Yucca Mountain would also permit easier access from the surface than would be possible at other sites. Access would be by two long, steep ramps tunneled in from the side of the mountain, instead of by vertical shafts. Furthermore, alone among the sites in the First Round inventory, the underground openings in the welded tuff are expected to be stable enough to make backfilling with crushed rock unnecessary; this offers an enormous advantage in terms of maintaining a capability for waste retrieval.

A principal technical disadvantage of Yucca Mountain has to do with the difficulty of predicting groundwater movement well enough to meet licensing requirements. Dependable predictive models for the unsaturated zone above the water table have not yet been made. Another disadvantage is that the region is seismically active. Natural earthquakes can be expected, not to mention the earthshaking from underground nuclear weapons shots conducted at Pahute Mesa and Yucca Flats, each some thirty miles from the Yucca Mountain site. But the repository would be at a depth ideal for attenuation of the effects of earthquakes.[46]

[46]Data have been compiled and analyzed on the effect on eighty different underground facilities of seventy major earthquakes in North and South America, Japan, India, and the Mediterranean. It was found that damage was slight at depths down to 900 meters, and was least between 200 to 900 meters (a Yucca Mountain repository could be at a depth of between 300 and 400 meters). H. R. Pratt, "Earthquake Damage to Underground Facilities and Earthquake Related Displacement Fields," in *Proceedings of the Workshop on Seismic Performance of Underground Facilities* (Aiken, S.C., Savannah River Laboratory, 1981) pp. 74, 370.

The principal concern would be to have all surface facilities for spent fuel handling built to resist groundshaking. The nearby E-MAD facility, built in the 1960s for testing nuclear rocket engines, has experienced earthshaking from hundreds of underground weapons tests without being damaged.[47]

As for land-use and environmental conflicts, a repository at Yucca Mountain would not threaten, nor be perceived as threatening, a town, a park, a farming region, or a major river. The one disturbing conflict that is present could be eliminated if the adjoining Nellis Air Force Base can be persuaded not to route its practice bombing runs over Yucca Mountain.

Nye County, which includes Yucca Mountain and the Nevada Test Site, is basically supportive of the repository siting effort, seeing an opportunity for some growth and jobs for people in small communities like Beatty and Amargosa Valley. "People here are used to it," says Robert Revert, chairman of the Nye County Commission, speaking of nuclear activities. "They have grown up with it and are educated to it. They are not frightened." According to Revert, the county's principal concern is to be allowed a voice in matters that affect its interests, such as avoiding a boom-and-bust cycle. "The Nuclear Waste Policy Act doesn't mention local government," he adds. "We are the ones who are going to get the impact, and we have no input."[48] Others in support of the siting effort are Nevada's professional engineers, labor unions in southern Nevada, and many business people, as evidenced by a resolution adopted by the North Las Vegas Chamber of Commerce.[49]

But Governor Richard H. Bryan has been strongly opposed to consideration of Yucca Mountain as a repository site, warning that a repository there might label Nevada "the country's nuclear wasteland" and ruin tourism. "If Nevada is to avoid this nuclear stigma we must unite in our opposition to it," Bryan told the Nevada legislature in January 1985. The Las Vegas and Clark County commissioners also have taken positions opposing the Yucca Mountain project. But the kind of deep, visceral public opposition that has been evident in places like southern Mississippi and west Texas appears to be lacking in Nevada. In a survey made by a Las Vegas councilman, almost half of the 2,400 respondents "did

[47]Interview by the author on June 6, 1986, with Don Vieth, DOE manager of the Yucca Mountain project. More recently, E-MAD was used in preparing spent fuel canisters for temporary emplacement deep in granite in the Nevada Test Site's experimental Climax facility.

[48]Interview by the author with Robert Revert, May 29, 1986.

[49]Department of Energy operations office, Las Vegas, "Outreach" document prepared September 1985.

not oppose establishing a high level nuclear waste dump on Yucca Mountain."[50]

The state legislature itself has dealt cautiously with the Yucca Mountain issue, having chosen to adopt no resolutions either favoring or condemning the project. Some key members believe that nuclear waste disposal could represent an important and advantageous new use for the Nevada Test Site (NTS). In 1978 Senator James I. Gibson, the Senate Democratic majority leader, chaired a special committee that reported to Governor Mike O'Callahan on alternative uses for the site in the event of a treaty banning nuclear testing. The committee concluded that nuclear waste disposal was a prime possibility deserving continued investigation, and suggested that, eventually, an "energy park" might be established on the NTS which would include reactors, spent fuel storage and reprocessing, and permanent waste isolation.[51]

If a nuclear test ban treaty should ultimately be negotiated, southern Nevada could lose the Nevada Test Site as one of its economic mainstays. This facility is one of the state's largest employers, providing some 6,800 jobs. Jeffrey Van Ee, an electrical engineer with the Environmental Protection Agency in Las Vegas, is a leader of the Toiyabe chapter of the Sierra Club in Nevada and eastern California. According to Van Ee, a number of Sierra Club members work at the NTS, many of whom would prefer to work on nuclear waste management than on nuclear weapons testing. Van Ee favors the Yucca Mountain investigation, but sees the siting process as suspect from the point of view of objectivity and fairness.[52]

Lack of fairness is an objection which a great many Nevadans cite not only with respect to the present national nuclear waste policy, but also with respect to how their state is treated generally. Nevadans feel that any time a site is needed for an activity or facility that no other state would tolerate, there will soon be a plan to put it in Nevada, whether it be nuclear weapons testing, MX missiles, "supersonic operating areas," bombing ranges, low-level waste, or, as now, high-level waste.

The day after the Department of Energy announced that the search for sites in the East was being suspended and that Nevada held one of the three sites selected in the West, the *Las Vegas Review-Journal* expressed indignation. It said the issue was not safety and that it really wasn't a problem of frightening tourists either. "If those underground atomic shots that ripple the upper floors of high-rise buildings in Las Vegas don't scare the tourists, then, sure as heck, a waste site isn't likely

---

[50]Ibid.

[51]Report of the Nevada Committee for the Utilization of State Resources to Meet National Needs, October 1978.

[52]Interview by the author with Van Ee, June 4, 1986.

to keep the folks away," the editorial observed. "What is at issue is the lack of fairness to Nevada, the disregard in Washington for the wishes of the people and the tendency of the technocrats and political forces in Washington to exploit Nevada's relative lack of national political power." The same day, the *Gazette-Journal* in Reno voiced the same complaint, but took a different tack, saying "Nevadans must begin to devise a strategy to exact some benefits in return [for hosting a repository]." "At long last," the editorial concluded, "Nevada deserves a break."[53]

The *Gazette-Journal* was no doubt speaking for many, and it is just this kind of sentiment to which Congress should respond in the event that Yucca Mountain is deemed to be a suitable primary candidate site. In principle, the government could impose a repository on Nevada and provide nothing beyond compensation for actual project effects. All of the land is federally owned and part of it is already dedicated to nuclear activities. Also, Nevada is indeed relatively weak politically, its congressional delegation being among the smallest. But for the nuclear industry as well as the State of Nevada it would be much better for Congress to strike a deal that leaves the Nevadans satisfied that they are getting a fair shake at last.

What would it take to accomplish this? If there are experts available to deal with this sort of thing, they are perhaps as likely to be found in Congress as anywhere. An effort should be made to reach an understanding with Nevada's governor, its senators and representatives, and possibly some of its key state legislators. It would consist of a quid pro quo, with Nevada to acquiesce in the siting activities at Yucca Mountain and in return receive substantial benefits, perhaps cash bonuses and generous payments in lieu of taxes, plus assurances that the state would be allowed a strong voice in certain matters of public concern, such as the way spent fuel would be shipped into Nevada. Obtaining the funds to back up an understanding of this kind should be no problem at all. Many hundreds of millions would be saved by abandoning the Deaf Smith and Hanford characterization projects. Furthermore, even a 5 percent increase in the present fee of one-tenth of a cent per kilowatt hour that is imposed on nuclear electricity would yield about $25 million a year.[54]

[53]"Feds Treat Nevada like Colony in Search for Nuke Waste Site," *Las Vegas Review-Journal*, May 29, 1986; "Nevadans Must Obtain Tradeoff for Waste Site," *Reno Gazette-Journal*, May 29, 1981.

[54]The DOE projection for total nuclear generation in fiscal year 1987 is 421 billion kilowatt hours, against which the fee of one-tenth of a cent per kWh will yield about $421 million; but the projection for fiscal 1989 is for 509 billion kWh and revenues of about $509 million (or, with a 5 percent increase in the fee, almost $535 million). Derived from Department of Energy, "Fiscal Year 1987 Congressional Budget Request," p. 430, table, a request by the Office of Civilian Radioactive Waste Management regarding use of the Nuclear Waste Fund.

426   TIME TO ACT

Some may call such dealings bribery, but the accusation is not easily
sustained. Generally speaking, bribery is intended to induce a betrayal
of trust by the offer of money or other favors. In the present context,
bribery could take the form of inducing state and local leaders to accept
short-term gains for their state at the expense of large long-term risks
to be carried by generations yet unborn. But as previously discussed,
the risks need not be large; at a properly chosen site, with a robust
overall system of containment, the risks can be very low, both for now
and for the many thousands of years that the waste remains dangerously
radioactive. However, the assurance of safety must be credible—hence
the importance of having a study, made under independent and respected auspices, to reexamine and reaffirm the potential that the multibarrier systems approach holds for geologic isolation and containment.

Successful negotiations leading to the designation of a primary candidate site—such as Yucca Mountain—could be important in arriving
at a workable, essentially voluntary, approach to obtaining a few additional sites to serve as backups should the primary candidate prove
unlicensable. The search for back-up sites could take advantage of a
recent U.S. Geological Survey study of the Basin and Range Province,
which includes most of the largely undeveloped and unoccupied desert
lands of the American Southwest. Done in cooperation with the state
geological surveys, this study found six large areas—the smallest of them
larger than Massachusetts—that were deemed promising for waste isolation.[55] But the study took only geohydrologic considerations into account and was not concerned with land-use or environmental conflicts
in the surface environment. The initial screening would have to be
completed, preferably again with state help, by redrawing the maps of
candidate areas to eliminate places where conflicts can be expected.
Next, the maps could be circulated to the state and county governments
concerned; they would be asked whether they would agree to siting
investigations, and under what conditions and with what benefits for the
hosts.

To sum up, the approach to repository siting and development out-
lined above emphasizes avoidance of conflicts and close questions. Also
to be avoided are procedural marathons inevitably associated with attempts to screen large numbers of potential sites and then compare and
rank several of them in an effort to pick the best. The preferred approach
is to select a primary candidate site for characterization from a modest
number that have been investigated from the surface. Characterization

[55]M. S. Bedinger, K. A. Sargent, and J. E. Reed, "Geologic and Hydrologic Characterization
and Evaluation of the Basin and Range Province Relative to the Disposal of High Level
Radioactive Waste," U.S. Geological Survey Circular 904-4 (1984).

of the primary site would be accompanied by development of a test and evaluation facility to optimize, for this specific site, a system of containment robust enough to overwhelm the uncertainties that the natural geologic barriers may present if taken alone. The repository system, built in hard rock, would allow ready retrieval of spent fuel or high-level waste canisters or casks for many decades, or perhaps for a century or longer.[56] The approach recommended also emphasizes independent peer review and generous benefits for the host state and locality.

This is the philosophy and policy proposed, but neither this nor any other philosophy will prove serviceable absent the will to embrace it and carry it out.

Common Ground

For a new nuclear waste policy to be adopted and successfully carried out, the interests with a stake in the outcome must find common ground. They must all agree that the problem is urgent. Although variously motivated, they must also want early progress enough to agree on the few available practical strategies. Elegant but impractical national site screening strategies must be seen as the prescription for political paralysis that they surely are.

But the struggle over nuclear waste policy has gone on so long that the mutual suspicions that divide the familiar players run deep and are likely to persist. These players include the nuclear industry, the potential host states, and the environmental and anti-nuclear groups. The consensus supposedly represented by the Nuclear Waste Policy Act of 1982 was illusory, and the environmentalists and anti-nuclear activists were never really a part of it anyway.

What is needed is a new, clearer, and broader consensus, with strong participation by certain importantly affected interests that were not much heard from in 1982. These new players would include individuals and groups concerned about the risks of nuclear weapons proliferation and nuclear terrorism; the governors and members of Congress from states with growing accumulations of spent fuel and military high-level waste; the utility ratepayers who are footing the bill for the waste disposal effort, together with the utility regulatory commissioners who represent them; and certain important elements of the environmental

---

[56]It should be noted that a credible system of retrieval must include provision for surface storage of any spent fuel removed from the repository; in this sense monitored retrievable surface storage is a necessary complement to the system of deep geologic isolation discussed here.

community never previously directly involved with radioactive waste or other nuclear issues, such as the National Wildlife Federation and the National Audubon Society.

These important interests would bring their own political weight to the waste policy deliberations, and something more besides. By the force of their example, they could perhaps draw both the nuclear industry and the environmental and anti-nuclear lobbyists into the circle of agreement over strategy. Left to their own inclinations, the industry and anti-nuclear lobbyists might well lock themselves into positions that obstruct agreement. Anti-nuclear lobbyists, for instance, might be sorely tempted to denounce any proposal to find a primary candidate site in Nevada and offer generous benefits to the state on the grounds that the proposal was a cynical attempt to follow the path of least political resistance and to bribe a politically weak western state into becoming the nation's nuclear waste dump. Similarly, utility lobbyists might dismiss any proposed new emphasis on artificial barriers as a ploy to load the nuclear industry with excessive and unnecessary costs.

Viewed overall, the mix of parties with reason to press for early progress in establishing a geologic repository is impressive, as will be seen below.

The nuclear control groups. Among those groups and individuals most interested in the risks of nuclear weapons proliferation and nuclear terrorism are the Federation of American Scientists (FAS) and such federation leaders as Frank von Hippel of Princeton University and John Holdren of the University of California, who are on record as opposing reprocessing and the use of plutonium fuel in the commercial nuclear fuel cycle because of the proliferation risks.[57] The Natural Resources Defense Council (NRDC) and the Union of Concerned Scientists (UCS), both widely known as critics of nuclear power, also have reason to join in an effort to make the geologic disposal program succeed. These groups have made control and reduction of nuclear weapons a major focus of their concerns and are aware of the advantages of direct geologic disposal of spent fuel for the nonproliferation regime. Moreover, Henry Kendall, professor of physics at MIT and chairman of the UCS, is impatient at the government's failure to move more expeditiously to the geologic disposal of nuclear waste. As he views it, this is a problem that presents no fundamental technical obstacles to its solution.[58]

The Federation of American Scientists, the Union of Concerned Scientists, and the Natural Resources Defense Council all have within their

[57]Interview by the author with FAS director Jeremy J. Stone, November 3, 1986.
[58]Kendall was interviewed by the author on May 25, 1982, and July 22, 1986.

ranks scientists who are highly knowledgeable about the relative risks of nuclear power. These scientists are much more likely to be trusted by an uneasy public than are industry experts suspected of a pro-nuclear bias. On some questions, moreover, their assessments can be reassuring. For instance, James MacKenzie, physicist and one of the early leaders of the UCS, believes (and will tell anyone who asks) that the transport of spent fuel that has been properly aged and is properly escorted can be quite safe[59]—contrary to the impression conveyed by the disaster scenarios long put forward, and often believed in, by some anti-nuclear activists.

Officials of states hosting accumulations of spent fuel. No one should have a stronger interest in seeing the geologic disposal problem resolved than the governors, the members of Congress, and the other state and local elected officials of the thirty-four states where nuclear reactors are operating or are under construction. Unless the spent fuel now accumulating at the reactor sites can be moved to a geologic repository or monitored retrievable storage facility, these officials will ultimately find themselves host to de facto nuclear waste storage depots that will have to be kept under constant custodial care. Two states, Washington and South Carolina, have both spent fuel from commercial power reactors and large and growing accumulations of high-level military waste. Whether officials of these numerous states actually will be an effective force in saving the geologic disposal effort is unclear. But certainly there is no reason for them to be complacent. The safety risks associated with at-reactor spent fuel pool storage appear low but are by no means zero, and the greater the density of the pool storage the greater the release of radioactivity may be in some accident scenarios.

The utility ratepayers and the public utility commissions. On the order of 20 to 30 billion dollars will be collected from utility ratepayers over the next three decades to support the nuclear waste management effort. In view of the enormous sums to be spent, the ratepayers—and more particularly the state public utility commissions who represent them—should expect something for their money. They could add substantial weight to efforts to bring about adoption of a practical, technically and politically defensible repository siting policy.

[59]MacKenzie, now with the World Resources Institute, was interviewed by the author on March 28, 1986. Mackenzie is persuaded that the casks used for the transport of spent fuel provide adequate containment, and he scoffs at those nuclear critics who see the casks as susceptible to catastrophic accidents. "The crash and fire tests are *quite* convincing," he says.

The ratepayers might find a useful precedent in the taxpayers' coalition against the Clinch River breeder reactor project, formed in 1982 by the National Taxpayers Union and sixteen other groups. The Clinch River coalition included labor unions, religious organizations, and environmental groups. This successful lobbying effort was built around the theme that the taxpayers, not the nuclear industry, were going to bear the soaring costs of the Clinch River breeder.[60] A ratepayers' coalition would not have the aim of killing the geologic disposal program but of saving it, and possibly saving money too. To protest the folly of spending $1 billion on characterization of a site as poorly chosen as the one at Hanford might be the way to start, but eventually the principal aim should be to reorient the siting effort away from its present tortuous course to one that is more promising and direct.[61]

The utilities and the rest of the nuclear industry. Collapse of the present controversial and complex siting effort should be welcomed by the utilities and the rest of the nuclear industry as good riddance. It makes sense to characterize only one especially promising primary candidate site and to concentrate on repository systems development of value to the nuclear industry everywhere. In principle, the concept of a repository system that provides a convincing hedge against uncertainty by emphasizing long-lived spent fuel canisters or casks could make siting possible in a variety of geographic and geologic environments.

But to gain this advantage such a system must first be developed, and the place for that first-of-a-kind endeavor is at a remote western site relatively free of real or perceived conflict. If the site permits ready retrieval of spent fuel after emplacement, this should be seen as a further advantage, especially by those who believe that some day reprocessing and use of plutonium fuel will be economically attractive and feasible at acceptable risk. Moreover, the sooner the problem of nuclear waste disposal is solved, perhaps the sooner adequate attention will be paid to matters even more critical to the industry's growth and survival, such as development of an economically competitive reactor that is inherently

---

[60]Interview by the author on July 22, 1986, with Jill Lancelot of the National Taxpayers Union.

[61]A subcommittee of the National Association of Utility Regulatory Commissioners has been keeping an eye on the nuclear waste disposal effort, but up through mid 1986 its main focus was on management efficiency and on seeing that the Department of Energy pays its fair share of waste management costs in light of the military high-level waste to be disposed of. According to its chairwoman, Edwina Anderson of the Michigan Public Service Commission, the subcommittee probably would continue to stay away from siting issues "unless those issues begin to impinge on costs." Interview by the author with Edwina Anderson, July 21, 1986.

less susceptible than are today's reactors to equipment failure and human error.[62]

The environmental and anti-nuclear groups. These groups constitute a broad category of rather disparate interests, ranging from some whose sole purpose is to oppose nuclear power to others whose aims have nothing whatever to do with nuclear issues. But for the environmental community as a whole a breakthrough in dealing technically and politically with spent fuel and high-level waste disposal could be relevant, directly and indirectly, to a variety of hazardous waste problems. On the merits, anti-nuclear groups such as Ralph Nader's Critical Mass and the environmental groups that have taken positions against nuclear power should push for early progress toward establishing the first repository. But whether they will do so is another matter.

The chairman of the Sierra Club, Michael McCloskey, believes personally that the idea of having environmental groups seek common ground with other interests on the geologic disposal problem is "statesmanlike and makes great sense." But he wonders whether he and other environmental-group leaders, were they to try to bring about such a common effort, might not find themselves without followers. He observes that among environmentalists, especially in the potential repository host states, the fact and appearance of mismanagement and political manipulation by the Department of Energy in repository siting have led to a steady erosion of confidence in geologic disposal as the best means of waste isolation. This perhaps could be overcome, he says, by an emphasis on reforming the way siting is carried out. But McCloskey sees another obstacle that would have to be overcome. "I suspect many environmentalists want to drive a final stake in the heart of the nuclear power industry before they will feel comfortable in cooperating fully in a common effort at solving the waste problem," he says. "Their concern would arise from the possibility that a workable solution for nuclear waste disposal would make continued operation of existing plants more feasible, and even provide some encouragement for new plants."[63]

As McCloskey suggests, this may indeed be the way that many environmentalists and anti-nuclear activists would react. But the pertinent

[62]A prescription for an "inherently safe" reactor that is immune to loss-of-coolant accidents and core meltdowns was set forth by Alvin M. Weinberg and others in *Second Nuclear Era*. Prepared by the Institute for Energy Analysis at Oak Ridge, this study does not find the present light water reactors to be unsafe, but notes that safety improvements achieved since the Three Mile Island accident have been gained at "the expense of greater complexity and cost."

[63]Interview by the author with Michael McCloskey, October 1986; letter from McCloskey to the author, June 16, 1986.

question today about the future of nuclear power is not—despite Chernobyl—whether there will be a future; rather, it is whether nuclear power will be made safer and more publicly acceptable. Nuclear power is not dead and is not dying, and the $150 billion investment in nuclear power that has been made in the United States certainly is not going to be abandoned for lack of permanent means of waste disposal. Spent fuel will just continue to accumulate at the reactor sites, most of it in the pools, possibly increasing the potential for large releases of radioactivity in the event of a severe accident.

Lack of permanent disposal facilities would tend to discourage further orders for nuclear reactors in the United States, though not necessarily everywhere abroad. But the extremely high capital cost of light water reactors of conventional design—much of this cost associated in one way or other with reactor safety—will itself probably suffice to discourage new orders, at least so long as coal-fired electricity generation is considered acceptable environmentally. This fundamental industry problem, which may never go away until a reactor inherently safer and more forgiving is designed, has been intensified by the Chernobyl accident.

A Time to Act

The consequences of allowing the waste problem to continue are anything but trivial. Just as nearly all stand to gain if the problem is solved, all stand to lose if it is not. Hundreds of millions of dollars have already been spent in the wrong places, and if billions more are not spent in some of those same places it will only be because the Department of Energy is deterred by the courts or by the Congress, and in Congress the fragile consensus represented by the Nuclear Waste Policy Act seems to have been shattered already.

The most serious consequence of not finally putting the repository siting effort on a more predictable and promising path will be the failure to come to grips with the nuclear imperatives of containment and safeguards. The dangerous residues of the fission process, all highly toxic and some having the potential to be made into nuclear explosives, are best contained and kept secure if left in the spent fuel and isolated in deep geologic formations. Of all nations, the United States has the best chance to perfect and demonstrate by the end of this century a technically and politically robust system for meeting these imperatives—and meeting them in a way persuasive to other countries, even those that see reprocessing and breeders as ultimately critical to their energy security.

The most urgent consideration is to discourage the economically premature and politically foolhardy use of plutonium fuel abroad. For plutonium fuel to enter routine use and commercial traffic in a world in which political instability and terrorist activity are rife presents risks that are quite beyond our powers to assess. To go along complacently in the face of such developments recalls the Joseph Conrad story *Typhoon* and the stolid Captain MacWhirr, who lacked the wit to imagine the force and ferocity of cyclonic winds. Despite a falling barometer and other ominous portents of a typhoon that would all but sink his ship, MacWhirr kept steady to his course, occasionally muttering "There's some dirty weather knocking about." There may or may not be typhoons ahead, and the risks should not be overstated. But neither should they be slighted or forgotten.

Glossary, Acronyms, and Abbreviations

ACDA   U.S. Arms Control and Disarmament Agency.

actinides   In the periodic table, the series of elements comprising nos. 89 through 105, which together occupy one position in the table, beginning with actinium. They include uranium and all of the transuranic elements, which are man-made (*see* transmutation).

activity   Short for radioactivity.

AEA   Atomic Energy Authority (Brit.).

AEC   Atomic Energy Commission (U.S.).

AFR   Away-from-reactor (surface storage facility).

AGNS   Allied-General Nuclear Fuel Services; Allied-General.

AGR   Advanced gas-cooled reactor.

ALARA   "As low as reasonably achievable," a radiation protection principle, held by national and international scientific and regulatory authorities and applied to radiation exposures, with costs and benefits taken into account.

ALATA   "As low as technically achievable," a radiation protection principle advocated by some critics of nuclear regulation who regard ALARA as insufficient.

alpha-emitter   A radioactive substance, such as plutonium, that emits alpha particles.

alpha particle   A helium atom (two protons and two neutrons) positively charged from the loss of its two electrons. Alpha radiation is much less penetrating than gamma rays or beta particles (a sheet of paper will stop an alpha particle), but is much more densely ionizing (*see* ionizing radiation). Significant cellular damage, possibly leading to cancer, may occur if an alpha-emitter such as plutonium enters the body by inhalation, ingestion, or through a cut or wound.

ANDRA   Agence Nationale pour la Gestion des Déchets Radioactif (National Agency for Radioactive Waste Management) (Fr.).

back end of the fuel cycle   *See* front end of the fuel cycle.

background radiation   The radiation in the natural environment, including cosmic rays and radiation from naturally radioactive elements, both inside and outside the bodies of humans and animals.

beta-emitter   A radioactive substance that emits beta particles.

beta particle   An elementary particle emitted from a nucleus during radioactive decay, having a single negative electrical charge and a mass equal to 1/1837 of a proton. A negatively charged beta particle is an electron. Beta radiation may cause skin burns, and a beta-emitter such as strontium is harmful if it enters the body.

BNFL   British Nuclear Fuels Limited.

boiling water reactor   A light water reactor, in which the cooling reactor water is allowed to boil and generate steam as it passes through the core. Boiling water reactors lack the secondary steam-generating circuit found in pressurized water reactors. (*See* light water reactor; pressurized water reactor.)

breeder reactor   A nuclear reactor designed to produce more fissionable atoms than it consumes. The new fuel material is created when fertile isotopes capture neutrons, with the transmutation of uranium-238 to plutonium-239 (*see* fertile isotopes; transmutation).

burnup   Fissioning of nuclear fuel. Measured in megawatt days (of heat) per ton of heavy metal. More burnup means more fission products and more radioactivity. (*See* fission; heavy metal; megawatt day per ton.)

BWR   Boiling water reactor.

CEA   Commissariat à l'Energie Atomique, French government nuclear energy agency.

CFDT   Confédération Française Démocratique du Travail, French trade union identified with the Socialist party.

CGT   Confédération Générale du Travail, French union identified with the Communist party.

chain reaction   In fission, a self-sustaining or repeating reaction that occurs when an atomic nucleus absorbs a neutron and is split, releasing additional neutrons, which in turn can be absorbed by other fissionable nuclei that release still more neutrons.

CLAB   Acronym for Sweden's central storage facility for spent fuel.

cladding   The tubular outer jacket for nuclear fuel pellets. Cladding prevents leaching of the fuel and the release of fission products into the reactor's coolant. Cladding materials include zirconium alloys, aluminum or its alloys, and stainless steel. Fuel of most modern light water reactors is zirconium-clad.

COGEMA   Compagnie Générale des Matières Nucléaire, nuclear fuel cycle company owned by the Commissariat à l'Energie Atomique.

CONAES   Committee on Nuclear and Alternative Energy Systems, National Academy of Sciences (U.S.); disbanded in 1980.

critical   Sustaining a chain reaction, used in reference to nuclear reactors.

curie   The basic quantitative unit used to describe intensity of radioactivity. A curie is equal to 37 billion disintegrations per second (approximately the rate of decay of 1 gram of radium).

deuterium   A nonradioactive isotope of hydrogen, having a nucleus about twice as heavy as the normal hydrogen nucleus; often called heavy hydrogen (*see* heavy water). The symbol is D.

DOE   Department of Energy (U.S.).

doubling time   The time required for a breeder reactor to produce enough plutonium in excess of its own needs to provide for start-up of a second reactor.

DWK   Nuclear fuel cycle firm owned by West German utilities.

EDF   Electricité de France, the French electric utility.

EEG   Environmental Evaluation Group, New Mexico state organ to oversee the federal Waste Isolation Pilot Plant (WIPP) project.

electrons   The orbital, negatively charged elementary particles surrounding the positively charged nucleus of an atom. The number and energies of these electrons determine the chemical properties of the atom, and the sharing of electrons creates chemical bonds between different elements and accounts for the formation of molecules and compounds. By displacing electrons from molecules, alpha, beta, or gamma radiation can break such chemical bonds (*see* beta particle; ion, ionization; ionizing radiation).

enrichment   An isotopic separation process used to increase the concentration of fissionable uranium-235 in uranium beyond the 0.7 percent found in nature (*see* uranium). High-enriched uranium (more than 90 percent fissile) is used in nuclear weapons, submarine reactor fuel, and some research reactors. The low-enriched uranium (2 to 4 percent fissile) used in commercial light water reactors cannot be used as a nuclear explosive.

EPA   Environmental Protection Agency (U.S.).

ERDA   Energy Research and Development Administration, predecessor of the Department of Energy (U.S.).

fertile isotopes   Isotopes, not fissionable in themselves, which can be converted or transmuted into fissile isotopes following irradiation by neutrons in a nuclear reactor (*see* isotope; transmutation).

fission   The splitting of an atomic nucleus into two or more fission products or fragments, accompanied by the release of relatively large amounts of energy and generally one or more neutrons. Fission can occur spontaneously, but usually it is caused by absorption of neutrons. (*See* fission products.)

fission products The fragments or radionuclides resulting from fission of an atomic nucleus, plus nuclides formed by radioactive decay of these fragments.

front end of the fuel cycle The mining, milling, and enrichment of uranium and the fabrication of uranium fuel used in nuclear reactors. By contrast, the "back end" of this fuel cycle involves the reprocessing of spent fuel, fabrication of plutonium fuel (a mixed oxide of plutonium and uranium), and disposal of radioactive waste; or, alternatively, disposal or storage of spent fuel.

fuel element A rod, tube, plate, or other form into which nuclear fuel is fabricated.

fuel pellets Small, ceramic-like pellets of uranium oxide or mixed (plutonium-uranium) oxide, which, together with the cladding, make up the fuel rods that constitute a fuel assembly.

fuel rod *See* fuel element; fuel pellets.

gamma-emitter A radioactive substance that emits gamma rays.

gamma rays High-energy, short-wavelength electromagnetic radiation, which frequently accompanies alpha and beta emissions and always accompanies fission. Although similar to X rays, gamma rays are more energetic than X rays and are nuclear in origin. Gamma rays are very penetrating (several feet of concrete can be required for effective shielding from gamma radiation).

GESMO General Environmental Statement on the Use of Recycled Plutonium in Mixed Oxide Fuel in Light Water Reactors.

gigawatt One billion watts.

gigawatt year The continuous generation of one billion watts of electric power over the period of one year.

GSIEN Groupe de Scientifiques pour l'Information sur l'Energie Nucléaire, French public interest scientific group.

half-life The time during which half of the atoms of a radioactive substance decay to another nuclear form. Half-lives vary from millionths of a second to billions of years.

heavy metal Any of the three elements—uranium, thorium, or plutonium—that can be used as nuclear fuel. Their atomic weight or mass, determined by the total number of protons and neutrons in the nucleus, is much greater than that of elements lower in the periodic table. For example, uranium, with an atomic weight of 235 or 238 (depending on the isotope), is more than four times heavier than iron, which has an atomic weight of about 56. (*See* plutonium, uranium.)

heavy water Water containing significantly more than the natural proportion of heavy hydrogen (deuterium) atoms to ordinary hydrogen atoms. Used in some nuclear reactors to moderate (slow down) high-velocity neutrons.

high burn-up fuel Fuel that has been left in a reactor much longer than fuel of low burnup and that consequently is more highly radioactive from its greater content of fission products.

high-level waste The highly radioactive waste from a reprocessing plant. It contains the fission products and actinides separated from the dissolved fuel. Some species are very long-lived. High-level waste also encompasses spent fuel when the fuel cycle does not include reprocessing.

IAEA International Atomic Energy Agency.

ICRP International Commission on Radiological Protection.

INFCE International Nuclear Fuel Cycle Evaluation.

intermediate-level waste The more highly radioactive waste associated with maintenance of nuclear facilities, such as the spent ion exchange resins from cleanup of reactor coolant. Transuranic or plutonium-contaminated wastes may be included. In the United States, transuranic wastes are a separate category, and those wastes called intermediate-level by the Europeans are regarded as a class of low-level waste.

ion, ionization An ion is an atom that has gained or lost one or more electrons and thus become electrically charged. In its normal state, the atom is electrically neutral, its positively charged nucleus balanced by its negatively charged orbital electrons. Ionization is the process of adding electrons to or removing them from atoms or molecules, thereby creating ions. (*See also* ionizing radiation.)

ion exchange The reversible interchange of various ions between a solution and a solid material. This can be an important mechanism for retarding the migration within a geologic formation of radionuclides that are in solution in the groundwater. In waste management, ion exchange is the chemical process routinely used in nuclear power plants to separate and remove cesium and other radionuclides that are in solution in the reactor cooling water. Resin filters are used.

ionizing radiation Radiation that produces ionization by removal of orbital electrons. Inasmuch as the characteristics of the electrons determine the atoms' chemical properties and link different elements in molecules and compounds, removal of these electrons disrupts chemical bonds and damages cells in living tissue. Densely ionizing radiation, especially characteristic of alpha particles, strips off electrons all along its path. The extent of ionization—that is, the number of ions created—varies with the kinetic energy of the particles or rays emitted, some species of alpha-, beta-, and gamma-emitters being much more potent in this respect than others. In the case of emitters that enter the body, their biological effect is partly a function of the emitters' chemical affinity for particular organs or body tissue. For instance, strontium mocks calcium and goes to the bone, where it may cause solid tumors by irradiating the bone cells, or leukemia by irradiating the marrow, generator of the blood cells.

IRG Interagency Review Group on Nuclear Waste Management (U.S.).

irradiate Expose nuclear fuel to neutron bombardment, in a reactor; more generally, to expose any material or living tissue to ionizing radiation.

ISFS International Spent Fuel Storage project.

isotope One of two or more atoms of the same element which chemically are identical, but which have slightly different atomic weights (mass relative to other atoms). Isotopic separation processes, as in uranium enrichment, take advantage of this difference in atomic weight.

JAIF Japan Atomic Industrial Forum, Japanese nuclear industry group.

JCAE Joint Committee on Atomic Energy, U.S. Congress; disbanded in 1975.

JNFS   Japan Nuclear Fuel Service.

KBS   Kärn-Bränsle Säkerhet (Nuclear Fuel Safety), Swedish project to develop data and planning to meet the nuclear waste disposal requirements of Sweden's Stipulation Act of 1977, as amended.

LDP   Liberal Democratic Party (Jap.).

light water   Ordinary water (H_2O), as distinguished from heavy water (D_2O).

light water reactor   A nuclear reactor that uses light (ordinary) water to moderate (slow down) high-velocity neutrons and remove heat from the reactor core.

low-level waste   In U.S. parlance, a broad category of wastes from the operation and maintenance of nuclear power facilities and from the use of radioisotopes in nuclear medicine and in industry. Some of these wastes are scarcely radioactive at all; others are radioactive enough to require heavy shielding. (*See also* high-level wastes; intermediate-level wastes.)

LWR   Light water reactor.

magnox reactor   A British nuclear reactor that uses magnesium-alloy-clad natural uranium fuel.

megacurie   One million curies.

megawatt day per ton   A measure of fuel burnup, megawatt day per ton can be thought of as the production of 1 million watts of heat in 1 day from 1 ton of fuel. Actually, the 1 million watts might be produced in less than a day or in more than a day, depending on the amount of fission taking place.

meltdown   The melting of nuclear reactor core materials, resulting from heat from fission or decay when effective cooling is lacking.

member   A minor stratigraphic unit of a geologic formation.

millirem   One-thousandth of a rem.

MITI   Ministry of International Trade and Industry (Jap.).

MRS Monitored retrievable storage.

MUF Material unaccounted for, used in this book in reference to plutonium unaccounted for in reprocessing-plant inventories.

nanocurie One-billionth of a curie.

natural radiation *See* background radiation.

NEA Nuclear Energy Agency, of the Organization of Economic Cooperation and Development.

NEPA National Environmental Policy Act of 1969 (U.S.).

neutron An uncharged elementary particle found in the nucleus of every atom heavier than hydrogen. The release of neutrons sustains the fission chain reaction in a nuclear reactor.

NFS Nuclear Fuel Services, Inc. (U.S.).

NIMBY Not-in-my-backyard.

NIREX Nuclear Industry Radioactive Waste Executive (Brit.).

NRC Nuclear Regulatory Commission (U.S.).

NRDC Natural Resources Defense Council, private U.S. environmental group.

NRPB National Radiological Protection Board (Brit.).

NRTS Nuclear Reactor Testing Station, Idaho; now the Idaho National Engineering Laboratory.

NTS Nevada Test Site.

nuclides Atomic forms of the elements.

NWPA Nuclear Waste Policy Act of 1982 (U.S.).

OECD Organization of Economic Cooperation and Development, international body.

OSTP   Office of Science and Technology Policy, of the White House.

picocurie   One-trillionth of a curie.

plutonium   A heavy fissionable element (atomic no. 94) created as a result of the capture of a neutron by the nonfissionable but abundant uranium-238. In its separated form, plutonium can be made into either nuclear fuel or a nuclear explosive.

PNC   Power Reactor and Nuclear Fuel Development Corp. (Jap.).

pressurized water reactor   A light water reactor having primary and secondary cooling circuits. In the primary circuit, heat is transferred from the reactor core to a heat exchanger by means of water kept under high pressure to achieve high temperature without boiling; in the secondary circuit, steam is generated.

psi   Pounds per square inch.

PTB   Physikalisch-Technische Bundesanstalt, West German technical agency.

PUREX   The standard chemical separation process for extracting plutonium and uranium from spent fuel. Residues from the process include high-level waste in the form of concentrated fission products, and transuranic waste in the form of plutonium-contaminated process trash.

PWR   Pressurized water reactor.

radioactive decay   The spontaneous decay or disintegration of an unstable atomic nucleus, generally accompanied by the emission of ionizing radiation.

radioactive decay product, or daughter   A nuclide resulting from the radioactive disintegration of a radionuclide. The decay product may be formed either directly or from successive transformations in a radioactive series.

radionuclide   A radioactive nuclide (*see* nuclides).

radwaste   Radioactive waste.

reactor core The central portion of a nuclear reactor containing the fuel elements.

recycling The reuse of fissionable material after recovering it from spent reactor fuel and processing it into new fuel elements.

rem Acronym for roentgen equivalent man, a dosage of ionizing radiation. The average background and medical dose in the United States is about 0.2–0.3 rem per year per person.

repository block The mass of rock that would contain the tunnels of a geologic disposal repository.

repository horizon The level in a rock formation at which a repository might be located.

reprocessing The chemical separation of irradiated nuclear fuel into uranium, plutonium, and fission products.

re-racking The replacement of existing fuel storage racks with modified racks designed to increase the amount of spent fuel that can be stored in pools at reactor sites.

RSSF Retrievable Surface Storage Facility.

RWMAC Radioactive Waste Management Advisory Committee (Brit.).

salt cake The wet crystalline solids left after liquids in a military high-level waste storage tank have been run through an evaporator. Cesium is found in the salt cake, but strontium and nearly all other fission products and actinides are in the underlying sludge. Even where much of the cesuim and strontium has been removed from the tanks, the salt cake and sludge remain highly radioactive.

saturated zone The portion of a geologic profile in which all pore spaces in the rock are filled with water. The water table marks the upper limit of the saturated zone.

scram To shut down a nuclear reactor suddenly.

SFR Acronym for the Swedish repository for intermediate-level (and all but the least radioactive low-level) wastes from reactor operation and maintenance. Being built in hard rock beneath the seabed.

shroud A major structural component of a light water reactor. Made of stainless steel. It becomes highly activated by neutron bombardment and, at reactor decommissioning, will require isolation as long-lived radioactive waste.

SIXEP Site Ion Exchange Plant (Brit.).

SKB, SKBF Swedish nuclear fuel and waste management company.

SKI Swedish Nuclear Power Inspectorate.

species A particular kind of atomic nucleus, atom, or molecule; a nuclide.

specific activity The radioactivity per unit weight of any radioactive material.

spent fuel Nuclear reactor fuel that has irradiated to the point that a chain reaction can no longer be efficiently sustained. The fission products absorb neutrons and thereby interfere with the fission process. If the spent fuel is not to be reprocessed, its disposal or storage represents the largest problem in radioactive waste management. If the fuel is reprocessed, the greatest problem is disposal of the high-level and transuranic wastes from reprocessing and plutonium recycling. (*See* PUREX; transuranic waste.)

thermal reactor In thermal reactors the fission chain reaction is sustained primarily by thermal neutrons, or neutrons slowed down by a moderator (such as water). The neutrons in a "fast" reactor or breeder, on the other hand, move at the speeds at which they are ejected from the fissioning nuclei, with little or no moderator present to slow them down.

THORP Thermal oxide reprocessing plant (Brit.).

TMI Three Mile Island (nuclear power plant, Pennsylvania).

transmutation The transformation in a reactor of one element into another by a nuclear reaction (or series of reactions) resulting from the absorption of a neutron. For example, the fertile uranium isotope U-238 is transmuted into the plutonium isotope Pu-239 by absorption of a neutron and the subsequent emission of two beta particles.

transplutonic element An element having a higher atomic number than plutonium.

transuranic element An element having a higher atomic number than uranium.

transuranic waste Waste material contaminated with plutonium and other elements having atomic numbers higher than 92. In the commercial fuel cycle, transuranic waste is produced primarily from the reprocessing of spent fuel and the manufacture of plutonium fuel.

TRU Transuranic waste.

tuff A rock formed of compacted volcanic ash and dust.

unsaturated zone The portion of a geologic profile above the water table. Water is present but does not fill all the pores in the rock (*see* saturated zone).

uranium A slightly radioactive element (atomic no. 92) that has become the basic raw material of nuclear energy. Uranium-235 (with a half-life of more than 700 million years) makes up 0.7 percent of the uranium in natural ore, and is fissionable; uranium-238 (with a half-life of more than 4 billion years) makes up the other 99.3 percent, and is fertile, meaning that by neutron capture it can be converted to the fissionable plutonium-239. Thorium, another heavy element (atomic no. 90), contains the fertile isotope thorium-232 which can be converted to the fissionable uranium isotope U-233; in principle, therefore, nuclear power could also have as its basis a thorium fuel cycle. (*See also* half-life.)

USGS United States Geological Survey.

vitrify In nuclear waste disposal, to solidify high-level waste by incorporating it into glass.

WIPP Waste Isolation Pilot Plant, New Mexico.

WISP Waste Isolation Systems Panel, National Academy of Sciences (U.S.).

yellowcake The semirefined uranium oxide (U_3O_8) produced in uranium mining and milling.

Name Index

Subject Index

455